The Disassembly Line

About the Authors

Seamus McGovern, Ph.D., is an Electronics Engineer at the Volpe National Transportation Systems Center. He concurrently holds a commission in the U.S. Navy Reserve as an Aerospace Engineering Duty Officer, as well as a part-time industrial engineering faculty appointment at Northeastern University. A peer reviewer for six scientific journals, Dr. McGovern's academic efforts have been recognized with the award of four competitive fellowships, 11 merit-based scholarships, election to three scholastic honor societies, and a national best dissertation prize.

Surendra M. Gupta, Ph.D., is a Professor of Mechanical and Industrial Engineering and the director of the Laboratory for Responsible Manufacturing, Northeastern University. He has authored or coauthored well over 400 technical papers published in books, journals and international conference proceedings. Dr. Gupta has taught over 100 courses in such areas as operations research, inventory theory, queuing theory, engineering economy, supply chain management, and production planning and control. Among the many recognitions received, he is the recipient of the Outstanding Research Award and the Outstanding Industrial Engineering Professor Award (in recognition of teaching excellence) from Northeastern University as well as a national Outstanding Doctoral Dissertation Advisor Award.

The Disassembly Line

Balancing and Modeling

Seamus M. McGovern

Surendra M. Gupta

McGraw Hill

New York Chicago San Francisco
Lisbon London Madrid Mexico City
Milan New Delhi San Juan
Seoul Singapore Sydney Toronto

The McGraw-Hill Companies

Cataloging-in-Publication Data is on file with the Library of Congress.

The Disassembly Line

1 2 3 4 5 6 7 8 9 0 DOC/DOC 1 9 8 7 6 5 4 3 2 1 0

ISBN 978-0-07-162287-5
MHID 0-07-162287-X

Sponsoring Editor
Michael Penn

Acquisitions Coordinator
Michael Mulcahy

Editorial Supervisor
David E. Fogarty

Project Manager
Manisha Singh, Glyph International

Copy Editor
Priyanka Sinha

Proofreader
Medha Joshi

Production Supervisor
Richard Ruzycka

Composition
Glyph International

Art Director, Cover
Jeff Weeks

Dedicated to our Families

Joyce, Conall and Rory, and James, Concetta and Seana

Seamus M. McGovern

Sharda, Monica and Neil

Surendra M. Gupta

Contents

Preface xv
Acknowledgments xix

Part I Disassembly Background

1 Introduction **3**
1.1 Overview 3
1.2 Motivation 5
1.3 Scope and Objectives 6
1.4 Format and Notation 7
1.4.1 General and Disassembly-Line Balancing Notation 8
1.4.2 Assembly-Line Balancing Notation 13
1.4.3 Disassembly-to-Order Notation 13
1.4.4 Disassembly Just-in-Time Notation 14
1.4.5 Disassembly Revenue Notation 15
1.4.6 Queueing Theory Notation 15
1.5 Outline 16

2 Assembly Lines **19**
2.1 Introduction 19
2.2 Production Background 19
2.3 History of the Assembly Line 22
2.4 Assembly-Line Balancing 23
2.5 Line Modeling 24
2.6 Summary 31

3 Disassembly Lines **33**
3.1 Introduction 33
3.2 Overview 33
3.3 History of the Disassembly Line 37
3.4 Disassembly-Specific Considerations 38
3.4.1 Product Considerations 38
3.4.2 Line Considerations 38
3.4.3 Part Considerations 39
3.4.4 Operational Considerations 41
3.4.5 Demand Considerations 43
3.4.6 Assignment Considerations 44
3.4.7 Other Considerations 45

4 **Related Research** **47**
- 4.1 Introduction 47
- 4.2 Environmentally Conscious Manufacturing and Product Recovery 47
- 4.3 Assembly-Line Balancing and Manufacturing Systems 49
- 4.4 Disassembly and Remanufacturing 52
- 4.5 Optimization and Algorithms 59

5 **Graphical Representations of Products to Be Disassembled** **61**
- 5.1 Introduction 61
- 5.2 General Product Disassembly Representations 61
  - 5.2.1 Product Case Study 62
  - 5.2.2 Connection Diagrams 63
  - 5.2.3 Assembly Trees, Reduced Disassembly Trees, Connection State Diagrams, and Subassembly State Diagrams 64
  - 5.2.4 Disassembly Precedence Graphs 66
  - 5.2.5 Constrained Connection Diagrams 67
- 5.3 Task-Based Precedence Diagrams 68
- 5.4 Disassembly Constraint Graphs 68
- 5.5 Schematic/Flowchart-Style Vertex and Arc Disassembly Representation 70
- 5.6 Other Representations 74

6 **Computational Complexity of Combinatorial Problems** **77**
- 6.1 Introduction 77
- 6.2 Complexity Theory Background 78
- 6.3 NP-Completeness 82
- 6.4 NP-Completeness in the Strong Sense 85
- 6.5 NP-Hardness 86
- 6.6 Overview of NP-Complete Problems 87
- 6.7 Solving NP-Complete Problems 88
- 6.8 Other Classes of Problems 89
- 6.9 Heuristic Performance Analysis 90
- 6.10 Hardware and Software Considerations 91

Part II Disassembly-Line Balancing

7 **Disassembly-Line Balancing Overview** **95**
- 7.1 Introduction 95
- 7.2 The Disassembly-Line Balancing Problem 95

7.3 Model Considerations 96
7.4 Balancing Objectives 99

8 Description of the Disassembly Line and the Mathematical Model 101
8.1 Introduction 101
8.2 Problem Overview 102
8.3 Balance Measure and Theoretical Bounds Formulation 103
8.4 Hazard Measure and Theoretical Bounds Formulation 105
8.5 Demand Measure and Theoretical Bounds Formulation 106
8.6 Direction Measure and Theoretical Bounds Formulation 107
8.7 Models and Measures as Prototypes 108
8.8 Matrices for Precedence Representation 109

9 Computational Complexity of DLBP 111
9.1 Introduction 111
9.2 DLBP NP-Completeness 111
9.3 DLBP NP-Completeness in the Strong Sense 113
9.4 DLBP NP-Hardness 114

10 Combinatorial Optimization Searches 117
10.1 Introduction 117
10.2 Combinatorial Optimization Methodologies 118
10.3 Exhaustive Search 121
10.4 Genetic Algorithm 122
10.5 Ant Colony Optimization 122
10.6 Greedy Algorithm 123
10.7 Adjacent Element Hill-Climbing Heuristic ... 124
10.8 *k*-Opt Heuristic 124
10.9 H-K General-Purpose Heuristic 125
10.9.1 Introduction 125
10.9.2 Heuristic Background and Motivation 125
10.9.3 Comparison to Other Methodologies 127
10.9.4 The H-K Process 129
10.9.5 Other H-K Formulations 131

11 Experimental Instances 133
11.1 Introduction 133
11.2 Personal Computer Instance 134

11.3 The 10-Part Instance 136
11.4 Cellular Telephone Instance 137
11.5 DLBP A Priori Optimal Solution Benchmark Instances 140
11.5.1 Background 140
11.5.2 Mathematical Formulation 140
11.5.3 Probabilistic Analysis of the Benchmark Data Set 143

12 Analytical Methodologies 149
12.1 Introduction 149
12.2 Graphical Analysis Tools 149
12.3 Multiple-Criteria Decision-Making Considerations 151
12.4 Normalization and Efficacy Index Equations 154
12.5 Simulation 156

13 Exhaustive Search 159
13.1 Introduction 159
13.2 Model Description 159
13.3 Numerical Results 161
13.3.1 Personal Computer Instance 161
13.3.2 The 10-Part Instance 162
13.3.3 Cellular Telephone Instance 162
13.3.4 DLBP A Priori Problem Instances . . . 162
13.4 Conclusions 164

14 Genetic Algorithm 165
14.1 Introduction 165
14.2 Model Description 165
14.3 DLBP-Specific Genetic Algorithm Architecture 166
14.4 DLBP-Specific Qualitative Modifications . . . 167
14.5 DLBP-Specific Quantitative Modifications 168
14.6 Numerical Results 170
14.6.1 Personal Computer Instance 170
14.6.2 The 10-Part Instance 170
14.6.3 Cellular Telephone Instance 171
14.6.4 DLBP A Priori Instances 173
14.7 Conclusions 180

15 Ant Colony Optimization 181
15.1 Introduction 181
15.2 Model Description 181

15.3 DLBP-Specific Qualitative Modifications and the Metaheuristic 182
15.4 Quantitative Assignments 187
15.5 Numerical Results 187
15.5.1 Personal Computer Instance 187
15.5.2 The 10-Part Instance 188
15.5.3 Cellular Telephone Instance 189
15.5.4 DLBP A Priori Instances 191
15.6 Conclusions 197

16 Greedy Algorithm 199
16.1 Introduction 199
16.2 Model Description 199
16.3 Numerical Results 203
16.3.1 Personal Computer Instance 203
16.3.2 The 10-Part Instance 204
16.3.3 Cellular Telephone Instance 204
16.3.4 DLBP A Priori Instances 206
16.4 Conclusions 211

17 Greedy/Adjacent Element Hill-Climbing Hybrid 213
17.1 Introduction 213
17.2 Model Description 214
17.3 Numerical Results 217
17.3.1 Personal Computer Instance 217
17.3.2 The 10-Part Instance 217
17.3.3 Cellular Telephone Instance 218
17.3.4 DLBP A Priori Problem Instances 220
17.4 Conclusions 226

18 Greedy/2-Opt Hybrid 227
18.1 Introduction 227
18.2 Model Description 228
18.3 Numerical Results 232
18.3.1 Personal Computer Instance 232
18.3.2 The 10-Part Instance 232
18.3.3 Cellular Telephone Instance 233
18.3.4 DLBP A Priori Problem Instances 233
18.4 Conclusions 241

19 H-K Heuristic 243
19.1 Introduction 243
19.2 DLBP Application 243
19.3 DLBP A Priori Numerical Analysis for Varying Skip Size 244

19.4 Numerical Results 252
19.4.1 Personal Computer Instance 252
19.4.2 The 10-Part Instance 252
19.4.3 Cellular Telephone Instance 253
19.4.4 DLBP A Priori Instances (Numerical Results for Varying *n*) 253
19.5 Conclusions 261

20 Quantitative and Qualitative Comparative Analysis 263
20.1 Introduction 263
20.2 Experimental Results 264
20.3 Conclusions 272

21 Other Disassembly-Line Balancing Research 275
21.1 Overview 275
21.2 Problem Extensions 275
21.3 Additional Solution Methodologies 276
21.4 Probabilistic Disassembly-Line Balancing 277
21.5 Probabilistic Disassembly-Line Data 278
21.6 Future Research Directions 278

Part III Further Disassembly-Line Considerations

22 Overview of Additional Disassembly-Line Related Problems 283
22.1 Introduction 283
22.2 Mathematical Models 284
22.3 Computational Complexity 285
22.4 Case Studies 286
22.5 Analytical and Solution Methodologies 287

23 Disassembly-Line Product Planning 289
23.1 Introduction 289
23.2 Sustainable Product Design Overview 289
23.3 New-Product Design Metrics for Demanufacturing 290
23.3.1 Design Metrics for End-of-Life Processing 290
23.3.2 Case Study 291
23.3.3 Results and Analysis 291
23.4 Additional Design-for-Disassembly Studies 293
23.5 Summary 294

24 Disassembly-Line Design 295
24.1 Introduction 295
24.2 Background 295
24.2.1 Facilities Engineering Planning 296
24.2.2 Real Estate 296
24.2.3 Engineering and Construction 296
24.2.4 Environmental Engineering 296
24.2.5 Support 296
24.2.6 Life Cycle Considerations 297
24.2.7 Equipment Planning 298
24.2.8 Scope and Funding Considerations 299
24.3 Facility Location and Layout Overview 300
24.4 Mixed-Model Disassembly-Line Design 303
24.5 Disassembly-Line Design Metrics 304

25 Disassembly-Line Sequencing and Scheduling 307
25.1 Introduction 307
25.2 Machine Sequencing and Scheduling Overview 307
25.3 Disassembly Sequencing and Scheduling ... 308
25.3.1 First Come First Served 310
25.3.2 Shortest Processing Time 311
25.3.3 Earliest Due Date 311
25.3.4 Critical Ratio 312
25.4 Stochastic Sequencing and Scheduling 313
25.5 Additional Disassembly-Sequencing Studies 314
25.6 Additional Disassembly-Scheduling Studies 315

26 Disassembly-Line Inventory 317
26.1 Introduction 317
26.2 Inventory Theory Background 317
26.3 Disassembly to Order 318
26.3.1 Single Period, Disassembly to Order (Deterministic Yields) 319
26.3.2 Single Period, Disassembly to Order (Stochastic Yields) 321
26.4 Additional Disassembly-Inventory Studies 323

27 Disassembly-Line Just-in-Time 325
27.1 Introduction 325
27.2 Just-in-Time Background 325

27.3 Just-in-Time Research 327
27.4 Multikanban System for End-of-Life Products 328
27.4.1 Arrival Pattern 329
27.4.2 Demand Fluctuation and Inventory Management 329
27.4.3 Disassembly-Line Multikanban Model Overview 330
27.4.4 Material Types 330
27.4.5 Kanban Types 331
27.4.6 Kanban Routing Mechanism 331
27.4.7 Selection of Products 332
27.4.8 Determining Kanban Level 333
27.4.9 Simulation Results 334

28 Disassembly-Line Revenue 337
28.1 Introduction 337
28.2 Disassembly-Line Revenue Background 337
28.3 Disassembly-Line Revenue Modeling 338
28.4 Additional Disassembly-Line Revenue Studies 341

29 Unbalanced Disassembly Lines 343
29.1 Introduction 343
29.2 Background 343
29.3 Stochastic Line Modeling and Unbalanced Lines 344
29.4 Queueing Theory Background 344
29.5 Benchmark Data for Measuring Unbalance 348

References 349

Appendix: Acronyms 363

Author Index 365

Subject Index 369

Preface

The growing amount of waste created by products reaching the end of their useful life poses challenges for the environment, governments, and manufacturers. As such, manufacturers are increasingly recovering and processing their postconsumer products. End-of-life processing of products has become desirable due to consumer preference and increased public awareness; new, more rigid environmental legislation and government regulation; and the economic attractiveness of reusing products, subassemblies, or parts instead of disposing of them. Alternatives for these products include reuse, remanufacturing, recycling, storage, and disposal. With disposal considered to be the least desirable, the first process required by the remaining alternatives is disassembly. Obtaining valuable components and materials, while minimizing the space required for landfills and reducing the amount of processed toxins introduced into the environment, are all compelling reasons that disassembly—and specifically the disassembly line—is of such importance and has garnered so much interest. However, disassembly is labor intensive and therefore costly. Ensuring that the disassembly process is as efficient as possible is essential to enable economic justification of an end-of-life option other than disposal.

Just as the assembly line is considered the most efficient way to assemble a product, the disassembly line is seen to be the most efficient way to disassemble a product. Efficient techniques are then required to solve these problems which involve the number of workstations required and the disassembly sequencing of end-of-life products on the disassembly line. The challenge lies in the fact that disassembly possesses unique characteristics. While having similarities to assembly, it is not the reverse of the assembly process. The difficulty in obtaining efficient disassembly-line part removal sequence or workstation selection solutions stems from the fact that any solution sequence would consist of a permutation of numbers. This permutation would contain as many elements as there are parts in the product. The problem therefore tends to have high calculation complexities due to its rapid growth in computational runtime with incremental increases in the number of parts. As such, the observation can be made that the DISASSEMBLY LINE BALANCING PROBLEM would appear to be NP-hard (and, of greater interest, the decision version would appear to be NP-complete). In addition, the added complications of disassembly typically foster multiple objectives including environmental and economic goals that can frequently be contradictory. Also, disassembly-line problems are a relatively recent problem area; for example, the DISASSEMBLY LINE BALANCING PROBLEM was first formally described in 2002. For these reasons, disassembly has gained a great deal of attention in the contemporary scientific literature.

The multiobjective DISASSEMBLY LINE BALANCING PROBLEM seeks to find a disassembly solution sequence which: provides a feasible disassembly sequence, minimizes the number of workstations, minimizes total idle time, ensures similar idle times at each workstation, along with other, disassembly-specific concerns. Finding the optimal line balance is computationally intensive due to exponential growth. With exhaustive search calculations quickly becoming prohibitively large, methodologies from the field of combinatorial optimization hold promise for providing solutions to the DISASSEMBLY LINE BALANCING PROBLEM. In this book, the DISASSEMBLY LINE BALANCING PROBLEM is described then defined mathematically and proven to belong to the class of unary NP-complete problems. Four case-study instances of the problem are introduced, then disassembly-line versions of depth-first exhaustive search, genetic algorithm and ant colony optimization metaheuristics, a greedy algorithm, and greedy/hill-climbing and greedy/2-optimal hybrid heuristics are presented and compared, along with an uninformed general-purpose search heuristic. The problem instances are then used in numerical studies performed to illustrate the implementations and to demonstrate quantitative and qualitative analysis and comparison. Observations about the solution methodologies include their consistent generation of optimal or near-optimal solutions, their ability to preserve precedence, the speed of the techniques, and their practicality for implementation.

While the focus of this book is on the problem of balancing the paced disassembly line, other issues and models related to disassembly in general are also considered. Disassembly is a process that interacts with all phases of product recovery including before life (the period of design and life cycle analysis), the useful period (the time when the product is actually being manufactured or is in use), and end-of-life (the period in which a product completes its useful life and is ready for further processing, recovery, or disposal). Therefore, the third part of the book deals with the remaining significant areas of disassembly research, including product planning, line and facility design, sequencing and scheduling, inventory, just-in-time, revenue, and unbalanced lines.

The various techniques presented in this book form a body of knowledge directed at addressing problems involving disassembly lines. The importance of the disassembly line's role in end-of-life processing, the contemporary nature of the problems, and the NP-complete characteristics of many of them make disassembly-line problems relevant, timely, academically interesting, and scientifically challenging, and hence provide much of the primary motivation behind the formulation of this book.

Acknowledgments

While the field of disassembly as a research area has been around for about two decades, the DISASSEMBLY LINE BALANCING PROBLEM was first described in a formal way relatively recently (in 2002). Since then, this problem has gained a great deal of attention in contemporary scientific literature. This book is an attempt to capture the state of the art in one volume. To the best of our knowledge, this is the first book in the world written solely on the DISASSEMBLY LINE BALANCING PROBLEM.

This book would not have been possible without the help and encouragement of numerous people. We would like to thank hundreds of researchers, colleagues, and former graduate students whose works we have read and benefited from and whom we have met and interacted with, at conferences and workshops around the world. In addition, we thank our current students and colleagues who continually help us drive toward future discoveries.

We also thank McGraw-Hill for their commitment to publish and encourage innovative ideas and express our appreciation to its staff for providing seamless support in making it possible to complete this timely and vital book.

Most importantly, we are indebted to our families, to whom this book is lovingly dedicated, for constantly giving us their unconditional support which made the challenging undertaking of writing this book much more gratifying.

Seamus M. McGovern, PhD
Cambridge, Massachusetts
Surendra M. Gupta, PhD
Boston, Massachusetts

The Disassembly Line

PART I

Disassembly Background

CHAPTER 1
Introduction

CHAPTER 2
Assembly Lines

CHAPTER 3
Disassembly Lines

CHAPTER 4
Related Research

CHAPTER 5
Graphical Representations of Products to Be Disassembled

CHAPTER 6
Computational Complexity of Combinatorial Problems

CHAPTER 1
Introduction

This chapter provides an introduction to the book. Section 1.1 presents an overview of current disassembly issues. Section 1.2 describes the motivation of the text. The scope and objectives of the book are then given in Sec. 1.3 while the notations used throughout the text are detailed in Sec. 1.4. Finally, Sec. 1.5 presents an outline of the book.

1.1 Overview

Manufacturers are increasingly recycling and remanufacturing their postconsumer products due to new, more rigid environmental legislation, increased public awareness, and extended manufacturer responsibility. In addition, the economic attractiveness of reusing products, subassemblies, or parts instead of disposing of them has helped to further energize this effort. *Recycling* is a process performed to retrieve the material content of used and nonfunctioning products. *Remanufacturing* on the other hand, is an industrial process in which worn-out products are restored to like-new conditions. Thus, remanufacturing provides the quality standards of new products with used parts.

Product recovery seeks to obtain materials and parts from old or outdated products through recycling and remanufacturing in order to minimize the amount of waste sent to landfills. This includes the reuse of parts and products. There are many attributes of a product that enhance product recovery; examples include ease of disassembly, modularity, type and compatibility of materials used, material identification markings, and efficient cross-industrial reuse of common parts/materials. The first crucial step of product recovery is disassembly.

Disassembly is defined as the methodical extraction of valuable parts/subassemblies and materials from discarded products through a series of operations. After disassembly, reusable parts/subassemblies are cleaned, refurbished, tested, and directed to the part/subassembly inventory for use in remanufacturing operations. The recyclable materials can be sold to raw-material suppliers, while the residuals are sent to landfills. Disassembly is a process that interacts with all phases of product recovery including *before life* (the period of design and life cycle analysis), the *useful period* (the time when the product is actually being manufactured or is in use), and *end-of-life* (the period in which a product completes its useful life and is ready for further processing, recovery, or disposal).

Disassembly has gained a great deal of attention in the literature due to its role in product recovery. A disassembly system faces many unique challenges; for example, it has significant inventory problems because of the disparity between the demands for certain parts or subassemblies and their yield from disassembly. The flow process is also different. As opposed to the normal "convergent" flow in regular assembly environment, in disassembly the flow process is "divergent" (a single product is broken down into many subassemblies and parts). There is also a high degree of uncertainty in the structure and the quality of the returned products. The condition of the products received is usually unknown and the reliability of the components is suspect. In addition, some parts of the product may cause pollution or may be hazardous. These parts may require special handling that can also influence the utilization of the disassembly workstations. For example, an automobile slated for disassembly contains a variety of parts that are dangerous to remove and/or present a hazard to the environment, such as the battery, airbags, fuel, and oil. Various demand sources may also lead to complications in disassembly-line balancing. The reusability of parts creates a demand for them; however, the demands and availability of the reusable parts are significantly less predicable than what is found in the assembly process. Most products contain parts that are installed (and must be removed) in different attitudes, from different areas of the main structure, or in different directions. Since any required directional change increases the setup time for the disassembly process, it is desirable to minimize the number of directional changes in the chosen disassembly sequence. Finally, disassembly-line balancing is critical in minimizing the use of valuable resources (such as time and money) invested in disassembly and maximizing the level of automation of the disassembly process and the quality of the parts (or materials) recovered.

This part of the book (Part I) provides a background to many of the issues and much of the current research, as well as an introduction to the problems and possible solutions. Included in this is a review of assembly lines, an introduction to disassembly lines, a survey of current research, graphical representations of products, an overview of computational complexity, and a description of combinatorial optimization searches.

In Part II of this book, the DISASSEMBLY LINE BALANCING PROBLEM (DLBP) is addressed using combinatorial optimization methodologies. While exhaustive search consistently provides the optimal solution, its exponential time complexity quickly reduces its practicality. Combinatorial optimization techniques are instrumental in obtaining optimal or near-optimal solutions to problems with intractably large solution spaces. *Combinatorial optimization* is an emerging field that combines techniques from applied mathematics, operations research, and computer science to solve optimization problems over discrete structures. Some of these techniques include greedy algorithms, integer and linear programming, branch-and-bound, divide-and-conquer, dynamic programming, local optimization, simulated annealing, genetic algorithms, and approximation algorithms.

The seven techniques selected for application in this text seek to provide a *feasible* disassembly sequence (i.e., one in which no *precedence constraints* are violated since some tasks cannot be performed until their predecessor tasks have been completed), minimize the number of workstations, minimize the total idle time, and minimize the variation in idle times between workstations, while attempting to remove hazardous and high-demand product components as early as possible and remove parts with similar part removal directions together. Four data sets are used as case studies. These instances are used with all of the solution techniques to illustrate implementation of the methodologies, measure performance, and enable comparisons.

In Part III of this book, other problems related to the disassembly line are explored. This part visits the bulk of the remaining disassembly-related areas. A background—much of it from traditional assembly line and production theory—is provided and then disassembly-specific issues and research are detailed. These research areas include product planning, facility and line layout, sequencing and scheduling, inventory, just-in-time, revenue, and unbalanced lines.

1.2 Motivation

End-of-life processing for products is becoming increasingly desirable due to consumer preference, government regulation, and corporate financial interests. Obtaining components and materials that have some value while minimizing the amount of waste sent to landfills and reducing the amount of processed toxins introduced into the environment are all compelling reasons that disassembly is of such importance and has garnered so much global interest.

Disassembly has unique characteristics. While possessing similarities to assembly, it is not the reverse of the assembly process (Brennan et al., 1994); therefore, new and efficient approaches and methodologies are needed to effectively perform disassembly-line operations. The difficulty in obtaining efficient disassembly-line sequence solutions stems from the fact that a solution sequence consists of a permutation of numbers. This permutation would contain as many elements as there are parts in the product. As such, the observation can be made that the DISASSEMBLY LINE BALANCING PROBLEM would appear to be

NP-hard and the decision version would appear to be NP-complete. Also of interest, the DISASSEMBLY LINE BALANCING PROBLEM is a recent problem, first formally described in this century (Güngör and Gupta, 2002).

The importance of the disassembly line's role in end-of-life processing, the contemporary nature of the problem, and its NP-complete characteristics makes the DISASSEMBLY LINE BALANCING PROBLEM—and the disassembly line in general—interesting and relevant.

1.3 Scope and Objectives

This book introduces the disassembly line (Part I) and details its primary focus—disassembly-line balancing (Part II)—and then considers other disassembly-line problems that do not directly address balancing. Various techniques are explored that involve multiple objectives to address disassembly-line balancing. Since the decision version of the DISASSEMBLY LINE BALANCING PROBLEM is NP-complete, this book considers a rigorous combinatorial optimization treatment of the problem. As part of this, the problem is mathematically defined and proven to be NP-complete (as well as unary NP-complete and NP-hard), quantitative and qualitative evaluation criteria are developed, and four problem instances are introduced. Next, seven different techniques from the realm of combinatorial optimization are employed to solve the four instances. The seven methodologies are then compared to each other.

The first methodology used is exhaustive search, which here employs a depth-first search using a recursive backtracking procedure to visit all permutations of an instance's n parts to consistently obtain the optimal solution sequence. The exhaustive search algorithm is presented for obtaining the optimal solution to small instances of the DLBP. While always optimal, it is limited in the size of the instance it can solve since the time to solve an instance grows exponentially with the size of the instance. The second technique used is a genetic algorithm (GA), a metaheuristic that recombines and mutates the best solutions over many generations. The genetic algorithm considered here involves either a randomly generated or a hot-started initial population with crossover, mutation, and fitness competition performed over many generations. The third technique used is ant colony optimization (ACO), another metaheuristic. The ant colony optimization metaheuristic applied here is an ant system algorithm known as the ant-cycle model that is enhanced for the DLBP. Ant colony optimization uses software agents referred to as ants that grow a solution from first to last part under greedy decision-making rules. Successful ants add pheromone (in proportion to the quality of their solution) to their paths, all paths slowly evaporate, and the process repeats for a given number of cycles. The fourth technique used is a deterministic first-fit-decreasing greedy algorithm, similar to that developed for the BIN-PACKING problem. Two 2-phase hybrids are considered next. The first hybrid process consists of the greedy sorting algorithm followed by a hill-climbing local search. The problem-specific hill-climbing algorithm only considers swapping parts that are in adjacent workstations. A second deterministic hybrid process is then shown. It consists of the same greedy sorting algorithm followed by a 2-optimal (2-opt) local search. The 2-opt search is modified for the DLBP by exchanging parts instead of the arcs connecting them. The final technique used is an uninformed, modified British Museum–type search that moves through the search space similarly to exhaustive search but only samples the space, methodically visiting equally spaced solutions in a deterministic manner. Influenced by the hunter-killer search tactics of military helicopters, this general-purpose heuristic algorithm easily lends itself to the DLBP. All of the techniques deliver optimal or near-optimal solutions while preserving the precedence relationships among the components.

Next (Part III), other wide-ranging disassembly-line problems are introduced. These problems encompass the areas of product design, facility location and line layout, sequencing and scheduling, removed-part inventory, just-in-time, revenue, and intentionally or unintentionally unbalanced lines.

1.4 Format and Notation

The following general guidelines are used throughout the book. Using the format of Garey and Johnson (1979), names of problems, including NP-complete problems, are capitalized when referring to the problem but not when referring to a methodology (e.g., "the LINEAR PROGRAMMING problem" is a different use than "modeling a problem using linear programming") or when the formal name of the problem is not appropriate. Since the search methodologies in this book are used for problems other than the DLBP, the versions used in this book make use of a title case format and are often prefaced by "DLBP," while generic references to the methodologies are made in lowercase (e.g., "DLBP Exhaustive Search" and "Exhaustive Search" refer to the problem-specific methodologies developed for use in this book while "exhaustive search" refers to any type or implementation of exhaustive search). The first time a proper mathematical or scientific term is defined (or if it is not defined, the first time it is used in the chapter that primarily references it, or lacking this, the first time it is used in the book) it is italicized. This is primarily the case when mathematical terms have a specific meaning but make use of a word common to the English language, an example being use of the word "language" in Chap. 6. Italics are also used to highlight some proper names. Most of the mathematical conventions are as found in Cormen et al. (2001), Rosen (1999), or Garey and Johnson (1979). Finally, part removal times may also refer to *virtual parts* (also referred to as *virtual components*; Lambert and Gupta, 2005); that is, a task performed that is required (or desired or possible) and takes a finite amount of time but does not result in the immediate removal of any part. Therefore, the terms "task time" and "part removal time" are both used here with the understanding that the two are potentially distinct. Unless otherwise specified, work element, job, part, component, subassembly, and task may be used interchangeably throughout this text.

1.4.1 General and Disassembly-Line Balancing Notation

The following notation is used in the remainder of the book:

$\langle 1, 2, \ldots, n \rangle$	ordered n-tuple
$\{1, 2, \ldots, n\}$	set (using the formal definition, i.e., list of n distinct items)
$(1, 2, \ldots, n)$	list of n items
(p, q)	arc (i.e., direct edge) pq
$[p, q]$	edge pq
A-B	subassembly made up of part A and part B
$\leq_p$	polynomial time reduction or polynomial transformation; read as: "can be converted to," "is easier than," "is a subset of," or "is a smaller problem than"
$\prec$	partial ordering, that is, x precedes y is written $x \prec y$
$\lvert X \rvert$	cardinality of the set X
$\lceil x \rceil$	ceiling function of x; assigns the smallest integer $\geq x$, for example, $\lceil 1.3 \rceil = 2$
$\max(x, y)$	maximum of x and y
$\min(x, y)$	minimum of x and y
$\forall$	"for all," "for every"
$\in$	"an element of," "is in"
$\exists$	"there exists," "there exists at least one," "for some"
$\propto$	"in proportion to"
$\cap$	intersection
$\cup$	union
$\subseteq$	subset
$\wedge$	conjunction (logical AND)
$\vee$	disjunction (logical OR)
!	factorial
:	"such that," used primarily to avoid confusion with the vertical bars used to define cardinality

$\vert$	"such that," used primarily to avoid confusion with the pseudo-code equal sign ":=" also used as a triplet separator in scheduling theory and to represent absolute value in the efficacy index
$\rightarrow$	"maps to"
$\Leftrightarrow$	"if and only if"
α	weight of existing pheromone (trail) in path selection; also refers to the machine environment in scheduling theory
β	weight of the edges in path selection; also refers to the processing characteristics and constraints in scheduling theory
ε	empty string
γ	objective to be minimized in scheduling theory
$\eta_{p,q}(t)$	visibility value of edge (arc for the DLBP) pq at time t
ρ	variable such that $1 - \rho$ represents the pheromone evaporation rate
$\tau_{p,q}(\text{NC})$	amount of trail on edge pq (arc for the DLBP) during cycle NC
ψ_k	kth element's skip measure (i.e., for the solution's third element, visit every second possible task for $\psi_3 = 2$)
$\Delta\psi_k$	kth element's delta skip measure; difference between problem size n and skip size ψ_k (i.e., for $\Delta\psi = 10$ and $n = 80$, $\psi = 70$)
O	"big-oh," $g(x)$ is $O(h(x))$ whenever $\exists\, y : \vert g(x)\vert \leq y \cdot \vert h(x)\vert\ \forall\, x \geq z$
Π	product; also refers to a decision problem (consisting of a set D of decision instances and a subset $Y \subseteq D$ of yes-instances) in complexity theory
Θ	"big-theta," $g(x)$ is $\Theta(h(x))$ whenever $g(x)$ is $O(h(x))$ and $g(x)$ is $\Omega(h(x))$
Σ	summation; also refers to an alphabet (a finite set of symbols, e.g., $\Sigma = \{0, 1\}$) in complexity theory
Σ^*	set containing all possible strings from Σ
Ω	"big-omega," $g(x)$ is $\Omega(h(x))$ whenever $\exists\, y : \vert g(x)\vert \geq y \cdot \vert h(x)\vert\ \forall\, x \geq z$
a	MULTIPROCESSOR SCHEDULING problem task variable; also refers to a function variable in complexity theory
A	MULTIPROCESSOR SCHEDULING problem task set
b	function variable in complexity theory

B	MULTIPROCESSOR SCHEDULING problem deadline bound
BST_k	identification of kth part in temporary best solution sequence during adjacent element hill-climbing (AEHC) and 2-Opt
c	initial amount of pheromone on all of the paths at time $t = 0$; also refers to a cost function that maps feasible points to the real numbers in complexity theory
C_k	completion time of task k in a flow shop in scheduling theory
C_{max}	completion time of the last task to be completed in the flow shop in scheduling theory
CT	cycle time; maximum time available at each workstation
d_k	demand; quantity of part k requested
D	demand rating for a given solution sequence; also demand bound for the decision version of DLBP; also refers to the set of all instances of a decision problem in complexity theory
D^*	optimal demand rating for a given instance; also used to refer to the set of solutions optimal in D
D_{lower}	lower demand bound for a given instance
D_{upper}	upper demand bound for a given instance
DP	set of demanded parts
e	encoding function in complexity theory
$E[x]$	expected value of x
EI_x	efficacy index of measure x; generates values between 0 and 100 percent
f	a feasible solution in complexity theory
F	measure of balance for a given solution sequence; also refers to the finite set of feasible points in complexity theory
F^*	optimal measure of balance for a given instance; also used to refer to the set of solutions optimal in F
F_{lower}	lower measure of balance bound for a given instance
F_{upper}	upper measure of balance bound for a given instance
$F_{t,r}$	measure of balance of ant r's sequence at time t
Fm	flow shop with m machines
FS	feasible sequence binary value; FS = 1 if feasible, 0 otherwise
$g(x)$	function in complexity theory

h_k	binary value; 1 if part k is hazardous, else 0
$h(x)$	function in complexity theory
H	hazard rating for a solution; also hazard bound for the decision version of DLBP
H^*	optimal hazard rating for a given instance; also used to refer to the set of solutions optimal in H
H_{lower}	lower hazard bound for a given instance
H_{upper}	upper hazard bound for a given instance
HP	set of hazardous parts
i	counter variable
I	total idle time for a given solution sequence; also refers to an instance of a problem in complexity theory
I^*	optimum idle time for a given instance
I_j	total idle time of workstation j
I_{lower}	lower idle time bound for a given instance
I_{upper}	upper idle time bound for a given instance
ISS_k	binary value; 1 if part k is in the solution sequence, else 0
j	workstation count (1, …, NWS)
k	counter variable (typically $k \in \{1, 2, \ldots, n\}$ and identifies a part or refers to a sequence position)
$l(a)$	MULTIPROCESSOR SCHEDULING problem task length
L	Language in complexity theory, that is, any set of strings over Σ (e.g., $L = \{1, 10, 100, 010, 1001, \ldots\}$)
L_r	ACO delta-trail divisor value; set equal to $F_{n,r}$ for the DLBP
m	number of processors in the MULTIPROCESSOR SCHEDULING problem; also number of ants; also number of machines
n	number of parts for removal
N	number of chromosomes (population)
$\mathbf{N}$	set of natural numbers, that is, $\{0, 1, 2, \ldots\}$
NC_{max}	maximum number of cycles for ACO
NPW_j	number of parts in workstation j
NWS	number of workstations required for a given solution sequence
NWS^*	optimal (minimum) number of workstations for n parts
$\text{NWS}_{\text{lower}}$	lower bound on the minimum possible number of workstations for n parts

NWS_{upper}	upper bound on the maximum possible number of workstations for n parts
p	counter variable; also edge/arc variable providing node/vertex identification
$p(x)$	polynomial function in complexity theory
P	set of n part removal tasks
$Pr^r_{p,q}(t)$	probability of ant r taking an edge $[p, q]$ [arc (p, q) for the DLBP] at time t during cycle NC
PRT	set of part removal times
PRT_k	part removal time required for part k (PRT_k is onto mapped to PRT though not necessarily one-to-one since multiple parts may have equal part removal times)
PS_k	identification of kth part in a solution sequence, that is, for solution $\langle 3, 1, 2\rangle$, $PS_2 = 1$
PSG_k	identification of kth part in the solution sequence after application of the Greedy algorithm
PSS_k	identification of kth part in solution sequence after sorting
PST_k	identification of kth part in solution sequence after AEHC and 2-Opt
q	counter variable; also edge/arc variable (node/vertex identification)
Q	amount of pheromone added if a path is selected
r	set of unique part removal directions; also ant counter
r_k	integer value corresponding to part k's part removal direction
R	direction rating for a given solution sequence; also direction bound for the decision version of DLBP
R^*	optimal direction rating for a given instance; also used to refer to the set of solutions optimal in R
R_k	binary value; 0 if part k can be removed in the same direction as part $(k + 1)$, else 1
R_{lower}	lower direction bound for a given instance
R_m	mutation rate
R_{upper}	upper direction bound for a given instance
R_x	crossover rate
$\mathbf{R}^z$	set of real numbers in z dimensions
s	MULTIPROCESSOR SCHEDULING task size
S	solution structure in complexity theory
ST_j	station time; total processing time requirement in workstation j

t	time within a cycle for ACO; ranges from 0 to n
$T(n)$	function describing an algorithm's running time on a computer processor
TMP_k	identification of kth part in temporary sequence during AEHC and 2-Opt
V	maximum range for a workstation's idle time
x	general variable; also refers to a part removal direction
$X_{k,j}$	binary decision variable; 1 if part k is assigned to workstation j, else 0
y	general variable; also refers to a part removal direction
Y	set of all "yes" instances in complexity theory
z	general variable; also refers to a part removal direction
$\mathbf{Z}$	set of integers, that is, $\{\ldots, -2, -1, 0, 1, 2, \ldots\}$
$\mathbf{Z}^+$	set of positive integers, that is, $\{1, 2, \ldots\}$
Z_p	pth objective to minimize or maximize

1.4.2 Assembly-Line Balancing Notation

Part I also makes use of the following notation:

DQ	total daily quantity required
DT	available daily production time
U	utilization rate
T_x	artificial task x
TT	total time (the sum of the task times)

1.4.3 Disassembly-to-Order Notation

In Part III, disassembly-to-order (DTO) notation includes the following:

CEP_g	end-of-life product cost (acquisition, transportation, disassembly, and part cleaning, inspection, sorting, etc.) for DTO product g
CNP_k	new-part acquisition cost for DTO part k
CPD_k	disposal cost for DTO part k
g	DTO product identification
G	total number of DTO products
k	DTO part identification
na	total number of DTO parts for all products
n_g	number of DTO parts in end-of-life product g
Pr(sc)	probability of scenario sc occurring
SC	set of DTO recourse model scenarios

w	number of possible outcomes in the DTO recourse model
X	decision variable
$X1_g$	decision variable indicating the amount of end-of-life product g to acquire for disassembly
$X2_k$	decision variable indicating the amount of new part k to acquire in order to meet the overall demand
$X2_{k,\,sc}$	decision variable indicating the amount of new part k to acquire in order to meet the overall demand in scenario sc
$X3_k$	decision variable indicating the number of excess removed parts k that need to be disposed of
$X3_{k,\,sc}$	decision variable indicating the number of excess removed parts k that need to be disposed of in scenario sc
$YP_{g,\,k}$	yield of part k from product g (the amount of part k obtained after disassembly of end-of-life product g)
$YP_{g,\,k,\,sc}$	yield of part k from product g (the amount of part k obtained after disassembly of end-of-life product g) in scenario sc

1.4.4 Disassembly Just-in-Time Notation

In Part III, just-in-time (JIT) notation includes:

a	kanban container capacity for JIT
ARS_x	arrival rate of subassembly x from an external source
$C(n)$	total number of possible combinations of a set of n parts
D_e	JIT expected demand per unit time (e.g., per day)
DRP_x	demand rate for part x
FRP_x	furnish rate of part x
FRS_x	furnish rate of subassembly x
K	number of kanbans for JIT
L	JIT production lead time (equal to the production time plus waiting time plus movement time)
$LRP_{x,\,z}$	removal rate of part x at workstation z
LRS_y	removal rate of subassembly y
NPK_x	number of part kanbans for part x at a given point on the disassembly line
NSK_x	number of subassembly kanbans for part x at a given point on the disassembly line
RRP_x	request rate of part x
RRS_x	request rate of subassembly x

W	JIT buffer stock (equal to zero ideally; often 10 percent of $D_e L$ is used)
x	part, part kanban, or workstation identification in the multikanban system
y	part, part kanban, or workstation identification in the multikanban system
y_z	subassembly or subassembly kanban identification in the multikanban system
z	part, part kanban, or workstation identification in the multikanban system

1.4.5 Disassembly Revenue Notation

Also in Part III, revenue notation includes the following (the artificial task notation T_x from Part I is not repeated here):

$\mathrm{AP}_{\mathrm{kt}}$	set of AND predecessors of task kt
AL	total number of artificial tasks
C_{kt}	cost of each disassembly task kt undertaken
$\mathrm{CT}_{\mathrm{upper}}$	decision-maker determined upper bound on the cycle time
d_{kp}	part kp's demand
kp	identifies any of the np parts
kt	identifies any of the nt tasks
np	number of parts
$\mathrm{nr}_{\mathrm{kt,\,kp}}$	total count of each part kp removed by task kt
nt	number of tasks
$\mathrm{OP}_{\mathrm{kt}}$	set of OR predecessors of task kt
$\mathrm{OS}_{\mathrm{kt}}$	set of OR successors of task kt
P^-	set of parts having a negative revenue
P^+	set of parts having a positive revenue
q_{kp}	number of demanded part kp that is removed
$\mathrm{REV}_{\mathrm{kp}}$	revenue per unit demanded
S	cost per unit time of keeping one station open
u_j	decision variable
$x_{\mathrm{kt},\,j}$	binary variable equal to one if task kt is assigned to workstation j

1.4.6 Queueing Theory Notation

Finally, in Part III additional queueing notation includes:

λ	mean number of arrivals per time period
λ_e	effective arrival rate

μ	service rate; the mean number of units that can be served per time period
ρ	traffic intensity; also known as the utilization factor
σ^2	variance
k	state of the queueing system; that is, number of customers in the waiting line plus the service facility
L	length of the queueing system (which is given by the expected number of customers in the queueing system)
L_q	length of the queue itself
n	maximum number of customers in the system
$\Pr_k$	probability of being in state k
s	number of servers
t	time
W	expected time in the queueing system
W_q	expected waiting time in line

1.5 Outline

Organized in three parts and 29 chapters, this book is primarily concerned with the complete disassembly of products on a *paced* disassembly line for component/material recovery purposes. It investigates the qualitative and quantitative aspects of the multicriteria solution sequences generated using the various combinatorial optimization techniques.

Chapter 2 provides a background in assembly lines, assembly-line balancing, and assembly-line modeling in order to provide a foundation prior to pursuing the disassembly line.

Chapter 3 gives an introduction to the disassembly line, detailing the many disassembly-specific considerations and providing some fundamental definitions.

Chapter 4 presents a survey of the research into environmentally conscious manufacturing and product recovery, assembly and manufacturing, disassembly, and various optimization and search techniques.

Chapter 5 provides some of the variety of product representations used in the disassembly field, including a graph-theory- and electrical-schematic-based arc and vertex representation for instances of the DLBP, which is then used throughout the book.

Chapter 6 gives an overview of complexity theory including the concepts of NP-completeness, unary NP-completeness, and NP-hardness, along with a listing of some of the specialized solution methodologies to address these problem classes. Also in this chapter, the computer processor hardware, software, language, software engineering, and analysis considerations are documented.

Part II begins with Chap. 7. This chapter presents the problem statement and research objectives of this book.

Chapter 8 provides a detailed description of the DISASSEMBLY LINE BALANCING PROBLEM including all objectives, mathematical formulae, and theoretical bounds.

Chapter 9 builds on the overview of complexity theory given in Chap. 6 and provides the proofs that the DISASSEMBLY LINE BALANCING PROBLEM is NP-complete, unary NP-complete, and NP-hard, necessitating specialized solution methodologies including those from the field of combinatorial optimization.

In Chap. 10, each of the combinatorial optimization searches used in Part II is introduced.

Chapter 11 introduces the four case-study problem instances. These include the personal computer instance, the 10-part instance, the cellular telephone instance, and the group of known-optimal instances. The personal computer and 10-part problem instances are slightly modified case studies from the literature, while the cellular telephone instance is a 25-part experimentally determined instance from an actual electronic consumer product. The known-optimal instances are a variable-dimension data set with known-optimal measures in all of the criteria under evaluation in this book. This chapter also includes a thorough statistical analysis of the number and percentage of optimal solutions with instance size using the known-optimal data set.

Chapter 12 introduces an analytical methodology for the qualitative and quantitative study of DLBP results. Also, the multicriteria decision-making format used in the remainder of the text is detailed. The concept of normalization is discussed and formulae for the efficacy indices are listed. Finally, simulation as an analysis tool is discussed.

Chapters 13 through 19 present the combinatorial optimization methodologies of exhaustive search, the genetic algorithm, ant colony optimization, the greedy algorithm, the greedy/hill-climbing hybrid, the greedy/2-opt hybrid, and the uninformed deterministic search heuristic, respectively. These chapters also include the numerical results of the methodologies' application to each of the four instances.

Chapter 20 compares all of the methodologies using the same qualitative and quantitative processes as used in Chaps. 13 through 19. These include graphical depictions of performance and calculated efficacy indices with growth in instance size, using the known-optimal instances.

Chapter 21 concludes Part II with a discussion of disassembly-line balancing extensions, additional issues, solution methodologies, probabilistic considerations, and areas for future research.

Chapter 22 introduces Part III where disassembly-line problems that are not directly related to balancing are reviewed.

Chapter 23 demonstrates application of disassembly considerations to the planning phase of a product, including the use of the

Chap. 9 mathematical formulae for use as indices in measuring the efficiency of future disassembly on multiple, competing product designs.

Chapter 24 provides a detailed background into facility design considerations and the field's associated definitions, then reviews traditional location and layout problems, and finally reuses the Chap. 9 formulae metrics for comparing different line designs.

In Chap. 25, classical sequencing and scheduling as used in production analysis is reviewed and applied using disassembly data.

Inventory theory is discussed in Chap. 26, and deterministic and stochastic disassembly-to-order systems are described.

Chapter 27 extends the disassembly-related inventory theory to encompass just-in-time.

In Chap. 28, a model is provided in which the revenue generated from disassembly is the primary consideration.

Chapter 29 provides a short summary of queueing theory as part of a disassembly application of the assembly-line concept of unbalancing lines.

Finally, the Appendix provides the acronyms that are referred to throughout the book.

CHAPTER 2

Assembly Lines

2.1 Introduction

In order to take advantage of similarities, as well as to appreciate the subtle differences between the assembly-line balancing problem and the DISASSEMBLY LINE BALANCING PROBLEM (DLBP), a footing in assembly-line theory is required. In this chapter, Sec. 2.2 provides a standard background in production and manufacturing and also introduces some common terms and issues. Section 2.3 provides a short history of the assembly line. Section 2.4 reviews specific assembly-line considerations. Section 2.5 covers a progression of some basic assembly-line models, while Sec. 2.6 summarizes the chapter.

2.2 Production Background

Traditional assembly has many inherent challenges (Defense Acquisition University, 2009a). For complex or large products it is not unusual for production and manufacturing to account for approximately 30 percent of the total life cycle costs. For these same types of products, production and manufacturing costs can accrue to 3 times the resources spent on all of the development. As a result, a great amount of money is spent during a relatively short time period. When problems occur in manufacturing, they affect the entire supply chain at a rapidly growing rate and cost. Also, problems encountered in this phase are often not able to be resolved quickly or inexpensively. Factors such as changing design requirements, late design releases, lack of production planning, and unstable financing can contribute to increased costs and production delays.

To mitigate these risks, production and manufacturing is sometimes integrated using a three-step process. This production and manufacturing integration process consists of influencing the design process, preparing for production, and executing the manufacturing plan. By *influencing the design process*, the company integrates manufacturing considerations during the development phase and gives early consideration to the procurement strategy, product design, and overall project risk management. In *preparing for production*, the customer's needs are distributed down to the manufacturing operations of the factory floor. The process of *executing the manufacturing plan* allows for a smooth transition to production with known risks. The result should be uniform, defect-free products, with consistent performance, and at the lowest cost.

Design is transformed into the finished product through a process made up of five basic elements: machinery, method, material, man, and measurement (5M). The 5Ms are used to ensure the design is capable of being produced into a uniform, defect-free, reproducible product. *Machinery* varies in type, particularly in terms of the volume of production. Robotics and automated machines differ from those requiring a dedicated operator. For example, manufacturing a satellite involves several highly trained personnel and results in the production of just one or two units per year. This is in contrast to the automated machinery used to manufacture, for example, millions of ball bearings each year. *Method* represents the way that raw materials are formed, shaped, and held together. Often there are many methods to accomplish the forming of a part. Both the materials and design requirements drive the selection of the methods. For example, boring a hole may be performed by drilling with a bit abrasive water jet, or industrial laser. The accuracy and precision needed usually dictates the method. *Material* includes all the raw materials that are needed to produce the parts/assemblies for the system and for the production equipment itself. Materials are often dictated by the functional requirements of the product. *Man* describes the utilization of personnel. *Measurement* (or metrics) is crucial to the overall production process; however, the systems needed to measure things—from the raw material to the testing of the final product—are often overlooked. Measurement systems provide for precision and accuracy in the manufacturing process. Examples of measurement include inspection, gauges, tolerances, and statistical process control.

Producibility is the relative ease of manufacturing an item. Producibility is a design accomplishment resulting from a coordinated effort by systems/design engineering and manufacturing/industrial engineering personnel. This coordinated effort creates a functional hardware design that optimizes ease and economy of component fabrication, assembly, inspection, test, and acceptance. A producible design should not sacrifice desired function, performance, or quality.

The five top-level design goals for a producible product include *design for ease of fabrication, design for ease of assembly, design for multi-use, design to minimize number of parts,* and *design to maximize number of common parts.* Designs that are complex to form or shape may result in excessive fabrication time and often cannot be fabricated at consistent quality levels. New processes may enhance the ease of fabrication of products. As assembly is the primary driver of labor costs, the use of computer-based design aids can assist in assessing how easy a subsystem is to assemble. Trade-offs exist between easing assembly and fabrication: The complexity of the fabrication may need to be increased in order to reduce the assembly time. In terms of multiuse, when possible the design should specify that the same part or assembly process be used several times, including use across product lines. Multiuse permits certain economies of scale and helps support the logistics aspects of the system. Another consideration is minimizing the number of parts. In addition to the cost of material and fabrication, each part has an overhead cost. By minimizing the number of parts, the overall cost of the product is reduced. The minimization of parts applies to both the number of total parts as well as to the number of different parts. Finally, another design goal is to optimize the number of common parts. For example, designers may be limited to items from an approved parts list. Common parts can potentially be produced by a large number of suppliers. Also, the use of common parts is preferred over more intricate or complex parts; tailored, complex parts are generally harder to machine, cast, or otherwise manufacture. The use of common parts also benefits after-sale support.

Even with adequate planning, production problems can still occur. Understanding common causes of production problems can help to mitigate their impact. The major causes of these are unstable rates and quantities, design instability, undue emphasis on schedule, inadequate configuration management, and inattention to environmental impact. Any time the rate of manufacturing and/or the overall quantity to be manufactured changes, production efficiency can suffer, leading to increased cost. The goal of an effectively designed production line is to produce the product at a cost-effective rate and quantity. Increases and decreases to the projected rate and quantity can result in under- (too little) or over- (too much) production capacity. Both situations can result in cost increases. Design instability is the second type of problem. When the design changes, the 5Ms often change, and the manufacturing planning is reset to zero. This impact is particularly significant during the fabrication of parts. It takes time to redevelop and/or replan for production when the design is changed. Often there may not be enough time or money to address the impacts of a design change adequately. Design instability during production may result from the design effort not being completed or from changes in the customer requirements. It is also recognized that holding firm to a contractual schedule or a product release date, when there is excessively high or unknown manufacturing risk, can be prone to undesirable results. Schedule delays are common when the manufacturing solutions require new materials, methods, machines, personnel, or measurement systems. It is desirable for the process to be event-driven with entry/exit criteria, and not driven solely by schedule or planned release dates. Related to design instability is the understanding of which design to produce. Inadequate configuration management can have as significant an impact as design changes themselves. If the wrong design is executed in manufacturing, it simply creates additional work in trying to correct the problem. Good configuration management is closely linked with good design control. Lastly, given the expectation of moderately frequent changes to environmental laws and their continued complexity, manufacturing efforts can be hampered by facility closure or heavy fines for violating environmental laws. Throughout the production process, managers need to be aware of the long-term impact that environmental laws have on the total life cycle costs for maintenance (if applicable), logistics, overhead, and disposal. As a result, design and process actions are often taken in order to eliminate environmental hazards.

2.3 History of the Assembly Line

The assembly line is often considered to be the most efficient way to produce many identical or near-identical products. One definition of an assembly line is that it is a manufacturing process where parts are sequentially added together by workers (including robots) until a final product is attained at the end of the line. The most familiar of these processes is a flow shop where the product being assembled moves at a constant rate on a paced line, possibly on a conveyor belt. While often attributed to the Ford Motor Company's assembly-line efforts from 1908 to 1915, the history of sequential mass production can be traced much further back.

The first evidence of an assembly line is attributed to the commissioning by Chinese Emperor Qin Shi Huangdi of the Terracotta Army in approximately 215 BC. The 8000 life-sized figures each had separate body parts which were manufactured at different locations/facilities (and individually labeled as such), collected, and then assembled into the final product.

In the 1500s, the Venetian Arsenal (an Italian shipyard, key to Venice's expansion at the time) employed almost 20,000 workers and reportedly delivered a ship each day. This was accomplished using standardized parts and an assembly-line type of manufacturing process.

Prior to the early 1800s, pulley-block (four-part, wooden ships' sail-rigging pulley mechanisms) production had been performed by a variety of contractors. This was expensive, manual-labor intensive, and provided inconsistent quality. Then Portsmouth Block Mills began producing pulley blocks for the British Royal Navy using what is considered to be one of the first linear, continuous assembly lines. These—now standardized—pulleys were mass produced using all-metal machine tools.

While not creating any of these assembly-line concepts, Eli Whitney is credited with introducing interchangeable parts and machine (as opposed to hand) tools and jigs (a fixture for holding parts in position during assembly), moving the skills from the worker to the machine, and making use of unskilled or semiskilled workers. This all took place in his manufacturing of firearms in the early 1800s.

The Chicago meatpacking industry of the 1860s is often considered the first assembly line in the United States (and, obviously, the first disassembly line). In these facilities, workers would be stationary at fixed workstations and the product would come to them. The line was an overhead trolley system that followed a given path at a relatively constant rate. Each worker on the line performed one task and repeated that task for each product that passed his workstation on the line.

The Industrial Revolution (1860s to 1890s) saw extensive improvements in mass production technologies and concepts throughout Western Europe and the United States (primarily New England). In 1901 the assembly-line concept was patented in the United States by Ransom Olds and used to mass produce automobiles in his Olds Motor Vehicle Company. Inspired by the idea of one person performing one job over and over again in the Chicago slaughterhouses, the assembly line was applied and refined by the Ford Motor Company, moving from a concept in 1908 to a finalized system by 1915 that increased production of a vehicle from taking almost 13 hours to just over 1.5 hours, used fewer workers, and increased worker safety (since assignment at one workstation prevented them from roaming about the facility). From this time, assembly-line use expanded rapidly; companies that did not adapt them could be expected to disappear.

2.4 Assembly-Line Balancing

Dating back to the mid-1950s and early-1960s, assembly-line-balancing research has enjoyed a great deal of significant study. While commonly treated as being from the field of manufacturing and production (more precisely, the field of scheduling), assembly lines are also regularly studied in the field of *facility layout*. Finch (2008) considers four types of facility layouts: process-oriented layouts, product-oriented layouts, cellular layouts, and service layouts.

Process-oriented layouts are characterized by functional departments and have a primary objective of locating the departments having the greatest interaction physically near to each other.

Product-oriented layouts consist of production or assembly lines, where products flow through a large number of workstations, each adding components and labor until a final product is achieved. While providing low costs at high volumes per unit, they typically have minimal flexibility.

Cellular layouts offer a compromise between the two, requiring that a family of products be produced in a *cell* containing all of the resources necessary to produce all of the products in the family.

Finally, *service layouts* may be similar in design to either a product-oriented layout (but having a service, rather than a product, flowing through the system) or a process-oriented layout. An example of a service layout would be the floor plan and product placement (e.g., placing high-frequency purchase items at the rear, requiring customer traffic through low-frequency areas) in a retail store.

Other authors consider additional configurations, such as *fixed-position layouts* and *layouts based on group technology* (Nahmias, 2009).

Assembly-line balancing is also considered a component of contemporary *lean manufacturing*. A concise definition and example tailored toward end users and line managers can be gleaned from Tapping (2003). Here, line balancing is described as a six-step process in order to evenly load all workers, define the order in which work elements should be performed, and define the number of line workers required. It is implied that the last step (i.e., evenly distribute work elements) in the process is performed either exhaustively or by trial and error. Either way, it is also implied that this step can be readily accomplished using either of these methods. In fact, for all but the simplest cases, this seemingly innocuous sequencing portion of the process is exceptionally difficult, necessitating many of the heuristics and metaheuristics that have been proposed for and used in assembly-line balancing.

The recent flexibility of plants and of plant layouts has moved assembly-line balancing more toward the dynamic-scheduling problem area. While various scheduling algorithms exist that address the challenge of sequencing, in general these attempt to minimize the number of workstations while often not addressing balance between workstations.

2.5 Line Modeling

When assembly-line balancing is considered as a component of the facilities planning and layout process in management science, the design of the line (i.e., placement and sequence of production tools and equipments) and of the product itself are considered in advance of the facility layout and subsequent production in order to maximize the *utilization rate* U (ratio of the sum of all task times to the total line time) and minimize the *balance delay* (lost resource utilization resulting from differences in processing times at each workstation). Finch (2008) provides an overview of these and other definitions, as well as the following five-step line-balancing algorithm and associated example. In this facilities-focused line-balancing method, the cycle time CT required to meet some given output requirement is calculated as part of the line's design; the greater the output rate, the shorter the cycle time must be; the shorter the cycle time, the greater the number of workstations required to complete all of the production. Another design value of interest is given as being the *production lead time* (also known as the *line time*), which is the amount of time a product spends in a production system in order to be completed. This is calculated as

$$\text{Production lead time} = \text{CT} \cdot \text{NWS} \tag{2.1}$$

where NWS is the number of workstations in the production line. This assembly-line balancing process is provided as

1. Identify the tasks and their associated task times and precedence relationships.
2. Determine the cycle time necessary to satisfy output requirements using

$$\text{CT} = \frac{\text{Production time available per day}}{\text{Units of output required per day}} \tag{2.2}$$

3. Determine the theoretical minimum number of workstations $\text{NWS}_{\text{lower}}$ using

$$\text{NWS}_{\text{lower}} = \left\lceil \frac{\text{TT}}{\text{CT}} \right\rceil \tag{2.3}$$

   where the total time TT is the sum of the task times.
4. Assign tasks to the workstations.
5. Evaluate the utilization of the line using

$$U = \frac{\text{TT}}{\text{CT} \cdot \text{NWS}} \tag{2.4}$$

The vague balancing process of step 4 is indicated in Finch (2008) by example where little more than observation and trial and error is used. The corresponding example consists of a product with 17 production tasks and having the goal of producing 120 units per day in a typical 8-hour workday with step 1 given in Table 2.1 and depicted in Fig. 2.1.

With 480 minutes (8 hours) per day available and desiring that 120 units be produced during that period, step 2 gives a cycle time of 4 minutes. Since the times in Table 2.1 sum to 19.5 minutes, step 3 calculates a theoretical number of workstations as being five. Though step 4 does not provide precise details, the sample's solution can be seen to be developed by working from left to right and top to bottom while starting from the first vertex and preserving precedence. That is, start at vertex one (precedence allowing), and add vertices to its right that are connected by arcs (to preserve precedence) until the next vertex—if selected—would exceed CT. At that point, attempt to continue in the

Task	Description	Time	Predecessors
1	Inspect top board	0.5	—
2	Sequence in jig	0.6	1
3	Insert top braces	0.3	2
4	Screw braces	3.6	3
5	Remove top	0.2	4
6	Inspect bottom boards	0.5	—
7	Insert left end boards	0.4	6
8	Drill, insert center bolt	1.25	7
9	Remove left end assy	0.4	8
10	Insert right end boards	0.4	6
11	Drill, insert center bolt	1.25	10
12	Remove right end assy	0.4	11
13	Attach center brace	2.3	9, 12
14	Attach hinges	3.6	5
15	Attach base	1.3	13, 14
16	Inspect	0.7	15
17	Package	1.8	16

TABLE 2.1 Assembly Tasks

current workstation by using vertices that are listed below the previous row of vertices, and then continuing to the right, selecting vertices sequentially (and repeated for the next row down when the next vertex selected would exceed CT). This process continues until a new workstation is required (at which point the process starts over again using vertices not yet assigned to any workstation) or until all tasks have been assigned.

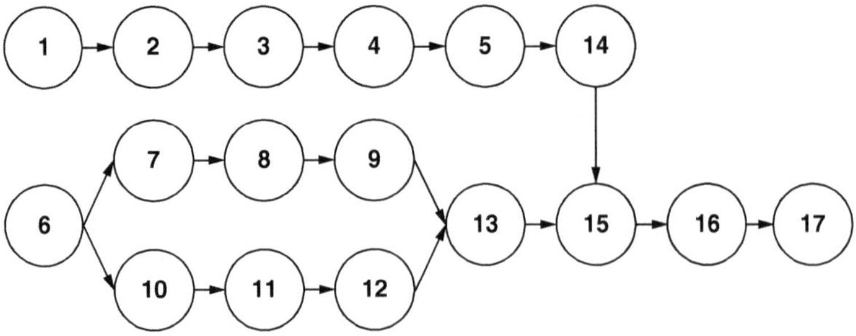

FIGURE 2.1 A 17-part assembly-line example precedence diagram.

Using this technique, the solution is given as 1, 2, 3, 6, 7, 8, and 10 in workstation one (3.95 minutes); 4 and 9 in workstation two (4.00 minutes); 5 and 14 in workstation three (3.80 minutes); 11, 12, and 13 in workstation four (3.95 minutes); and 15, 16, and 17 in workstation five (3.80 minutes). Overall utilization is seen to be 97.5 percent using Eq. (2.4).

The lean-manufacturing–based line-balancing process described by Tapping (2003) for evenly loading all workers, defining the sequence in which work elements are to be performed, and determining the number of workers required, consists of the following six steps:

1. Choose a process or work area (i.e., select the product, in the case of an assembly line).
2. Obtain individual task times for each work element (deterministic as given in this process and in its example).
3. Sum these task times to obtain the total time TT.
4. Create an *operator balance chart* (a visual depiction of each work station and its associated tasks, all of which are sized in proportion to time).
5. Determine the ideal number of line workers (the minimum number of workstations NWS_{lower}).
6. Evenly distribute work elements among workers as calculated in step 5.

Step 5 is calculated using

$$NWS_{lower} = \left\lceil \frac{TT}{takt} \right\rceil \tag{2.5}$$

where

$$takt = \frac{DT}{DQ} \tag{2.6}$$

with *takt* (the German word for "rhythm" or "beat") time being the rate needed to produce one part or product, DT representing the available daily production time, and DQ being the total daily quantity required [Eq. (2.5) is then the same as Eq. (2.3)]. The balancing portion of this algorithm (i.e., step 6) is not clarified, indicating that it is performed manually—either exhaustively or by trial and error—which is impractical for all but the smallest products.

While these algorithms—rooted in facilities planning, layout design, and lean manufacturing—do not reconcile the potentially daunting challenge in minimizing workstations and balancing workloads, they do provide an essential service in exposing line workers, manufacturing managers, and management science students to the importance of minimizing the size of the line and equalizing efforts at the workstations. The complexity of balancing is ultimately addressed in the fields of scheduling theory and of production analysis.

Some of the first researchers to provide a solution to this problem were Helgeson and Birnie (1961) whose *ranked positional weight technique* assigns weights to tasks based on the total time required by all subsequent tasks. Tasks are then assigned to workstations sequentially based on these weights, while not exceeding the cycle time or violating precedence constraints. As is the case with many assembly-line balancing processes, this algorithm works to minimize the number of workstations but with no process provided to equalize station times (i.e., balancing the workload is not considered). Also common to many assembly-line balancing techniques is its practical application of assisting the decision maker in adjusting cycle time in order to reduce the number of workstations or in reducing the cycle time while maintaining the number of workstations in order to reduce the total line time. Nahmias (2009) provides an example (Table 2.2) consisting of a product with 12 production tasks and having a cycle time of 15.

Since the installation times sum to 70 and with the cycle time of 15, Eq. (2.3) then gives a lower bound of five on the number of workstations. Figure 2.2 graphically depicts the precedence relationships.

The tasks' positional weights are then calculated by summing the time to complete these tasks and all subsequent tasks (e.g., task five's positional weight is seen to be $2 + 1 + 4 + 6 + 7 = 20$) and the tasks are sorted by decreasing positional weights (Table 2.3).

Task	Time	Predecessors
1	12	—
2	6	1
3	6	2
4	2	2
5	2	2
6	12	2
7	7	3, 4
8	5	7
9	1	5
10	4	9, 6
11	6	8, 10
12	7	11

Table 2.2 Assembly Tasks for Ranked Positional Weight Example

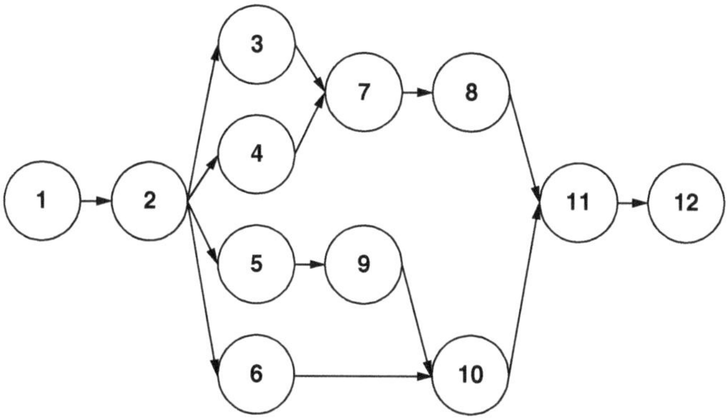

FIGURE 2.2 A 12-part assembly-line example precedence diagram.

Task	Positional weight	Decreasing positional weight	Positional weight order
1	70	70	1
2	58	58	2
3	31	31	3
4	27	29	6
5	20	27	4
6	29	25	7
7	25	20	5
8	18	18	8
9	18	18	9
10	17	17	10
11	13	13	11
12	7	7	12

TABLE 2.3 Calculated Ranked Positional Weights

Assigning tasks to workstations in positional weight order sees task 1 (task time of 12) assigned to workstation 1. Tasks 2 and 3 are considered next but are both too large (with each having times of 6, either would exceed the cycle time of 15); each of these tasks is subsequently assigned to workstation 2. Task 6 is considered next but is too large (task time of 12) and is therefore assigned to workstation 3. Task 4 (task time of 2) is considered and is seen to fit in workstation 1 but would result in violating task 2's precedence constraint. Task 4 does fit into workstation 2 and without violating task 6's precedence constraint. This process continues, resulting in the solution as shown in Table 2.4. The result

Workstation	1	2	3	4	5	6
Tasks	1	2, 3, 4	6, 5, 9	7, 8	10, 11	12
Station time	12	14	15	12	10	7
Idle time	3	1	0	3	5	8

TABLE 2.4 Initial Ranked Positional Weight Solution

is six workstations with a total idle time of 20 minutes and a total line time of 1 hour 30 minutes.

The ranked positional weight technique then allows the determination of the minimum cycle time that would result in, for example, five (versus six) workstations. Incrementing CT to 16 is attempted—and successful—giving the five-workstation solution shown in Table 2.5 (if CT = 16 were not successful, the cycle time would be incremented to 17, 18, and so on until a solution with five workstations was obtained). The result is five workstations with a total idle time of 10 minutes and a total line time of 1 hour 20 minutes.

Finally, the ranked positional weight technique can alternatively be used to determine the minimum cycle time while maintaining, for example, six workstations. Decrementing CT to 14 and then to 13 are both successful with CT = 13 giving the six-workstation solution shown in Table 2.6. The result is six workstations as originally found, but this time with a total idle time of only 8 minutes and a total line time of just 1 hour 18 minutes.

While assembly-line balancing primarily considers deterministic part-installation times, probabilistic times have been considered as well. More precisely, stochastic scheduling of lines can be broken into

Workstation	1	2	3	4	5
Tasks	1	2, 3, 4, 5	6, 9	7, 8, 10	11, 12
Station time	12	16	13	16	13
Idle time	4	0	3	0	3

TABLE 2.5 Reduced Workstation Ranked Positional Weight Solution

Workstation	1	2	3	4	5	6
Tasks	1	2, 3	6	4, 7, 5, 9	8, 10	11, 12
Station time	12	12	12	12	9	13
Idle time	1	1	1	1	4	0

TABLE 2.6 Reduced Cycle Time Ranked Positional Weight Solution

static (task times are stochastic, but tasks arrive for processing simultaneously) and dynamic (task arrival rates are stochastic). These are typically addressed using the well-established field of queueing theory (see Hillier and Lieberman, 2005, for a standardized overview of this field of study, and Pinedo, 2002, for a detailed summary of line-specific scheduling). It should be noted that the application of queueing theory necessitates a job shop (versus a flow shop) model of the line or the use of buffers (for storage) between workstations. It is for these reasons that lines with stochastic times are not further detailed in this chapter or to any significant level in the remainder of the book (see Secs. 21.4, 21.5, 25.4, 26.3, and 29.3 for exceptions to this).

2.6 Summary

Though not precisely the same as disassembly lines, the quantitative study of assembly lines provides a sound basis for the description and mathematical study of the disassembly line, as well as an appreciation of the computational complexity involved in finding ideal sequences.

While some algorithms that are made available to the practitioner are too simplistic for all but the smallest of lines, much of the breadth and depth of the half century of work in sequencing assembly lines can be leveraged in the study and analysis of disassembly lines. The wealth of research into assembly-line sequencing and balancing will not be reconstituted with disassembly terms in this text. Where applicable, these are appropriate techniques to be used, but areas where assembly-line algorithms and techniques can be directly applied or applied with slight modification to disassembly lines will not be discussed and are left to the reader. This text will primarily focus on significant extensions of some assembly line (and other) algorithms and new, disassembly-specific efforts.

Ironically, disassembly lines—which are only becoming a topic of rigorous research and analysis at the beginning of the twenty-first century—led to the creation of assembly lines at the beginning of the twentieth century: The disassembly lines of the Chicago slaughterhouses inspired Henry Ford to legitimize the assembly line, spurring manufacturing and the Industrial Revolution.

CHAPTER 3

Disassembly Lines

3.1 Introduction

Disassembly can be defined as the systematic separation of an assembly into its components, subassemblies or other groupings (Moore et al., 2001; Pan and Zeid, 2001). Disassembly is an important process in material and product recovery since it allows for the selective separation of desired parts and materials. In this chapter, Sec. 3.2 provides an overview of the state of the disassembly industry, while Sec. 3.3 provides a history of the disassembly line. Section 3.4 reviews disassembly-line considerations and terminology.

3.2 Overview

Popular press examples of why and how end-of-life processing is performed abound. Brown (2009) provides an overview of different techniques used by U.S. manufacturers, explaining how economics is a significant driver for these. With corporations looking to minimize costs, and with many of the obvious or easy ways to achieve this reaching practical limits, reducing inputs per unit of output is often seen as the next source of cost reduction. An environmental sense of urgency is listed as another powerful motivator, with the National Academy of Sciences, American Association for the Advancement of Science, and American Geophysical Union all described as having found compelling evidence of a link between human activity and climate concerns. Examples given in Brown's paper include a success story from IBM Corporation of how the company has created a $2 billion electronic equipment recycling business at 22 sites globally that is able to find uses for more than 99 percent of what is brought in.

Well-intended end-of-life solutions can have unintended consequences. According to a report by Walsh (2009), 80 percent of the U.S. population simply tosses obsolete electronic components directly into the trash, amounting to more than 350,000 cell phones and 130,000 computers every day. Unfortunately, of the electronic waste that is recycled, a portion of it ends up overseas, where valuable metals may be removed with little or no oversight or regulation, and using polluting mechanical removal methods such as chipping and scraping, burning away plastics and other nonmetals over open flames, or using hazardous chemicals such as acids in uncontrolled environments—often homes. An August 2009 Government Accountability Office report is described as listing the Environmental Protection Agency's (EPA) enforcement as "lacking" and described a formal investigation that found that 43 U.S. recycling firms were willing to ship broken computer monitors having lead-laden cathode-ray tubes to buyers in foreign countries without getting the required permissions from either the EPA or the receiving countries. However, recent successes include Apple's laptop design for easier recycling, Sony's pledge to only work with recyclers who do not export waste, Dell's take-back program and environmental audits of partners, and a general reduction, across the electronics-manufacturing industry, in the use of toxic metals (such as mercury) and small pieces of glass and aluminum. Still, the exporting of electronic waste remains a concern and is yet another reason why efficient, effective dismantling is essential, with a productive, balanced, disassembly line being a critical component of this.

Hindo (2006) claims that remanufacturing provides companies with another way to compete in the marketplace and then, as a case study uses Caterpillar Incorporated. With a general rule being given as 70 percent of the cost to build something new is in the materials and 30 percent is in the labor, it may often make sense to reuse the larger expense component of the two. Remanufacturing almost reverses these percentages (according to survey data of 270 manufacturing companies credited to Robert Lund and William Hauser) with materials' cost shrinking to 40 percent. In addition, remanufacturing gives an additional benefit of a lower total product cost, typically being half of that of a new product. (The fiscal performance of Caterpillar's remanufacturing operations led the company to create a separate division in 2005, with over $1 billion in sales and an expected growth rate of 15 percent annually.) Hindo lists Rochester Institute to Technology and Boston University reports as determining three requirements for a successful remanufacturing operation: design, logistics, and "low fashion." The product to be remanufactured must be designed for disassembly to allow it to be taken apart, remachined or refinished, and rebuilt for two, three, four, or more life cycles. The product also must have a reverse-logistics network in place to prevent the company from expending too many resources in determining where their end-of-life products are and how to collect them (examples given of how this has been done successfully include establishing robust dealer relationships and providing financial incentives to clients). Caterpillar relies on a process where, for example, a customer is charged full price for a remanufactured part that is sold usually for half price, and is refunded the difference only after the customer's end-of-life part is turned in and accessed as being able to be remanufactured. This process effectively incents customers to become partners. (A similar process—but one that is more regulatory than profit oriented in nature—is familiar to most consumers, where an end-of-life automobile battery has some monetary turn-in value.) Finally, it is claimed that fashion, popularity, and style cannot be remanufactured; when the level of desire a consumer has for a product extends beyond the product's performance, the product is only worth remanufacturing as a warranty replacement.

The aerospace industry has been on the forefront of remanufacturing for decades. Aircraft components are not only expensive due to high manufacturing costs (as a result of rigorous design, laborious test, precision machining, materials that are often rare or possess a high level of purity, and low rates of production), they also must be government–aviation-agency certified for use. Both of these elements easily justify remanufacturing (in addition to the obvious durability of the parts). However, only a fraction of the resources expended to develop a new aircraft is invested in allowing for disposal or recycling at its end of life (Ripley, 2008). A 2007 initiative by a European aircraft consortium has established an aircraft-dismantling facility in France. The Tarbes Advanced Recycling & Maintenance Aircraft Company (TARMAC AEROSAVE) expects to dismantle 300 surplus airliners every year and indicates that 85 percent of these aircrafts' parts can be reused, recycled, or recovered in a secure and environmentally friendly manner. Prior to this effort, most retired airliners were stored out in the open at desert locations, allowing toxic materials to enter the environment as the aircraft decayed. In addition, newer aircraft contain many computer components as well as exotic structural materials—such as titanium and carbon fiber as opposed to the almost exclusive use of aluminum and steel in older designs—further necessitating environmentally aware disposal.

Hazardous materials are of significant concern and are a factor in the processing of many end-of-life products. Different governments, government agencies, and organizations define hazardous materials in a variety of ways. Examples using the hazardous materials lists of the U.S. Army Material Command, Environmental Protection Agency, and Air Force Aeronautical Systems Center are demonstrated in Table 3.1 (Defense Acquisition University, 2010).

These materials can often be relatively expensive to dispose of as seen in Table 3.2, where the bulk rates paid by the U.S. Defense Logistics Agency are detailed (Defense Acquisition University, 2010). An individual company's costs can be expected to differ depending on quantities and services available.

Since all manufactured products may not be fully reusable or may have remaining materials that are not completely recyclable, some amount of waste can often be expected. Pearce (2009) lists three

Army Material Command "Big 3"	Environmental Protection Agency List of 17	Aeronautical Systems Center Priority List
Ozone depletors	Benzene	**Group I: highest priority**
Firefighting	Cadmium and compounds	Banned material: asbestos, PCBs, etc.
Refrigerants	Carbon tetrachloride	
Solvents	Chloroform	Last resort: hydrazine
Heavy-metal plating	Chromium and compounds	**Group II: reduce/ eliminate**
Cadmium	Cyanides	
Chromium	Dichloromethane	Chlorinated
VOC	Lead and compounds	Hydrocarbons
	Mercury and compounds	VOCs (MEK, benzene, etc.)
Stripping	MEK	Heavy metals and compounds
Cleaning	MIK	
Painting	Nickel and compounds	Sensitizers
	Tetrachloroethylene	**Group III: consider toxicity**
	Toluene	
	Tichchloroethyene	Advanced composite material components
	Xylenes	

Note: PCB = polychlorinated biphenyls,VOC = volatile organic compounds, MEK = methyl ethyl ketone, MIK = methyl isobutyl ketone

TABLE 3.1 Examples of Hazardous Materials Listings

Chemical	Cost to buy	Cost to dispose of
Freon (nonrefrigerant; cleaning)	$9.95	$4.19
Polyurethane paint	$8.52	$4.50
Zinc chromate primer	$8.61	$7.00–$15.00
Battery acid	$1.82	$3.40–$25.00
Trichloroethane (cleaning)	$4.83	$3.60
Hydraulic fluid	$4.00	$2.79–$9.00
Motor oil	$0.50–$1.00	$2.36
Acetone (cleaning)	$2.90	$4.07

TABLE 3.2 Hazardous Materials Cost Factor Chart (Costs are per Gallon)

ways to finance waste disposal: extended producer responsibility (the producer is responsible for disposal, either through assumption of disposal costs or via their own take-back program; e.g., the European Union's *Proposal for a Directive on Waste Electrical and Electronic Equipment*), advanced recovery fees (the producer collects a disposal fee at the time of purchase; e.g., the state of California's Electronic Waste Recycling Act), and end-of-life fees (consumers pay at the time of disposal; e.g., Hewlett-Packard Company's mail-back option).

Whether for environmental, remanufacturing, or regulatory reasons, efficient disassembly of an end-of-life product is critical.

3.3 History of the Disassembly Line

The disassembly line has existed for over a century, attributed to the Chicago slaughterhouses of the 19th century. The practices at these slaughterhouses are recognized as being the first production lines and they certainly provided the impetus that led to the first true assembly lines, attributed to the Ford Motor Company in the early 1900s. Assembly lines improved dramatically over the next 100 years, yielding improvements in speed, precision, flexibility, and product cost. Manufacturing improvements resulting from assembly lines motivated the creation or evolution of the fields of industrial engineering, human factors, operations research, and later, computer processor task scheduling. Assembly-line-specific research flourished as well, with many significant and fundamental technical problems being addressed in the literature of the 1960s. The success of assembly lines would ultimately lead to a need to remove the multitude of low-cost products (no longer able to be justifiably repaired when new replacements could be acquired at a lower cost) with the environmental movement of the 1960s and 1970s (and often dated to the 1962 publication of the Rachael Carson book *Silent Spring*). This, and the revolutionary idea that assembly problems could be most efficiently studied by considering them as reverse disassembly problems, led to the study of disassembly in the 1980s (Bourjault, 1984). The subsequent stream of disassembly research has steadily increased with continued motivation from government regulation, corporate profit, and consumer desires. By the 2000s the disassembly line had received its first disassembly-unique description as having features that separate it from treatment as an assembly line (Güngör and Gupta, 2002), and balancing the disassembly line was mathematically modeled (McGovern and Gupta, 2003a) and formally defined (McGovern and Gupta, 2006c). The success of assembly lines—inspired by the first disassembly line—has led full circle to the need for disassembly lines and associated disassembly-line-specific research (see Lambert and Gupta, 2005 for additional historical perspectives).

3.4 Disassembly-Specific Considerations

Güngör and Gupta (2002) were the first to formally propose the disassembly line. The following is a summary of some of the various considerations in a disassembly-line setting due to their observation of a disassembly line as being fraught with a variety of disassembly-unique complications.

3.4.1 Product Considerations

The number of different products disassembled on the same line is an important characteristic of a disassembly line. The line may deal with only one type of product whose original configuration is the same for every product received; for example, only personal computers (PCs) with certain specifications. The line may also be utilized to disassemble products belonging to the same family; that is, whose original configurations are only slightly different from each other. For example, there may be different models of PCs on the same line. The line may receive several types of products, partially disassembled products and subassemblies whose configurations are significantly or completely different from one another such as PCs, printers, digital cameras, motherboards, and monitors.

Changing the characteristics of the products complicates the disassembly operations on the line. Intuitively, balancing a disassembly line used for the disassembly of several types of products can become very complex; such a line may be balanced for a group of products, yet its status may become unbalanced when a new type of product is received.

3.4.2 Line Considerations

Various disassembly-line configurations may be possible. Some layouts will be inspired by assembly-line layouts. For example, layouts such as serial, parallel, circular, U-shaped, cellular, and two-sided lines (Finch, 2008) also find their way onto disassembly lines. Even so, new layout configurations may still be required to enable more efficient disassembly lines.

One of the most important considerations of a disassembly line is the line speed. Disassembly lines can be either paced or unpaced. Paced lines may be used to regulate the flow of parts on disassembly lines. However, if there is too much variability in the task times (which depends upon the conditions of the products being disassembled and the variety of the products processed on the line), it might be preferable to have an unpaced line. In such lines, each station works at its own pace and advances the product to the next workstation after it completes the assigned tasks. The advantages of paced lines over unpaced lines include less work in process, less space requirements, more predictable and higher throughput, and—if properly handled—less chance of bottlenecks. To take advantage of the positive aspects of a paced line, its speed can be dynamically modified throughout the entire disassembly process to minimize the negative effects of variability (including variability in demand).

3.4.3 Part Considerations

3.4.3.1 Quality of Incoming Products

When a disassembly system receives the returned products, their conditions are usually unknown: sometimes they are in good condition and relatively new, while at other times they are obsolete, nonfunctioning items. Therefore, there is a high level of uncertainty in the quality of the products and their constituent parts. There is a trade-off between the level of uncertainty and the efficiency of the disassembly line. When the level of uncertainty increases, the disassembly efficiency worsens, especially if the disassembly line has not been designed to cope with the uncertainty in product/part quality.

A part is considered to be *defective* if the part has a different structure and/or different operational specifications than its original structure and/or specifications. A part can become defective in various ways, such as suffering physical damage or being exposed to extreme operating environments. There are primarily two types of defects:

- **Physical defect** The geometric specifications (i.e., dimensions and shape) of the part are different from its original design. For example, the cathode-ray tube of a computer monitor can be broken, or the cover of a personal computer can be dented.
- **Functional defect** The part does not function as it was originally designed to. For example, the central processing unit (CPU) of a PC may be physically perfect but it may not function properly due to an internal problem.

There are three considerations with these two types of defects:

- **Removable defective parts (type A defect)** Some of the physically defective parts in the product can be disassembled even though they sustain some level of damage. This type of physical defect is referred to as a *type A defect*. When a part is physically defective and yet removable, it might take longer to remove that part. However, it does not affect the removal of other parts since the precedence relationship is not affected by the longer disassembly time.
- **Nonremovable defective parts (type B defect)** Sometimes a physically defective part in a product cannot be disassembled because it is either badly damaged (and thus gets stuck in

position) or its connection is of the type that must be physically broken. This is a *type B defect* and it has a tremendous impact on the efficiency of the disassembly line. For example, even if the nonremovable part does not have a demand, it may still precede other demanded parts. Also, a part with a type B defect may result in what is known as the *disappearing work-pieces phenomena* (Sec. 3.4.4.5).

- **Parts with functional defects** Assume that part k does not have a physical defect (type A or B), but rather sustains a functional defect. If this fact were known in advance, then the disassembler may or may not have an incentive to remove the part (unless, of course, it precedes other demanded parts). Therefore, in addition to concerns related to physical defects, the possibility of functional defects in parts on incoming products should also be incorporated in operating and balancing a disassembly line.

3.4.3.2 Quantity of Parts in Incoming Products

Another complication is related to the quantity of parts in the incoming products. Due to upgrading, downgrading, or cannibalization of the product during its useful life, the number of parts in the product may not match with its original configuration. The actual number of parts may be more or less than expected when the product is received. The following should be considered:

- The demanded part of the product may be absent or its quantity may be less than expected. This may require a provision in the calculation to ensure that the demand for the missing part is met. Another approach would be to carry out a preliminary evaluation of the product to make sure that all the parts of interest exist in the product before it is pushed down the disassembly line. This evaluation may be used to determine the route of the product on the disassembly line.
- The number of demanded parts may be more than expected. One example would be that of a PC disassembly line. Assume that there is a demand for the memory modules of old PCs. Further assume that PCs being disassembled were originally configured to have one memory module. However, if the owner of the PC upgraded the computer's memory using another module, two memory modules would be retrieved as a result of the disassembly of the PC, which is more than expected. This situation would affect the memory demand constraints and the workstation time (since, using this example, the disassembly of two memory modules could be expected to take longer than the removal of just one).

3.4.4 Operational Considerations

3.4.4.1 Variability of Disassembly Task Times

Similar to the assembly-line case, the disassembly task times may vary depending on several factors that are related to the condition of the product and the state of the disassembly workstation (or worker). Task times may be considered as deterministic, stochastic, or dynamic (dynamic task times are possible due to learning effects, which allow for a systematic reduction in disassembly times). For example, in their case-based reasoning approach, Zeid et al. (1997) demonstrated that the disassembly of products can be assisted by the reuse of solutions to similar problems encountered in the past, thus reducing disassembly times.

3.4.4.2 Early Leaving Work-Pieces

If one or more, but not all tasks of a work-piece that has been assigned to the current workstation cannot be completed due to some defect, the work-piece could leave the workstation early. This phenomenon is termed as the *early leaving work-piece*.

Due to an early leaving work-piece, the workstation experiences an unscheduled idle time. Note that the cost of unscheduled idle time is high because, even though the demand for parts associated with failed tasks is unfulfilled, the disassembly cost of failed tasks has been incurred. For example, assume that disassembly of the hard disk and the media drive of a PC is carried out at workstation 1. After the removal of the hard disk, if the holding screws of the media drive are stuck or damaged, it may not be possible to remove the media drive. Therefore, the work-piece (the PC being disassembled) can leave workstation 1 early. When this happens, workstation 1 remains idle for the duration of the normal disassembly time of media drive.

3.4.4.3 Self-Skipping Work-Pieces

If all of the tasks of a work-piece that have been assigned to the current workstation are not performed due to some defect they posses and/or precedence relationships, the work-piece leaves the workstation early without being worked on. This is known as a *self-skipping work-piece*.

Consider the example of the PC as given earlier. Assume that the disassembly of the hard disk must be carried out before the disassembly of the media drive. If the hard disk of the PC cannot be removed because its holding screws are damaged, neither of the tasks assigned to workstation 1 (disassembly of the hard disk and of the media drive) can be performed. Therefore, the work-piece can leave the workstation without being worked on; in other words, it self-skips workstation 1.

3.4.4.4 Skipping Work-Pieces

At workstation j, if one or more defective tasks of a work-piece directly or indirectly precede *all* the tasks of workstation $j + 1$ (i.e., the workstation immediately succeeding workstation j), the work-piece skips workstation $j + 1$ and moves on to workstation $j + 2$. This is known as a *skipping work-piece.*

A work-piece can skip one workstation, two workstations, and so on. In addition to unscheduled idle time, both the self-skipping work-piece and the skipping work-piece contribute added complexities in material handling (e.g., how to transfer the out-of-turn work-piece to the downstream workstation) and the status of the downstream workstation (e.g., it may be busy working on other work-pieces and therefore may require some sort of buffer allocation procedure to hold the skipped work-piece until the machine becomes available). Using the PC example, assume that the disassembly of the media drive precedes the removal of the memory modules and that these tasks are scheduled to be carried out at workstation 2. If the media drive cannot be removed at workstation 1 due to its damaged holding screws, the work-piece leaves workstation 1 early and may need to skip workstation 2 because neither the memory modules nor the sound card can be removed from the work-piece due to precedence constraints. In this case, workstation 2 remains idle for the duration of its originally scheduled tasks; the removal of the memory modules and the sound card in this example.

3.4.4.5 Disappearing Work-Pieces

If a defective task disables the completion of all the remaining tasks on a work-piece, the work-piece may simply be taken off the disassembly line before it reaches any downstream workstation. In other words, the work-piece effectively disappears, which is referred to as a *disappearing work-piece.*

A disappearing work-piece may result in starvation of subsequent workstations leading to a higher overall idle time, highly undesirable in a disassembly line. It is, in a way, a special case of a skipping work-piece where the work-piece skips all succeeding workstations. The consequences of the disappearing work-piece are similar to those of the skipping work-piece, but to a greater extent. For example, assume that the removal of the PC cover is carried out at workstation 1 in addition to the removal of hard disk and the media drive. If the top cover cannot be removed due to damage, the PC can be taken off the line since none of the parts inside the PC can be reached. In a way, it disappears and the idle times propagate throughout the remaining workstations.

3.4.4.6 Revisiting Work-Pieces

A work-piece currently at workstation j may revisit a preceding workstation $(j - p)$, where $(j - p) \geq 1$ and $p \geq 1$ and integer, to perform task x if the completion of current task y enables work on task x which was originally assigned to workstation $(j - p)$ and was, however, disabled due to the failure of another preceding task. These are termed *revisiting work-pieces.*

A revisiting work-piece results in the overloading one of the previous workstations. As a consequence, it may lead to complications in the material-handling system because of reverse flow, decoupling from the revisited workstation, or introduction of a buffer to hold the revisiting work-piece until the workstation becomes available. This would obviously have a financial impact as well.

As an example of a revisiting work-piece, assume that at workstation 1 the hard disk and the media drive are removed, at workstation 2 the memory modules and the sound card are removed, and at workstation 3 the power supply of the PC is removed. Also assume that memory modules can be removed after the removal of either the media drive or the power supply. Further assume that disassembly of the media drive precedes the removal of the sound card. If the media drive cannot be removed due to preexisting damage, the PC skips workstation 2 and goes to workstation 3 where the power supply is removed. Since the power supply has been taken out of the PC, the memory modules can be removed. Thus, the PC is sent back to the workstation 2 for disassembly of the memory modules. Complicated revisiting work-pieces that could make the material handling and flow control more difficult may simply exit the line.

3.4.4.7 Exploding Work-Pieces

A work-piece may split into two or more work-pieces (subassemblies) as it moves on the disassembly line. Each of these subassemblies acts as an individual work-piece on the disassembly line. This phenomenon is known as *exploding work-pieces*.

An exploding work-piece complicates the flow mechanism of the disassembly line; however, it can be planned for in advance since it is known which part's removal would result in the exploding work-piece. In a disassembly line, a complication occurs when the part that would normally result in an exploding work-piece cannot be removed due to some defect.

3.4.5 Demand Considerations

Demand is one of the most crucial issues in disassembly-line design and optimization since it is desirable to maximize the utilization of the disassembly line while meeting the demand for parts in associated planning periods. In disassembly, the following demand scenarios are possible: demand for one part only, demand for multiple parts, and demand for all parts.

Parts with physical or functional defects may influence the performance of the disassembly line. If part x is not demanded and it directly or indirectly precedes a demanded part y, then part x must be disassembled before the removal of part y. (Note that when part a precedes part b, which precedes part c, then part a is said to be a *direct precedent* of part b and an *indirect precedent* of part c.) Removal of part x may require additional time due to the presence of a defect. This may cause the processing time to exceed the cycle time, which could result in the starvation of subsequent workstations and blockage at these previous workstations. Placing buffers at the workstations may be necessary to avoid starvation, blockage, and incomplete disassembly.

If part x has a demand, then the various types of demand affect the number of products to be disassembled, and eventually the disassembly line's balance. There are three types of demand.

In the first type, the demand source may accept part x as is. This is possible, for example, when part x is demanded for its material content (in which case the defect in a part may not be important).

In the second type of demand, the demand source may not accept parts with any type of defect. This occurs when a part is used as is (e.g., in remanufacturing or in the repair of other products). Thus, if a demanded part has a defect, its disassembly does not satisfy the requirement. Since an objective of a disassembly line may be to meet the demand, the objective function and demand constraints must cope with this type of complication. If the part has this second type of demand and it does not directly or indirectly precede another demanded part, the disassembly of the part is redundant. This can have an effect on the efficiency of the disassembly line since the workstation responsible for removing the defective part will remain idle for an extended time.

In the third type of demand, the demand source may accept certain defective parts depending on the seriousness of the defect. This type of demand may be received from a refurbishing environment where the parts undergo several correction processes (e.g., cleaning and repair) before reuse. This type of demand introduces a further complication: the requirement for some sort of tracing mechanism to identify the type of defect and the associated demand constraint(s).

3.4.6 Assignment Considerations

In addition to precedence relationships, several other restrictions limit the assignment of tasks to workstations. While similar restrictions are also present in assembly lines, the following are some restrictions related specifically to disassembly.

Certain tasks must be grouped and assigned to a specific workstation (e.g., if the removed parts are to be sorted and packaged together for shipment to the demand source). Thus, assigning the disassembly of these components (parts or joining elements) to the same workstation minimizes the distance that the components travel in the disassembly system. Moreover, the tasks requiring similar operating conditions (e.g., temperature and lighting) can be restricted to certain workstations as well.

The availability of special machining and tooling at certain workstations or at a limited number of workstations may necessitate the assignment of certain tasks to these workstations. For example, disassembly of hazardous parts may be assigned to highly specialized workstations in order to minimize the possibility of contamination of the rest of the system. Similarly, the skills of human workers can be a factor in the task-assignment restrictions.

Tasks may be assigned to workstations so that the amount of repositioning of the work-pieces on the disassembly line (e.g., the product orientation changes) is minimized. Similarly, tasks may need to be assigned in order to minimize the number of tool changes in the disassembly process.

3.4.7 Other Considerations

There are additional uncertainty factors associated with the reliability of the workstations themselves. Some parts may cause pollution or nuisance due to the nature of their contents (e.g., oil and fuel), which may increase the chance of breakdowns or workstation downtime. Furthermore, hazardous parts may require special handling, which can also influence the utilization of the workstations.

CHAPTER 4

Related Research

4.1 Introduction

This chapter presents a survey of the literature related to the issues considered in this text. As this book focuses on the processing of end-of-life products on a disassembly line, in obtaining perspective on this, the chapter is organized as follows: Section 4.2 presents an overview of the literature addressing environmental concerns and environmentally conscious manufacturing. Section 4.3 reviews work that has been done on the issues associated with assembly-line balancing as well as with manufacturing systems in general. The next two sections provide a background on a great deal of the quantitative efforts that are covered in this book, including the mathematical aspects of disassembly and coverage of optimization concepts and methodologies. Specifically, Sec. 4.4 discusses the variety of approaches that have been taken to date to solve disassembly-related problems, and Sec. 4.5 focuses on the technical aspects needed to conduct the efforts in this book: from discrete mathematics to complexity theory to the design and history of various algorithmic approaches.

4.2 Environmentally Conscious Manufacturing and Product Recovery

With the U.S. Environmental Protection Agency estimating that 4 million tons of electronic waste was dumped in landfills in 2000, the U.S. National Safety Council estimating that by 2007 approximately 500 million personal computers will have been scrapped, while less than 10 percent of electronic waste was recycled in 2000, environmentally conscious manufacturing and product recovery are relevant and timely. Gualtieri (2005) provides these and other facts in a concise overview of end-of-life issues and environmental consequences. Included in this is the observation that Japan requires personal computer manufacturers to accept machines back at their end-of-life and Germany requires equipment manufacturers to recycle their products when returned by consumers. This is now the case for all European Union (EU) members with the implementation of the

EU's *Proposal for a Directive on Waste Electrical and Electronic Equipment* (WEEE). Once these products are collected, they fall under the area of product recovery.

Many papers have discussed the different aspects of product recovery. Güngör and Gupta (1999c) provide a especially thorough review of environmentally conscious manufacturing and product recovery. This significant paper (thoroughly updated by Ilgin and Gupta, 2010) is a survey of the available literature on the newly emerged research area of *environmentally conscious manufacturing and product recovery* (ECMPRO). ECMPRO is enforced primarily by governmental regulations and by customer perspective on environmental issues due to the perceived escalating deterioration of the environment (e.g., diminishing raw material resources, overflowing waste sites, and increasing levels of pollution). ECMPRO involves integrating environmental thinking into new-product development, including design, material selection, manufacturing processes, and delivery of the product to the consumers, plus its end-of-life management. Diverse problems include: product life cycle, disassembly, material recovery, remanufacturing, and pollution prevention. This survey presents the development of research in ECMPRO and provides a state-of-the-art survey of published work. Brennan et al. (1994) provide a similarly detailed study involving planning issues in the assembly and disassembly environments.

Toffel (2002) focuses on academic research performed in the area of product recovery. This paper primarily considers the management literature and proposes some areas for further research. Details include coverage not just of WEEE, but of the EU's *Directive on End-of-Life Vehicles* as well, which requires automakers to reuse or recycle 85 percent of an automobile's weight in 2006 and 95 percent by 2015, along with setting aside funds to process cars built before 2002 (cars built after this date include an end-of-life processing tax).

Sodhi and Reimer (2001) present a variety of mathematical models for addressing the reverse supply chain for electronic components. Useful to any reverse supply chain researcher, Sodhi and Reimer provide a detailed breakdown of the typical material composition of electronic scrap as well as a table with 1999 values for 11 of the principle materials having some economic value that are found in electronic components. The authors conclude that 1 ton of electronic waste could yield up to $9193.46 if sold at 1999 market prices, providing economic impetus for further research.

While many of the papers in the literature consider electronic components at their end-of-life, Schultmann and Rentz (2001) provide a novel case study of the environmentally conscious disassembly and recycling of buildings. Along with the case study, they were also able to show that their methodology had both environmental and economic benefits. In a separate but related paper, Bader (2004) considers the growth in *construction and demolition* debris due to an increase in building and provides an overview of the specialized recycling equipment (e.g., one contemporary large shredder is equipped with a 20,000-lb rotor holding forty-two 95-lb hammers and is capable of generating 250,000 ft-lb of torque) and processes being used, which are significantly different than those in, for example, electronic product recycling.

Kotera and Sato (1997) propose an automated disassembly and recycling process for home electronic appliances (refrigerators, central air conditioners, window air conditioners, washing machines, and televisions). This is another paper that provided detailed case-study data about the products considered and their material breakdown, including valuable parts and materials, and hazardous materials.

4.3 Assembly-Line Balancing and Manufacturing Systems

A major part of manufacturing and assembly operations, the *assembly line* is a production line where material moves continuously at a uniform rate through a sequence of workstations where assembly work is performed. With research papers going back to the 1950s, the ASSEMBLY LINE BALANCING problem is well defined and fairly well understood. While having significant differences from assembly-line balancing, the recent development of the DISASSEMBLY LINE BALANCING PROBLEM (DLBP) requires that related problems be fully investigated and understood in order to better define the DLBP and to obtain guidance in the search for appropriate methodologies to solve it.

Elsayed and Boucher (1994) include an overview of assembly-line balancing in their text. They provide a description of the problem, definitions of the *simple assembly-line balancing type I* (SALB-I) and *simple assembly-line balancing type II* (SALB-II) models, and heuristics including: Kilbridge-Wester (assign numbers to each task depending on number of predecessors; low-number tasks are assigned to workstations first), Moodie-Young (two phase; first assign tasks with no predecessors to workstations in order of declining time value and repeat for all tasks as more predecessors are completed, then redistribute idle time equally to all workstations), Helgenson-Birnie (assign tasks to workstations based on highest positional weight, i.e., time of the longest path from the task through the remainder of the precedence network), *immediate update first fit* (use one of eight score functions from the literature, place the task with the highest score at the first workstation where capacity and precedence constraints will not be violated, update tasks, and repeat), and *rank-and-assign* (similar to immediate update first fit but tasks are placed solely on score).

Pinedo (2002) provides a thorough overview of scheduling theory, algorithms, and systems, with the first third of the text dealing primarily with deterministic problems in scheduling and covering both the theoretical and applied aspects. Scholl (1995) visits the balancing and sequencing of assembly lines.

Gutjahr and Nemhauser (1964) were the first to describe a solution to the ASSEMBLY LINE BALANCING problem with an algorithm developed to minimize the delay times at each workstation. The heuristic accounts for precedence relationships and seeks to find the shortest path through a network with the resulting technique being similar to *dynamic programming*.

Erel and Gokcen (1999) develop a modified version of Gutjahr and Nemhauser's line-balancing problem algorithm by allowing for *mixed-model* lines (assembly lines used to assemble different models of the same product) by allowing multiple state times and then, during construction of the solution network, considering all state times before categorizing a given set of completed tasks to a workstation.

Suresh et al. (1996) were the first to present a genetic algorithm to provide a near-optimal solution to the ASSEMBLY LINE BALANCING problem. The genetic algorithm was designed to minimize idle times and minimize the probability of line stoppage; that is, minimize the probability that the station time exceeds the cycle time.

Silverman and Carter (2003) compare the relative effectiveness of two station-loading rules (the "probability rule" and the "percent rule") in assembly-line design when considering high and low task-time variability, high and low costs when the cycle time is exceeded, and using task instances of $n = 45$ and $n = 70$. They also consider two methods of introducing slack time for the case where the cycle time is exceeded.

Tabu search is used in balancing assembly lines in Lapierre et al. (2006) using SALB-I with instances from the literature (*Arcus 1* and *2*) and a case study from industry (a home appliance manufacturer with a product having 162 tasks and 264 precedence constraints).

Hackman et al. (1989) propose a *branch-and-bound* heuristic for the SALB-I problem. Ponnambalam et al. (1999) compare line-balancing heuristics with a quantitative evaluation of six assembly-line-balancing techniques where the evaluation criteria is the number of workstations, line efficiency, smoothness index, and runtime using 20 instances.

Thilakawardana et al. (2003a) propose a line-balancing algorithm based on *forward-loading theory* (a stochastic version of the Hoffman precedence matrix approach). Based on this work, a front-loading genetic algorithm for SALB-I was then developed (Thilakawardana et al., 2003b).

Simulated annealing is the approach to SALB-I used by Hong and Cho (1997). Case studies for the process include an electrical relay and an automobile alternator. McMullen and Frazier (1998) use simulated annealing for solving a multiobjective, parallel workstation version of SALB-I.

Lapierre and Ruiz (2004) present a case study of an appliance instance with 248 tasks, 400 precedence constraints, and 4 task attributes for balancing a two-sided assembly line that also had two heights. The case study is programmed in database software and solved using an enhanced priority-based heuristic. The study is useful as both an actual application and in demonstrating how the use of logic and randomness in the algorithm enables the heuristic to find good solutions.

While the work in this book primarily makes use of deterministic part removal times, on an actual disassembly (or assembly) line, these times could be stochastic. However, the work-sampling operations required to define the standard deviation of stochastic task times on an assembly line can be time consuming (Tiacci et al., 2003). Tiacci et al. (2003) study the significance of the task times' standard deviation on the performance of the system model in order to determine if it is worth collecting enough data to calculate the standard deviation. They then propose a methodology for performing assembly-line balancing with a reduced set of data. This is done using the means of the stochastic task times as inputs alone, that is, without their standard deviation. An application of the methodology to a case study of a trailer assembly line having 53 tasks and a cycle time of 240 minutes is then considered.

Bautista and Pereira (2002) use ant colony optimization to solve the ASSEMBLY LINE BALANCING problem and compared these to their hill-climbing heuristics. Their research uses a greedy heuristic, similar to immediate update first fit, for the ASSEMBLY LINE BALANCING problem to be used in comparison to a developed ant algorithm. It also demonstrates two novel hill-climbing local search techniques. McMullen and Tarasewich (2003) also use ant colony optimization to solve the ASSEMBLY LINE BALANCING problem. The paper gives an excellent and concise explanation of the ant colony optimization metaheuristic. The authors interestingly make use of parallel workstations, stochastic task durations, and mixed-models; however, the ant colony optimization metaheuristic is given an extremely long time to solve the instances, including 50 hours each for instance sizes of $n = 29$, $n = 40$, $n = 45$, and $n = 74$.

Moore and Gupta (1996) provide an in-depth survey of the state-of-the-art and the work to date on the use of *Petri nets* (a graphical and mathematical modeling technique developed in the 1960s to model concurrent computer system operations) for representing flexible and automated manufacturing systems. Areas covered in the paper include: flow shops, automatic transfer lines, job shops, flexible manufacturing systems, and assembly, as well as qualitative and quantitative analysis. Bukchin and Tzur (2000) consider the design of flexible assembly lines for the purpose of minimizing equipment costs. They formulate the problem of selecting equipment and assigning tasks to workstations with the objective of minimizing total equipment costs, given some fixed cycle time as an integer programming model. They then generate lower bounds on this model and apply branch-and-bound before developing their own problem-specific heuristic algorithm.

Boysen et al. (2008) provide guidance on selection of assembly-line balancing models for different situations. They provide a proposed notation to identify the type of line with characteristics that include number of models (single, mixed, or multiple), line control (paced, unpaced asynchronous, or unpaced synchronous), frequency (new installation or reconfiguration), automation (manual or automated), and industry (automobile or all other).

4.4 Disassembly and Remanufacturing

The disassembly-line problems under consideration in this book require a study of the available literature on remanufacturing in general and on disassembly specifically.

Lambert (2003) provides a survey of the available literature on *disassembly sequencing*, that is, the search for all possible disassembly sequences and selection of the optimum solution out of these (commonly used in assembly analysis, maintenance, and in-processing of end-of-life products). The author relates that in each application a slightly different approach is used including level of detail, degrees of freedom, and the role of uncertainty. The author covers the current state-of-the art at both a high-level (part level) and low-level (process level, including logistics) of detail.

Lee et al. (2001) present a survey of the available literature on planning and scheduling problems in disassembly systems. *Disassembly planning* includes product representation, disassembly sequencing with disassembly level and end-of-life options, and related product design/redesign issues. *Disassembly scheduling* is the problem of determining the order quantity of the used product to fulfill the demand of disassembled parts and subassemblies. This paper presents a review of the state-of-the-art and suggests further research directions including (a) algorithms to generate all feasible disassembly sequences, (b) methods to treat uncertainty in estimating recovery values, product states, and disassembly operations, and (c) disassembly scheduling models to consider system capacity and uncertainty. In particular, the integration of disassembly planning and scheduling problems is suggested as a fundamental research direction to develop an integrated framework to solve operational problems in disassembly systems.

Gupta and Taleb (1994) have written what is considered the founding document in the science of disassembly planning and scheduling. It presents an algorithm for scheduling the disassembly of a product made up of subassemblies and parts. Although it is similar to *material requirements planning* (MRP) in concept and in the structure of the information, it is not the reverse of MRP primarily because there exist multiple demand sources in the case of disassembly, as opposed to a single demand source in the case of assembly. Taleb et al. (1997) then present a follow-up paper where the issue of parts and materials commonality when scheduling disassembly is addressed. Commonality introduces additional complexity by creating alternative procurement sources for the common component items. Since a certain part can now have more than one parent, it can be procured by dismantling any of its parents and, therefore, there is a decision to be made. It is also desired to determine the procurement strategy that fulfills all of the demand while minimizing the ordering of the root item. It is preferred to keep the demand and inventory levels in a certain proportion at the beginning of each period so the net requirements will be in the chosen proportion when proceeding from one level to the next in the product structure. Therefore, the modules are no longer independent and cannot be processed separately for all time periods but rather are processed in parallel, proceeding from the highest level to the root level. Since the on-hand inventory account is now shared by all the items that represent the same part, the process cannot go into the next period without considering the inventories of all other common items for that period, so an inventory update has to be performed at every time period for all subassemblies at a certain level before moving to the next period. Also, since division is used in proceeding from one level to the next, the resulting fractions are rounded up to ensure the quantity needed of that item (in assembly this problem is not present due to the use of multiplication of integers).

Tang et al. (2001b) focus on demanufacturing from a systems perspective since the allocation of resources in a demanufacturing system is seen as being so critical to efficient operations. The authors presented a disassembly planning and demanufacturing scheduling method for an integrated, flexible demanufacturing system having the ability to account for the high uncertainty expected to be found in demanufactured products' structures and components and the fact that the disassembly termination goals are not necessarily fixed. The system is modeled using workstation, product, and scheduling Petri nets. The algorithms provide for scheduling and disassembly planning. The model and algorithms are demonstrated using personal computers as the product to be demanufactured. Tang et al. (2001a) also present an algorithm to facilitate disassembly-line design and optimization.

Moore et al. (2001) present a Petri-net-based approach to automatically generate *disassembly process plans* from the disassembly precedence matrix (see Sec. 8.8), which is automatically generated from computer-aided design (CAD) drawings of the product using a specialized algorithm. Zussman and Zhou (1999, 2000) and develop disassembly Petri nets for the design and implementation of adaptive disassembly systems. Zha and Lim (2000) integrate expert systems and ordinary Petri nets to develop an expert Petri net model for disassembly planning. Rai et al. (2002) develop a Petri-net-based heuristic approach for disassembly sequence generation. Kumar et al. (2003) and Singh et al. (2003) deal with the complexity of Petri nets by proposing a stochastic Petri net that consists of a knowledge base, graphic characteristics, and artificial intelligence. Gao et al. (2004) propose a fuzzy reasoning Petri net to deal with the uncertainty associated with the disassembly process. Tang et al. (2006) consider the uncertainty associated with human factors in disassembly planning and propose a fuzzy-attributed Petri net to deal with this uncertainty. Grochowski and Tang (2009) integrate a disassembly Petri net and a hybrid Bayesian network to develop an expert system capable of determining the optimal disassembly action without human assistance.

Tiwari et al. (2001) develop a cost-based methodology to determine the disassembly process for a particular product to maximize profits by using a Petri net model. In the model, one component at a time is removed from a product in a downward cascading Petri net graph. Each removed component can be sent to one of five buffers: servicing (restored for reuse), disassembly (nondestructive extraction of desired and hazardous components), dismantling (destructive materials separation prior to recycling), recycling (reuse of materials), or dumping (costs only, no revenues generated). Using product data, five indices (servicing, disassembly, dismantling, recycling, and dumping) for each component are calculated. The largest positive index for each component is selected and then the Petri net representation is updated appropriately (i.e., with the selection of the single destination for each removed component) providing the disassembly sequence for a given product.

Kongar and Gupta (2002a) present a genetic algorithm to provide a near-optimal solution to the multicriteria *disassembly sequence plan* problem. Solution fitness was calculated based on three criteria: part demand, directional change, and disassembly method change. The top four parents were kept along with all of the children to make up the next generation. The process was repeated for 100 generations or less than a 1.0005 average fitness difference between generations, whichever came first.

Das and Naik (2002) propose a *disassembly bill of materials* to be provided by manufacturers to enable the recycler to determine an efficient disassembly sequence for a product at its end of life. The authors also provide a model, with an example, for calculating the percent return after disassembly to determine if disassembly is worth performing on a particular item (when the percent return is negative, disassembly is a money-losing proposition; when positive, the ratio gives the disassembly-return-on-investment). The most valuable portions of this paper may be insights into the disassembly process provided by the authors as a result of their interaction with disassembly workers, along with the well-thought-out and detailed classification and quantification of disassembly processes and their associated levels of effort.

Güngör and Gupta (2001b) present a technique for automatically generating a disassembly sequence plan from a CAD file. The disassembly precedence matrix contains all zeros except where the row-part precedes removal of the column-part (in that case the direction of blocking is entered in the element; if blocked in more than one direction, 1 is entered). Various penalty factors are assigned (in this case: tool change, direction change, and hazardous material content) along with two growth factors (the tree analysis depth size and the number of nodes to be selected at each analysis depth).

Güngör and Gupta (2001a) present an algorithm for solving the DISASSEMBLY LINE BALANCING PROBLEM in the presence of failures with the goal of assigning tasks to workstations in a way that probabilistically minimizes the cost of defective parts. This is done by first finding all task assignments that minimize the number of workstations for a given cycle time, then comparing them to find the set with the lowest likelihood of complications due to failure. Initially, the incomplete state network is generated from the disassembly precedence matrix and the cycle time, and then the state network is generated in a fashion similar to that of Gutjahr and Nemhauser (1964). A weighted state network is developed with edges weighted by the idle times. The shortest directed path of the network is determined using Dijkstra's shortest path algorithm. The cost of each of the shortest directed paths is calculated based on known costs and given probabilities of failure (penalizing for high costs and/or probabilities associated with leaving the workstation early, skipping the next workstation, self-skipping the workstation, and/or disappearing) with the lowest-cost, shortest, directed path being selected as the near-optimum solution.

Güngör and Gupta (2002) provide a first description of disassembly lines and the unique complications associated with them (including AND, OR, and complex AND/OR precedence relationships). An algorithm is also developed that solves the general, paced DISASSEMBLY LINE BALANCING PROBLEM. The algorithm works as follows. Once the disassembly precedence matrix is generated, candidate tasks (initially the components without any predecessors) are ranked (by time-to-remove, demand, number of other parts blocked, hazardous material content, and direction the work-piece faces) and then put into the first workstation. This is performed until all of that workstation's cycle time is filled, and then the process is repeated on a second workstation for the remaining components, and so on. This process continues until all the components have been removed.

Güngör and Gupta (1998) develop an algorithm that, after an initial disassembly sequence plan is generated, modifies the plan in real-time to deal with product uncertainty (defective, modified, or hazardous parts) that may render the initial plan infeasible or require destructive disassembly. A four-part flashlight is used as a model product. The disassembly sequence plan is generated based on a prior paper's technique and then disassembly is carried out using the new plan until an unexpected situation arises. In that situation, the sequence is modified using a reordering algorithm and the process continues. In the example given, the optimum disassembly sequence plan of ⟨cap, battery, receptacle, handle⟩ becomes ⟨handle, battery, receptacle, cap⟩ due to the damaged cap-to-receptacle joint that disables the unscrewing of one from the other.

Güngör and Gupta (1997) use time-to-disassemble (the sum of the individual component removal times with penalties assigned for changes in removal direction and changes in removal tool type) as the performance evaluation criteria to compare disassembly paths. A heuristic algorithm is developed to determine the highest performance, feasible disassembly sequence. A numerical value is assigned to each subassembly to be removed (with a higher number indicating that the subassembly is more difficult to remove), as is a direction of disassembly reference (plus or minus x, y, or z), a list of predecessor subassemblies, and a list of joint types. Each subassembly is then placed into a list based upon how many different joints restrain it. The lists are ranked from least number of joints to most, and internally from lowest difficulty in removal to highest. Items are removed using these lists, precedence allowing. From this, a feasible near-optimal disassembly sequence is generated.

Lambert (1999) develops a method for optimally solving general disassembly sequence generation problems using linear programming. It is based upon earlier research on optimal disassembly sequence generation, specifically graphical search methods. The technique is expanded to solve the combined disassembly/clustering problem (*clustering* is the combining of different disassembled products into desirable categories, e.g., steel parts). A product is first graphically represented as a *disassembly graph* (frequently used in assembly/disassembly, this is a graph that proceeds from left—the initial product—to right—each of the individual components—depicting all feasible disassembly paths). This graphical information is then listed in a transition matrix that depicts all disassembly actions in the columns and all feasible subassemblies (from the initial product to the individual components and all subassemblies in between) as rows; all entries are zero except those created by an action (1) and those destroyed by an action (–1). The transition matrix is then used to select the equivalent feasible revenues and costs for each action/subassembly combination. The objective function seeks to maximize revenue and minimize cost. The problem is then modeled using linear programming and solved to optimality. Clustering is demonstrated through a modification where no further disassembly is performed once a given subassembly is composed of components consisting of a single material.

O'Shea et al. (1999) demonstrates an application of dynamic programming to disassembly. In this paper, a method is described for selecting automatic tools to be procured for use in the disassembly of a product component, specifically a commercial washing machine valve. The method uses a graphical representation (*cluster graph*) to symbolize the product in order to allow for the simple identification of the relationships between parts in the product. The graphical model is then converted into a dynamic programming model that includes cost data. The solution of the dynamic programming model provides an optimum determination of the tools that should be employed. Using a different approach, Torres et al. (2004) reports a study for nondestructive automatic disassembly of personal computers. Duta et al. (2002) presents some methods for modeling automated disassembly.

Kongar and Gupta (2002b) present the first paper to use multicriteria decision-making techniques—goal programming—to solve a disassembly-planning problem. A multicriteria optimization model of a disassembly-to-order system is developed to determine the best combination of each product type to be taken back at their end of life and then disassembled to meet the demand for its subcomponents and raw materials under a variety of physical, financial, and environmental constraints in order to achieve six preemptive goals. These goals include maximum total profit, maximum sales from materials, minimum number of disposed items, minimum number stored items, minimum cost of disposal, and minimum cost of preparation. When solved using goal programming, the model provides the number of reused, recycled, stored, and disposed items, as well as the values of various other performance measures.

Lambert (2002) develops an algorithm to solve disassembly sequence generation optimally using mathematical programming. Starting with the assembly diagram of the finished product, a connection diagram is created. Logical relations establishing precedence for each individual component are then formed. From this point on, the algorithm can be automated. *Coherent subassemblies* (i.e., touching parts) are generated from all connection diagram combinations.

Detachable subassemblies (i.e., subassemblies obtained through disassembly actions) are generated from the precedence relations. The intersection of the sets of coherent and detachable subassemblies provides a listing of feasible subassemblies. From this, feasible actions can be assigned (1 if a child subassembly is created, −1 if a parent subassembly is destroyed). A matrix can then be formed with these 1s and −1s (all other entries 0). To obtain the optimal disassembly sequence, the action cost and subassembly revenue are used with the matrix to maximize the net revenue using linear programming (which includes a constraint designed to check each feasible action individually). This technique is applied first to simple examples for demonstration and then to a more complex problem (an automatic transmission assembly frequently used in the literature). Finally, it is applied to electronic equipment having a hierarchical modular structure (a television). Since a good deal of the process can be fully automated, it is a significantly simpler and more importantly, less error-prone method of determining all feasible disassembly sequences.

Veerakamolmal and Gupta (1998) present a quantitative mathematical programming model for product disassembly and recycling. The model aims to recover common components by computing the number of products to be disassembled in order to meet the demand while minimizing disassembly and disposal costs. A subsequent paper (Veerakamolmal and Gupta, 1999) proposes an index to analyze the design efficiency of electronic products at their end of life. The design efficiency is measured using a *design for disassembly index*. The design for disassembly index uses a disassembly tree that relies on the product's structure. The disassembly tree can be used to identify the precedence relationships that define the hierarchy of the product's structure (which in turn, represents the order in which components can be retrieved). The design for disassembly index can be used to analyze the merits and drawbacks of different product designs. The index offers designers a measure to help improve future products. A comprehensive procedure for developing the index is provided and demonstrated using an example.

Lambert and Gupta (2002) build on the disassembly work by Veerakamolmal and Gupta (1998) with a comparison being made between different modeling methods for the disassembly of electronic appliances, with their typically modular design and hierarchical assembly/disassembly tree structure. Two traditional approaches are compared: disassembly graphs and *component disassembly optimization* (created by developing an inverse MRP). A new technique is then demonstrated which combines the advantages of both.

With cellular telephone–replacement cycles averaging 18 months in Western Europe, more than 100 new models being introduced every year, and noting that 70 percent of phones recovered in the United States and the European Union are beyond economical reuse (i.e., the remanufacturing costs are too high or the subsequent demand for the remanufactured product is too low), Franke et al. (2006) developed a remanufacturing plan for cellular telephones using a linear optimization model and discrete-event simulation.

Kekre et al. (2003) develops a simulation-based line-configuration model for an actual remanufacturing line that collects reusable automobile component modules and refurbishes them for sale in the aftermarket. The objective is to minimize the number of workstations and seek similar idle times between workstations. While the implementation of the improvements is of interest (simulation indicated throughput could be increased by 30 percent), the case-study portion of the paper should provide other researchers with a well-written overview of an actual remanufacturing process as well as some of the associated concerns and limitations.

Additional disassembly studies can be found in Altekín (2005), Duta et al. (2005), Güngör and Gupta (1999a, 1999b), Gupta and Güngör (2001), Kongar et al. (2003), McGovern and Gupta (2005a, 2006a, 2006c, 2006d, 2007a, 2007c, 2008a, 2008b, 2008c, in press), McGovern et al. (2004), and Prakash and Tiwari (2005).

4.5 Optimization and Algorithms

Key to addressing any engineering problem is to understand how complex or easy it is, what it shares with similar problems, and appropriate methods to obtain reasonable solutions. For these reasons, a background in optimization and algorithms is helpful.

Tovey (2002) provides a well-structured review of complexity, NP-hardness, NP-hardness proofs (including the concise style of Garey and Johnson), typical NP-hard problems, the techniques of specialization, forcing, padding, and gadgets, mathematical programming versus heuristics, and other complexity classifications.

Rosen (1999) provides a useful text in the general area of discrete mathematics including set theory, logic, algorithms, graph theory, counting, set theory, and proofs.

Lambert and Gupta's (2005) text reviews various aspects of disassembly. It reviews general and disassembly- or assembly-specific aspects of number theory, complexity, set theory, and graph theory. Product examples of varying complexity are presented and used effectively throughout the text to demonstrate techniques and limitations. The examples chosen are very appropriate and should prove helpful to any reader. In general, all terms are well defined and the mathematical notation used is appropriate and consistent throughout. The text provides detailed, valuable background for any quantitative study in disassembly.

The work of Cormen et al. (2001) is a standard reference for algorithm design, development, and complexity evaluation while Miller and Freund (1985) provide an engineering-based approach to probability and statistics. Papadimitriou and Steiglitz's (1998) work is the de facto text on combinatorial optimization as is Garey and Johnson's (1979) work in the area of NP-completeness.

Holland (1975) is credited with developing the genetic algorithm. Koza's (1992) work deals with *genetic programming* but this text provides a wealth of information relevant to genetic algorithms as well. One of the best known texts on genetic algorithms is by Goldberg (1989).

Dorigo et al. (1996) were the first to describe ant colony optimization. Dorigo et al. (1999) provide a summary and review of ant algorithms.

Hybrids of genetic algorithms, simulated annealing, naïve evolution, and hill climbing with two different packing heuristics can be found in a paper by Hopper and Turton (2000). The various hybrids are applied to the TWO-DIMENSIONAL RECTANGULAR PACKING problem (a problem from the categories of *cutting* and *packing*), which is concerned with how to fit differently shaped objects into a container in a way that minimizes the open space.

Osman and Laporte (1996) provide a remarkably well-researched paper on all forms of metaheuristics, the basic concepts of each, and references to applications. While this paper would appear to be somewhat dated considering how recent many of these approaches are, it provides a massive reference of the work performed up to its publication date and is well organized and very in-depth. A follow-on paper by Osman (2004) is more compact and also more current. Iori (2003) also provides an in-depth study and comparison of metaheuristics.

CHAPTER 5

Graphical Representations of Products to Be Disassembled

5.1 Introduction

This chapter reviews traditional graph-theory-based methodologies for depicting a product and its precedence relationships, and then demonstrates several slightly different treatments of vertices and arcs that address disassembly-specific data presentation requirements.

This chapter is organized as follows: Section 5.2 reviews traditional product representations and provides a historical perspective to disassembly representations. Section 5.3 introduces task-based precedence diagrams. Section 5.4 reviews disassembly constraints graphs, while Sec. 5.5 presents the electronics schematic/flowchart-type vertex and arc representation along with examples. Section 5.6 discusses some other representation used.

While often used interchangeably, for consistency and for ease of conceptualizing, where applicable and appropriate in this chapter and throughout the book, *undirected graphs* will consist of *nodes* and *edges* and *directed graphs* will consist of *vertices* and *arcs*.

5.2 General Product Disassembly Representations

The bulk of product disassembly representation methods (e.g., Fig. 5.1) are based on graph theory and rooted in product assembly representations. While this information is found contained is a variety of literature sources, it is well summarized and detailed by Lambert and Gupta (2005), where they also include a historical perspective and worked examples.

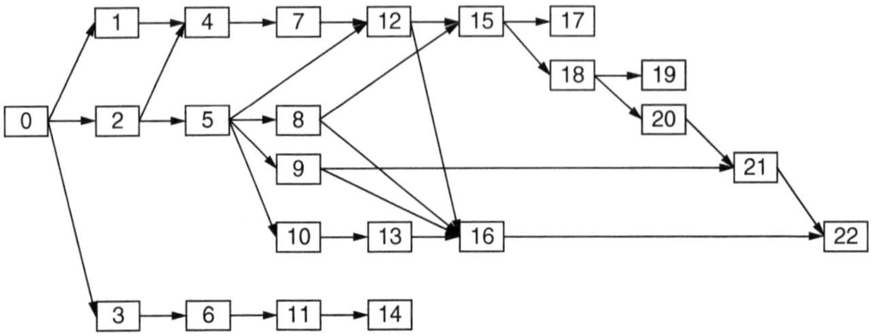

FIGURE 5.1 Example of a disassembly representation.

Some of these methods reviewed by Lambert and Gupta and discussed in this section include a case study (referred to as "Bourjault's pen" or the ballpoint pen), Bourjault's assembly tree, connection diagrams, reduced disassembly trees, connection state diagrams, subassembly state diagrams, disassembly precedence graphs, and constrained connection diagrams.

5.2.1 Product Case Study

With the variety of methods used to graphically represent end-of-life products as part of their disassembly planning, a prototypical case study enables a demonstration of the methods. The case study used here is the pen example by Bourjault (1984). Bourjault's pen is a component of his Ph.D. dissertation research and has been used as the basis of various researchers' assembly and disassembly developments. (While the original 1984 doctoral thesis was published in French, an English-language summary can be found in Bourjault, 1987.) An early study of disassembly, most of the analysis in this study was driven by assembly. In assembly, as long as precedence constraints are not violated, there exist many degrees of freedom in the ordering of operations. As a result, many sequences may be possible. In selecting a sequence, typically all feasible sequences are considered. Unfortunately, if the investigation of all possible sequences is performed by starting at some initial part and then progressively adding parts in order to attain the final product, many of these potential sequences will ultimately reach a point at which no further progress can be made. This is due to blocking by a previously installed part and it results in an infeasible sequence. In addition, this blocking may not be recognized for some time due to the time-consuming process of building a sequence from a single part. This effect is readily demonstrated with the ballpoint pen used by Bourjault.

The pen consists of six components: body (A), head (B), cartridge (C), ink (D), button (E), and cap (F); and five connections: 1 (body A to head B), 2 (button E to body A), 3 (head B to cartridge C), 4 (cartridge C to ink D), and 5 (cap F to body A) as seen in Fig. 5.2 and Fig. 5.3.

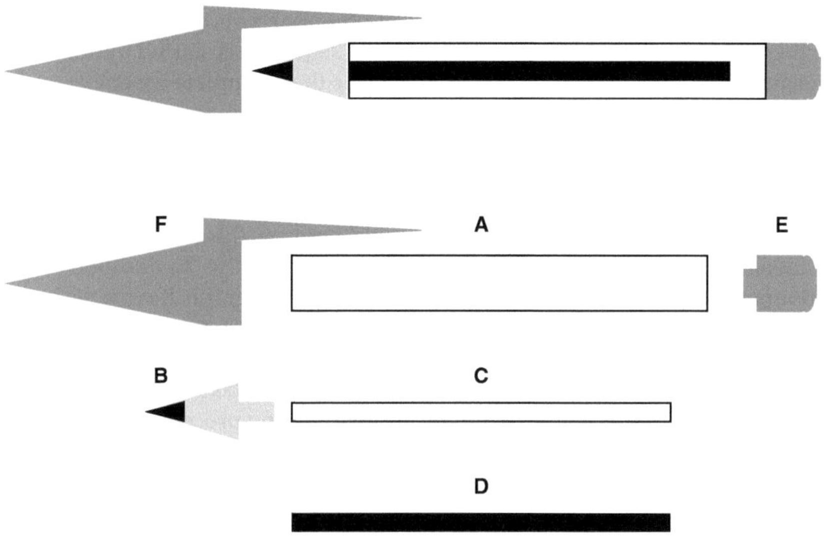

FIGURE 5.2 Cross section of Bourjault's ballpoint pen.

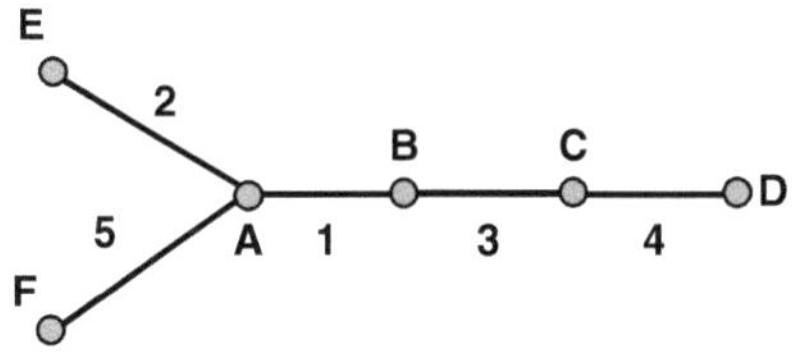

FIGURE 5.3 Connection diagram for Bourjault's ballpoint pen.

5.2.2 Connection Diagrams

A *connection diagram* is an undirected graph where the nodes represent the parts and the edges represent the connections. The diagram indicates all of the connections between parts and represents all of the topological relationships (or *topological constraints*; there are also geometric constraints and technical constraints as detailed later in the chapter). Topological constraints consist of all connections including mating (i.e., physical contact only between two adjacent parts; in other words, where there is contact, but no connection or bond). Lambert and Gupta (2005) also describe the optional inclusion of *virtual components* (fasteners such as welds, glue, etc.) and *quasi components* (fasteners such as screws, or connectors such as straps, where the part or material is insignificant in terms of recovery value, disposal cost, etc.), neither of which are seen in the pen example (the connection diagram for which is depicted in Fig. 5.3).

With the connection diagram example of Fig. 5.3, it can be seen that the part-ordering information is incomplete; it is not entirely clear what parts are true predecessors of other parts and what all feasible sequences would be using Fig. 5.3. If a sequence is started by, for example, attaching E and B to A, it is clear that complete assembly will be obstructed because C and D can never be installed (see Fig. 5.2). Precedence constraints (i.e., an obstruction that prevents the removal of a part) are referred to as *geometric constraints.*

To prevent the generation of these types of infeasible sequences, Bourjault proposed studying the disassembly process (his research considered disassembly as simply the reverse of assembly). In addition, Bourjault made use of formal reasoning including: graph theory, set theory, and the formulation of logical relations. The application of formal reasoning was partially motivated by the presumed replacement of human labor by robots. An additional consideration was the desire for automation of the assembly planning and sequencing process.

5.2.3 Assembly Trees, Reduced Disassembly Trees, Connection State Diagrams, and Subassembly State Diagrams

The use of formal reasoning by Bourjault included his introduction of his *assembly tree* (*graphe de choix directs* in Bourjault's original paper). It is a directed graph (although arrows are not depicted) with the process starting from the top and proceeding downward. Each assembly sequence is represented in the tree by a linear subgraph. Vertices correspond with a connection state and arcs correspond with an assembly (and later on, disassembly) task. The assembly tree is demonstrated in Fig. 5.4 using the pen example.

It can be seen from Fig. 5.4 that 12 assembly sequences are possible using the pen case study. It is also noted that many subsequences in the tree appear repeatedly. For example, connection state 1-3 (describing the subassembly consisting of connections 1 and 3) appears twice. The complete subtree that is associated with connection

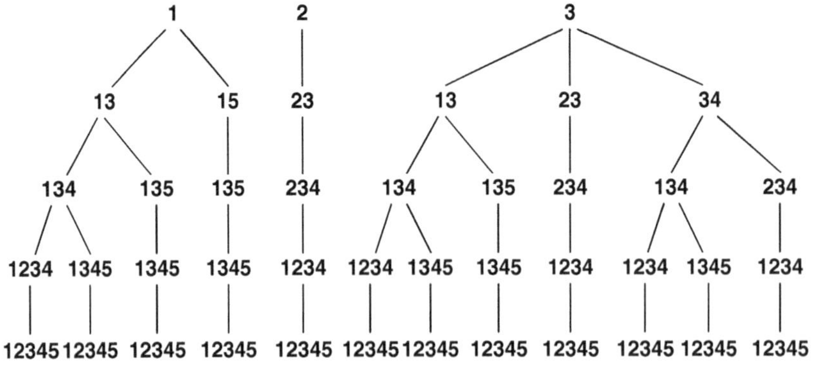

FIGURE 5.4 Assembly tree for the ballpoint pen model.

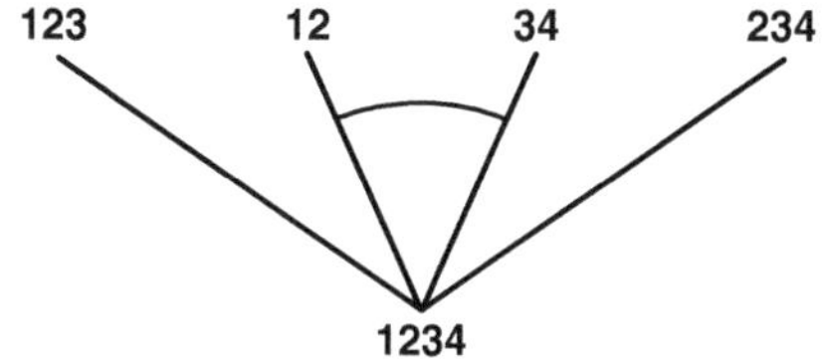

Figure 5.5 Assembly tree logical AND example.

state 1-3 therefore also appears twice. However, this feature can be exploited for use in significantly reducing the size of the tree.

As the tree represents the construction of subassemblies rather than simply indicating precedence relationships, there could exist a need to demonstrate the mating of two subassemblies rather than simply the addition of a single part to a subassembly. While, from top to bottom, each branch represents a logical OR (i.e., there are multiple, optional ways to build the product), a semicircle is often used to represent a logical AND. In the example of Fig. 5.5, there are three ways (additional permutations are not allowed in order to simplify the example) to build the product (or subassembly) made up of parts 1, 2, 3, and 4. These include: adding part 4 to subassembly 1-2-3 OR adding part 1 to subassembly 2-3-4 OR adding subassembly 1-2 AND subassembly 3-4 to each other (as indicated by the connecting semicircle).

The *disassembly tree* proceeds from the finished product (at level 1) down to the individual parts. Again using the pen example, at level 1 two results are possible: remove connection 2 or remove connection 5. This results in two branches. Continuing in this fashion, a disassembly tree is developed. The notation here is modified slightly from Bourjault's in accordance with that of Lambert and Gupta (2005). Specifically, the connection state Ø (the *empty state*, which corresponds with the set of disjoint components) is included, and the connection states that are found to appear previously in the tree are encircled and investigated no further giving the *reduced tree* (Fig. 5.6*a*). Combining any

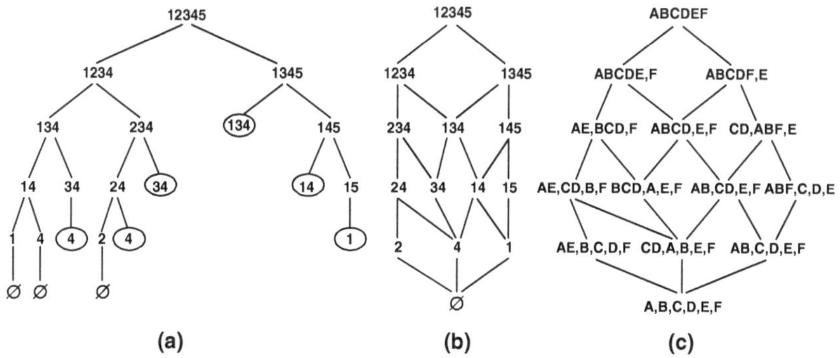

Figure 5.6 Disassembly of the ballpoint example using: (a) reduced tree diagram, (b) connection state diagram, and (c) subassembly state diagram.

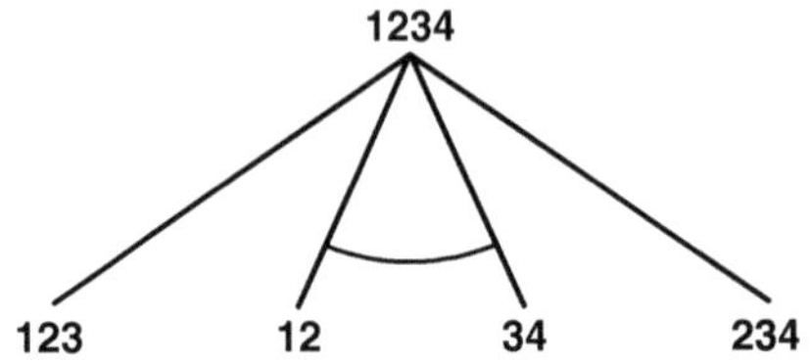

FIGURE 5.7 Disassembly tree logical AND example.

of the encircled multiple connection states results in the *connection state diagram* (Fig. 5.6*b*), which still contains the full information of the original version. This can also be depicted as a *subassembly state diagram* as seen in Fig. 5.6*c* and displaying all parts instead of connections.

Similar to the case with the assembly tree, the disassembly tree represents the removal of parts from the final product, and then from the remaining subassemblies. While precedence relationships are inferred, this is not the main purpose of the tree representation. There could, however, be a requirement to show the breaking of the main subassembly into two, smaller subassemblies (as opposed to the removal of a single part). In the disassembly tree moving from top to bottom, each branch represents a logical OR (i.e., there are multiple, optional ways to disassemble the product), a semicircle can be used to represent a logical AND. In the example of Fig. 5.7, there are three ways (other permutations are not allowed in order to simplify the example) to disassemble the product (or subassembly) made up of parts 1, 2, 3, and 4. These include: remove part 4 from subassembly 1-2-3 OR remove part 1 from subassembly 2-3-4 OR remove subassembly 1-2 AND subassembly 3-4 from each other (as indicated by the connecting semicircle).

5.2.4 Disassembly Precedence Graphs

Disassembly precedence graphs represent the ordering of parts according to their immediate predecessors, providing a visual representation of precedence constraints and of possible feasible disassembly sequences. This is accomplished primarily through the use of directed arcs. These graphs are closely related to Bourjault's tree and include the product's parts and any tasks (e.g., inspection, cleaning, and testing, in addition to the disassembly tasks). Figure 5.8 demonstrates a typical layout, notation, shapes, and flows using the ballpoint pen

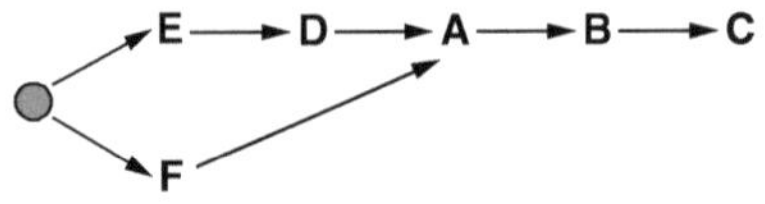

FIGURE 5.8 Disassembly precedence graph of the ballpoint pen (also known as partial ordering).

example. The graph goes from left (the fully assembled end-of-life product) to right (the last part or parts for removal based upon precedence constraints).

A logical OR is implied when more than one arc leaves a vertex (e.g., remove E OR F after the start in Fig. 5.8) while a logical AND is implied when more than one arc enters a vertex (e.g., remove D AND F before A in Fig. 5.8); however, some researchers choose to further label each of these relationships throughout a graph. In the example of Fig. 5.8, there are three possible part removal solution sequences: ⟨E, D, F, A, B, C⟩, ⟨E, F, D, A, B, C⟩, and ⟨F, E, D, A, B, C⟩. Finally, representing any of these three possibilities as the disassembly precedence graph is often referred to as a *totally ordered* disassembly precedence graph, while the depiction in Fig. 5.8 is referred to as a *partially ordered* disassembly precedence graph.

5.2.5 Constrained Connection Diagrams

One of the most widely used ways of representing precedence relations is by *constrained connection diagrams*. These hybrid diagrams are built using a connection diagram and then adding the directed edges of the disassembly precedence graphs in order to represent the precedence constraints. The final structure consists of undirected edges (the connections), nodes/vertices (the parts), and directed edges/arcs (representing the precedence constraints).

While authors follow varying conventions, the example of Fig. 5.9 (i.e., the ballpoint pen case study) follows the work of Lambert and Gupta (2005). In this format, a precedence relation is represented by an arrow pointing from a part (i.e., vertex) to those parts that are obstructing its removal. Multiple arcs sharing the same tail (directed edges have a *tail* and a *head*, where the direction is from the tail to the head) indicate a logical OR relationship, while multiple tails emanating from the same vertex indicate a logical AND relationship. As an example using Fig. 5.9, the arcs (D, B) and (D, E) indicate that (B∨E) must be removed prior to part D. Similarly, (B∨C) must be removed as well in order to access part D. As a result, ((E∨B)∧(B∨C)) must be

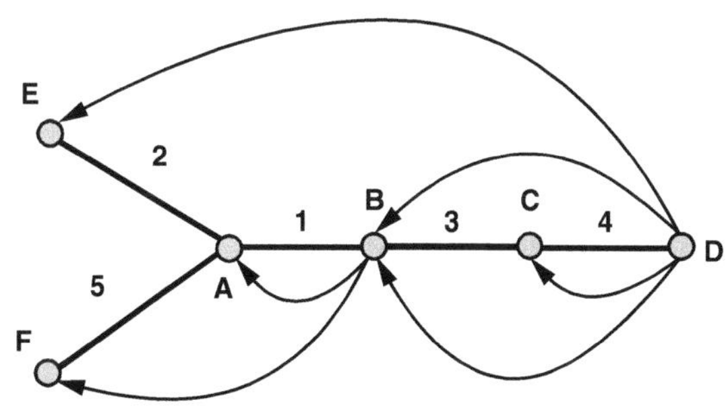

FIGURE 5.9 Constrained connection diagram for Bourjault's ballpoint.

removed prior to part D. In Fig. 5.9, the (D, E), (D, B), (B, A), and (B, F) arcs represent the geometric constraints in this product, while the (D, B) and (D, C) arcs represent *technical constraints* (constraints that are dependent on tools, costs, procedures, etc.). The technical constraint in the pen example is due to recognition that the ink D cannot be removed from the sealed cartridge C until the head B is removed.

Figure 5.9 indicates that disassembly must start at either E or F (A cannot be removed due to topological constraints). If E is removed first, this would delete arcs (D, E) and (D, B). The next part to be removed would need to be F (again due to A's topological constraints), which deletes arcs (B, A) and (B, F). Since arcs (D, B) and (D, C) remain, the part removal sequence proceeds with the removal of A, then B, and finally C. If these two arcs did not remain, part D could alternatively be removed after the removal of part F.

Additional information on Bourjault's pen, Bourjault's assembly tree, connection diagrams, reduced disassembly trees, connection state diagrams, subassembly state diagrams, disassembly precedence graphs, and constrained connection diagrams, as well as other representations such as the *cut-set method*, can be found summarized in Lambert and Gupta (2005).

5.3 Task-Based Precedence Diagrams

While tasks can be represented in part-based diagrams with the use of virtual parts, it may be of interest to exclusively address the precedence relationships between tasks. Product disassembly task precedence relations and complex relationships between tasks can be addressed using a format described by Altekín et al. (2008). Consisting of the familiar application of vertices and arcs and making use of the concept of depicting OR relations with a semicircle connecting the affected arcs, this diagram also adds the notion of *artificial* parts.

OR predecessors are indicated by a semicircle attached to all of the predecessor tasks' arcs as is found with other disassembly precedence diagrams, AND predecessors are implied by the presence of the predecessor tasks' arcs ending at a common task vertex (Figs. 5.10*a* and 5.10*b*, respectively). Artificial tasks (indicated by T_1, e.g., in Fig. 5.10*c*) are used to address the case where there are two or more groups of predecessor tasks (with all but one of the groups allowed to be as small as a single task), all of which are related by a logical OR.

5.4 Disassembly Constraint Graphs

It is possible to graphically indicate both precedence constraints and attaching bonds in a single diagram (Li et al., 2010). This representation simultaneously makes use of both directed and undirected edges, while the nodes/vertices continue to represent parts (or tasks). Two nodes in this *disassembly constraint graph* can exhibit four different configurations:

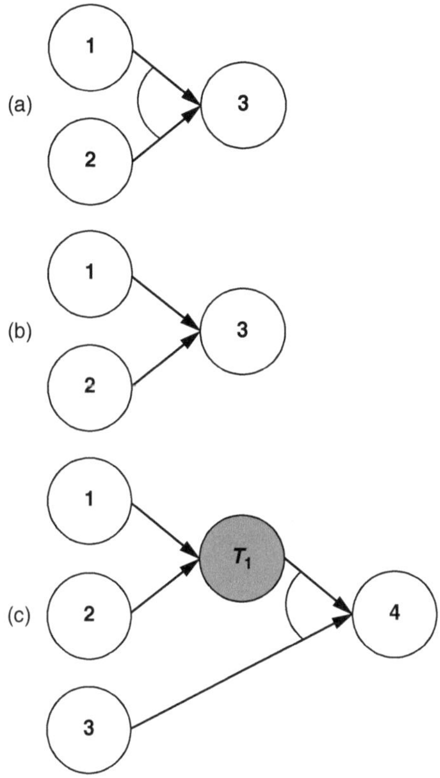

Figure 5.10 Disassembly-task-based precedence diagrams where: (a) tasks 1 and 2 are OR predecessors of task 3, (b) tasks 1 and 2 are AND predecessors of task 3, and (c) tasks $(1 \wedge 2) \vee 3$ are/is a predecessor of task 4.

no relationship, contacting constraints, precedence constraints, or both contacting and precedence constraints. No relationship is indicated by the lack of an edge or arc between two nodes and is interpreted as there being no physical attachment between the two parts and no immediate blocking. An edge indicates some type of bond between the two parts that must be broken before one or the other can be removed. An arc indicates blocking by an immediate predecessor (though others may exist further in/out) that must be removed before one or the other can be removed (the node at the tail of the arc is the immediately blocking part, while the node at the arc's head is the part that is blocked). An edge and an arc between the two parts indicates there is a physical bond between the two parts that must be broken and that once the bond is broken, one part immediately blocks the other's removal. (Note that these directions are the reverse of those given in Sec. 5.2.5.)

As demonstrated in Fig. 5.11*a*, part 1 immediately blocks the removal of part 2. Figure 5.11*b* shows that part 1 and part 2 share a mechanical bond (e.g., glue, screws, nuts and bolts, and welds) that

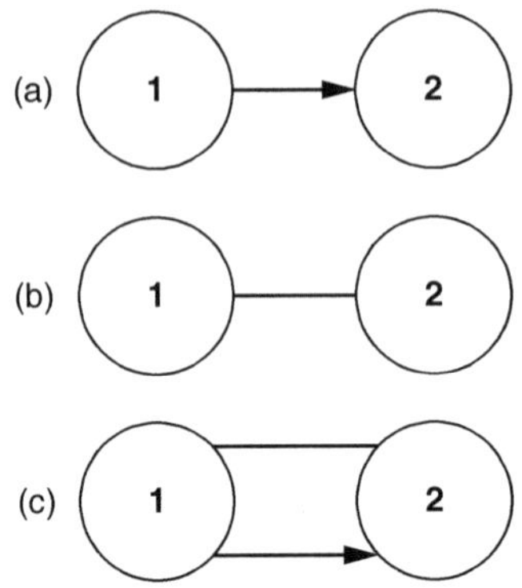

FIGURE 5.11 Disassembly constraint graph where (a) part 1 is an immediate predecessor of part 2, (b) part 1 and part 2 are connected, and (c) both conditions are true.

must be broken prior to the removal of one or the other. Finally, Fig. 5.11*c* demonstrates that part 1 immediately blocks the removal of part 2 and part 1 is connected to part 2.

Finally, in the disassembly constraint graph, only the immediate relationships are indicated. For example, other blocking parts that can be addressed with parts that are more immediately blocked are not given an arc connecting it to the later-blocked part.

5.5 Schematic/Flowchart-Style Vertex and Arc Disassembly Representation

The DISASSEMBLY LINE BALANCING PROBLEM (DLBP) can be represented by a directed graph (*digraph*). A modified digraph-based disassembly diagram is effective for displaying the additional and relevant information in a disassembly problem's precedence diagram (McGovern and Gupta, 2005c). It is based on the familiar format found in assembly-line balancing problems (as depicted in Elsayed and Boucher, 1994; see Lambert, 2003 for other representations), which consists of a circular symbol (the vertex) with the task identification number in the center and the part removal time listed outside and above. This is selectively enhanced to provide greater usability in application to disassembly problems. The part removal time (PRT_k) of part k is kept outside of the vertex, but in the upper-right corner. Working around in a clockwise manner, the additional considerations found in disassembly are listed using, as a default, the order of priority as given in this book, that is, "H" (only if the part is labeled as hazardous, else blank), the part's demand (d_k), and the part's removal direction (r_k); see Fig. 5.12.

This representation provides a reference to all of a given instance's disassembly-unique quantitative data and concentrates this data in one location, while still enabling all of the graphical

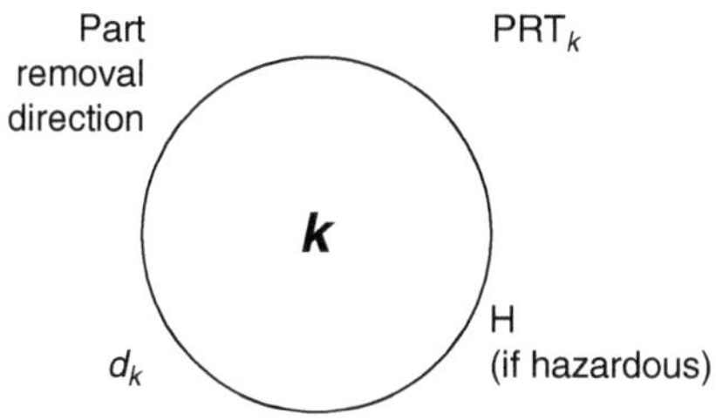

FIGURE 5.12 Disassembly task/part vertex representation.

relationship information that a traditional precedence diagram portrays. Arcs continue to be depicted as solid lines terminated with arrows. The arrow's direction points (top to bottom as found in this book or alternatively, left to right) toward subsequent disassembly tasks; that is, the arrows leave the predecessor parts. This format tends to be more angular, similar to an electronics schematic or a flowchart.

Another modification is the use of broken lines to express OR relations by connecting them to each other. The OR relationship is depicted in Fig. 5.13; that is, remove (1∨2) prior to 3 (note that, e.g., part 1 takes 2 seconds to remove, is hazardous, has a demand of 7, and can only be removed in direction $-y$).

Although all parts should be included in the diagram, only parts with demands and/or parts that are predecessors to subsequently demanded parts are required to be listed if incomplete disassembly is possible (since it is always desirable to remove demanded parts and it is required to remove their predecessors). Parts without demands that do not precede other parts having demands can be deleted from the diagram to conserve space—however, all of any deleted part's additional information (PRT_k, "H" if applicable, and r_k) will no longer be available to the reader. Since, in this case, the diagram would be incomplete, the beginning and end of the disassembly precedence diagram can optionally be depicted using flowchart-type terminator boxes, with the start box

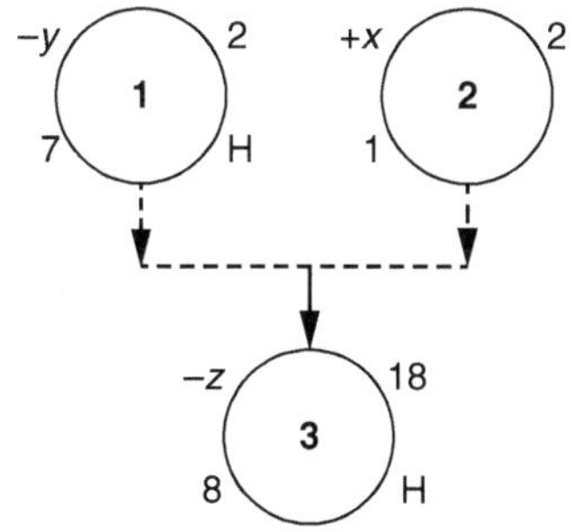

FIGURE 5.13 OR example depicting the requirement to remove 1 OR 2 prior to 3.

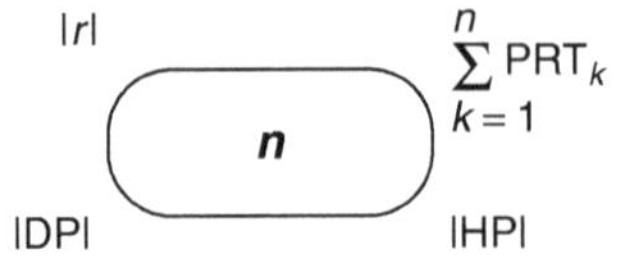

FIGURE 5.14 Disassembly precedence diagram *start* terminator box (optional).

used to contain summary information about the product being disassembled. The total number of parts n is listed in the center and, working around the outside in a clockwise manner, the total of the part removal times for the entire product is listed in the upper right corner, then the number of hazardous parts HP, the number of demanded parts DP, and the number of part removal directions r (Fig. 5.14).

The set of demanded parts is defined as

$$\text{DP} = \{k : d_k \neq 0 \ \forall \ k \in P\} \tag{5.1}$$

where P is set of n part removal tasks.

By taking advantage of the established benefits of electrical wiring diagrams, this design lends itself to readily representing products with many parts. As a result, one of the benefits of this wiring-diagram-type representation—in addition to containing disassembly-specific details in the diagram itself—is that it can be easier to read in the case of products with many parts. This is due to the straight lines of the wiring diagram being more amenable to easily and clearly crossing one or more pages in a document, as well as not being restricted by the legibility of multiple angled lines converging on a single part as found in other diagramming formats; the straight lines are easier than angled lines to trace on products with many parts.

Another benefit is its ability to represent multiple disassembly paths (as does, e.g., the disassembly tree of Sec. 5.2.3) as well as unusual or complex logical part relationships. Prior to demonstrating more complex logical part relationships, some examples of the more basic relationships can be seen in Figs. 5.15 through 5.20 (in order to focus on the part's relationships alone, in the interest of clarity, the part's characteristics are left off the examples).

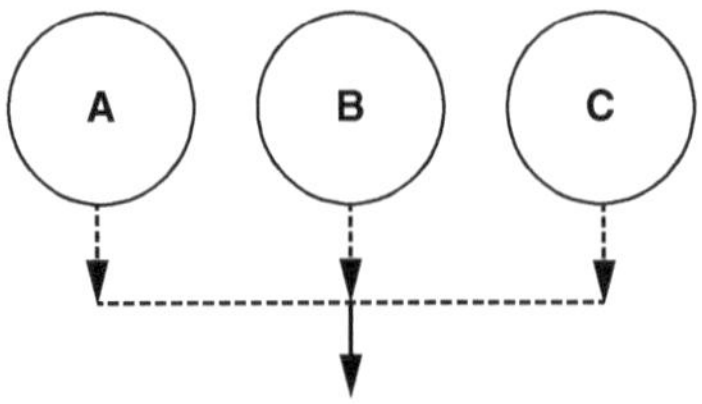

FIGURE 5.15 Disassembly precedence relationship (A∨B∨C).

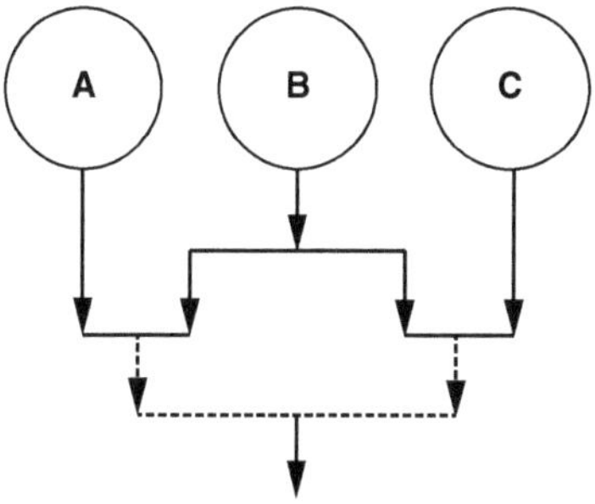

FIGURE 5.16 Disassembly precedence relationship (A∧B)∨(B∧C).

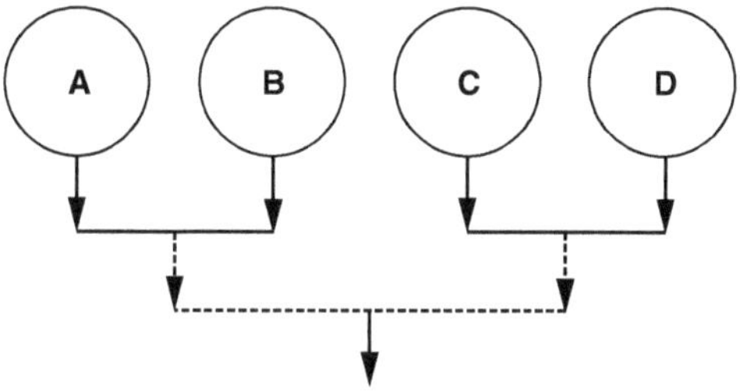

FIGURE 5.17 Disassembly precedence relationship (A∧B)∨(C∧D).

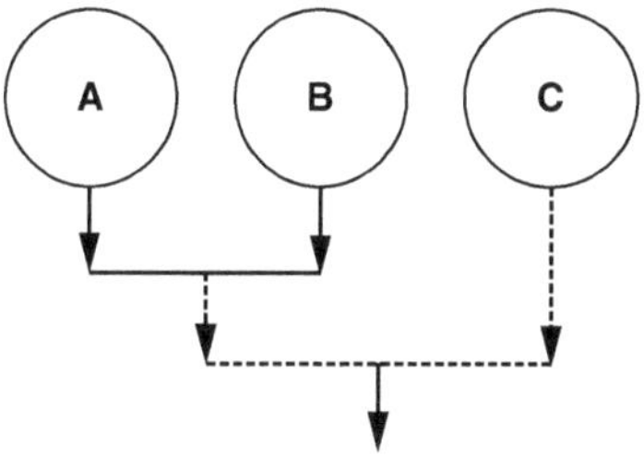

FIGURE 5.18 Disassembly precedence relationship (A∧B)∨C.

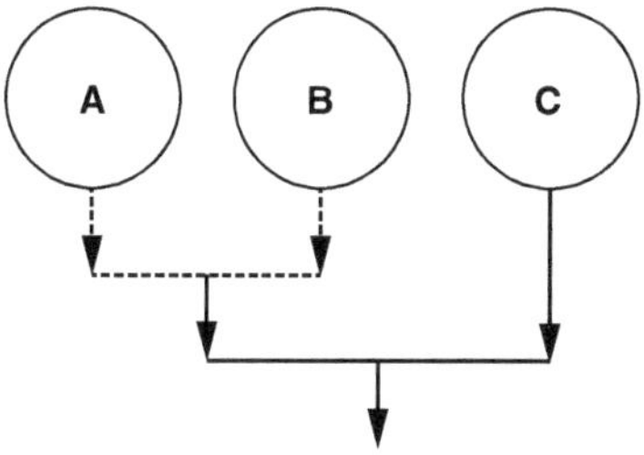

FIGURE 5.19 Disassembly precedence relationship (A∨B)∧C.

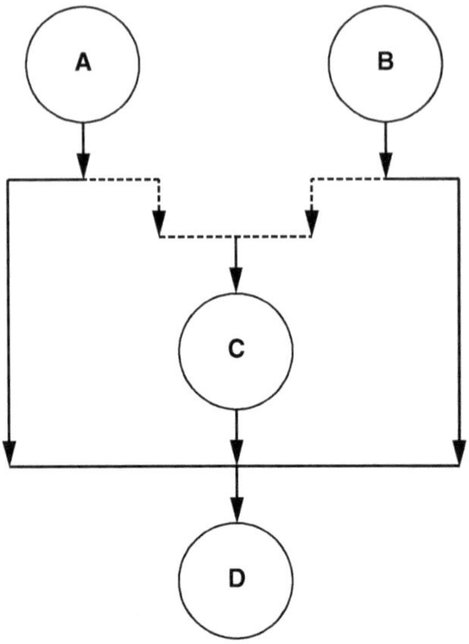

FIGURE 5.20 Disassembly precedence relationships: remove (A∨B) before C, remove (A∧B∧C) before D.

Examples of relationships that are more complex can be seen in Figs. 5.21 through 5.23 and include (A∨B∨C)∧D, ((A∧B)∨(B∧C))∧D, and (A∧B)∨(A∧C)∨(A∧C). Figure 5.23 also demonstrates lines crossing that are not in contact. A necessary feature of wiring diagrams (sometimes represented with a semicircle to indicate a wire's motion away from the page into a third dimension), this is helpful in representing disassembly paths in products with intricate precedence relationships.

5.6 Other Representations

Various graphical approaches have been developed to present disassembly sequencing. Kaebernick et al. (2000) use a cluster graph that is created by sorting the components of a product into different levels based on their accessibility for disassembly. Lambert (1997) presents an AND/OR-graph-based graphical method for the generation of the optimum disassembly sequence. Li et al. (2006) present a disassembly

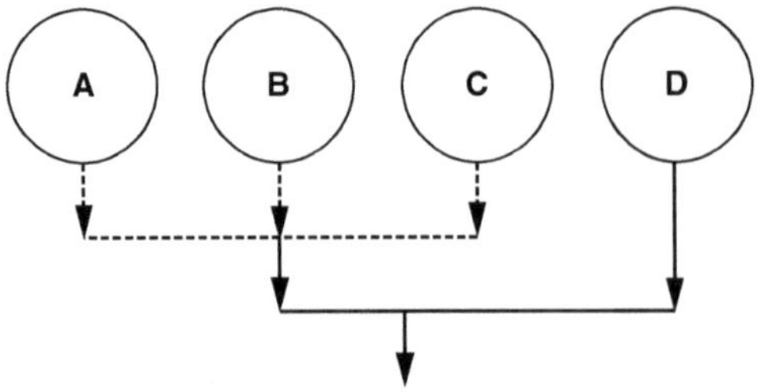

FIGURE 5.21 Disassembly precedence relationship (A∨B∨C)∧D.

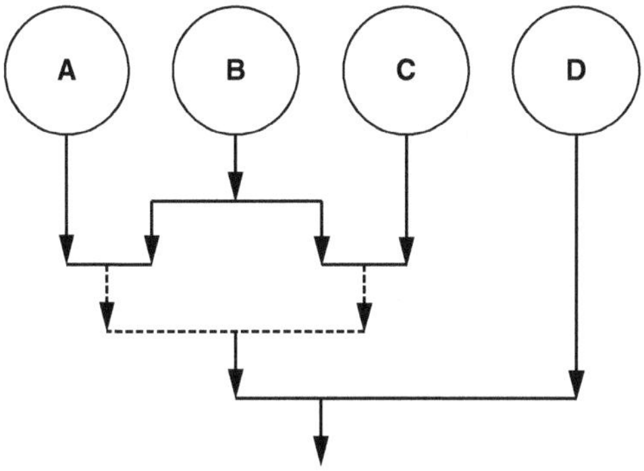

FIGURE 5.22 Disassembly precedence relationship ((A∧B)∨(B∧C))∧D.

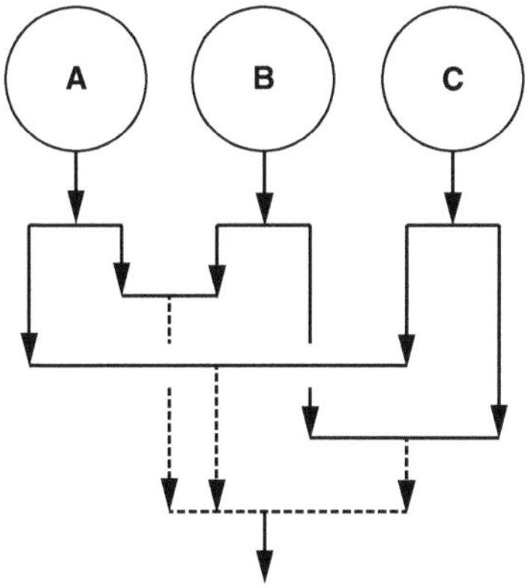

FIGURE 5.23 Disassembly precedence relationship (A∧B)∨(A∧C)∨(B∧C).

constraint graph to generate possible disassembly sequences for maintenance. Dong et al. (2006) propose a method for the automatic generation of disassembly sequences from a hierarchical attributed liaison graph. O'Shea et al. (1999) provide an example of a cluster graph.

Petri nets represent a popular alternative for disassembly modeling. Petri nets are a graphical and mathematical modeling technique developed in the 1960s to model concurrent computer system operations. The work of Moore and Gupta (1996) indicate they also have application in disassembly. Moore et al. (1998, 2001) present a Petri-net-based approach for the automatic generation of disassembly process plans for products with complex AND/OR precedence relationships.

Building on disassembly trees and disassembly precedence graphs, the *transformed AND/OR graph* is a form of a digraph that can be used to represent products to be disassembled (Koc et al., 2009). A disassembly tree represents all possible subassemblies (but represents precedence relations only indirectly). Disassembly precedence graphs contain all of the task information needed to completely disassemble a product. The transformed AND/OR graph is a version of the disassembly tree, which is modified to include disassembly precedence graph information. The graph is formed as follows. Each arc in the disassembly tree is represented in the transformed AND/OR graph using a vertex, while each vertex in the disassembly tree corresponding to a subassembly is represented in the graph using an *artificial vertex* (or artificial task; see Sec. 5.3). Precedence relationships between the tasks (now vertices in the transformed AND/OR graph) and subassemblies (now artificial vertices) are maintained in the new diagram. The artificial vertices may be preceded or succeeded by more than one vertex; however, only one of the predecessors and one of the successors should be processed during disassembly. Arcs in the transformed AND/OR graph are of two types: the first type of arc represents the normal precedence relation case, while the second type of arc indicates alternative paths that are indicated using a connecting semicircle.

CHAPTER 6

Computational Complexity of Combinatorial Problems

6.1 Introduction

In order to attempt to solve the DISASSEMBLY LINE BALANCING PROBLEM (DLBP), its structure and complexity first must be established and understood. This chapter focuses on providing an introduction to complexity theory as well as discussing hardware, software, and analysis considerations, including a review of the hardware used for generating the case-study results as well as discussing software engineering, software reusability, and data analysis.

Section 6.2 describes computational complexity and combinatorial optimization along with "easy" and "hard" problems and their associated polynomial-time and nondeterministic-polynomial-time solutions. Section 6.3 reviews the class of NP-complete problems. Section 6.4 covers the class of unary NP-complete problems and Sec. 6.5 addresses the class of NP-hard problems. Section 6.6 provides a synopsis of the details in the previous sections. Section 6.7 notes the currently accepted general concepts for attempting to solve NP-complete problems. Section 6.8 discusses other classes of problems to enable a complete discussion of complexity theory within this chapter. Section 6.9 discusses the analysis of suboptimal solutions generated by these techniques, while Sec. 6.10 reviews computer-related considerations and the specifics of the hardware and software used in the analysis performed in this book.

Note that a great deal of the explanations in this chapter can be attributed to Garey and Johnson (1979), Karp (1972), Papadimitriou and Steiglitz (1998), and Tovey (2002). Because of the importance of

NP-completeness theory, its occasional misrepresentations in texts and papers, and the terse and seemingly cryptic illustrations traditional to the precise field of mathematics, the bulk of the information presented in this chapter is offered here exclusively as supplementary information to allow a more full understanding of the DLBP complexity proofs that follow. This is especially necessary since the three proofs presented here are structured in that same succinct format, requiring some degree of interpretation of various subtle details that are often considered to be "obvious" and are typically left to the reader. In fact, this book uses the notoriously compact proof format used by Garey and Johnson (1979), which is actually longer and more detailed than that used by Karp (1972) in his seminal paper. Readers familiar with NP-completeness are invited to skip this chapter and should have little difficulty in appreciating the proofs of Theorem 9.1 in Sec. 9.2, Theorem 9.2 in Sec. 9.3, and Theorem 9.3 in Sec. 9.4.

6.2 Complexity Theory Background

Computational complexity is the measure of how much work—typically measured in time or computer memory—is required to solve a given problem. If the problem is easy, it can probably be solved as a linear program (LP), network model, or with some other similar method. If a problem is hard, finding an exact solution is apt to be time intensive or even impractical, requiring the use of enumerative methods or accepting an approximate solution obtained with heuristics.

Though similar in name to the ASSEMBLY LINE BALANCING problem [and more specifically the simple assembly line balancing type I (SALB-I) problem, a claimed NP-hard combinatorial optimization problem], the DISASSEMBLY LINE BALANCING PROBLEM contains a variety of complexity-increasing enhancements. The general ASSEMBLY LINE BALANCING problem can be formulated as a scheduling problem, specifically as a *flow shop*. In its most basic form, the goal of DLBP is to minimize the idle times found at each flow-shop machine (workstation). Since finding the optimal solution requires investigating all permutations of the sequence of n part removal times, there are $n!$ possible solutions. The time complexity of searching this space is $O(n!)$. This is known as *factorial complexity* and referred to as *exponential time*. Though a possible solution to an instance of this problem can be verified in polynomial time, it cannot always be optimally solved. This problem is therefore classified as *nondeterministic polynomial* (NP). If the problem could be solved in polynomial time it would then be classified as *polynomial* (P). [More precisely, Karp (1972) defines P to be the class of languages *recognizable*—recognition in languages is equivalent to solving in decision problems—in polynomial time by *one-tape deterministic Turing machines* and NP to be the class of languages recognizable in polynomial time by *one-tape nondeterministic*—i.e., guessing—*Turing machines*.]

Combinatorial optimization is an emerging field that combines techniques from applied mathematics, operations research, and computer science to solve optimization problems over discrete structures. Combinatorial optimization problems are optimization problems where some c (a cost function that maps feasible points to the real numbers) is optimized over a finite set F (the set of feasible points). The goal is to find $f \in F$ such that $c(f)$ is maximized, minimized, or equal to some value; that is, find an optimal f. Note that there can also be more than one optimal solution, or none. Usually, F is considered to be the set of solutions, and in all cases F is finite. Combinatorial optimization problems are those that can be put in form (F, c), for example, the TRAVELING SALESPERSON PROBLEM (TSP; given a finite set of cities, edges, and weights, the lowest-cost *tour* is desired) is a combinatorial optimization problem where F is the set of the Hamiltonian cycles (i.e., cycles that visit every node exactly once) and c is the total cost of a cycle (which is to be minimized). To solve one of these problems, one must consider every solution, compute their cost functions, and select the best one. This technique is known as the *exhaustive* approach to solving a problem. However, exhaustive search takes $n!$ time, which makes this solution technique impractical for all but the smallest problems.

The basic distinction made between solution techniques is whether they are easy [they take polynomial time $O(x^a)$, where x is indicative of the problem size and a is some constant; this group includes n, $n \log n$, n^2, n^3, $10^8 n^4$, etc.] or they are hard [taking exponential time $O(a^x)$; this group includes 2^n, 10^n, $n^{\log n}$, $n!$, n^n, etc.; Papadimitriou and Steiglitz, 1998]. [Using *asymptotic notation*, a function $g(x)$ is $O(h(x))$ whenever there exists a constant a such that $|g(x)| \leq a \cdot |h(x)|$ for all $x \geq 0$.] Complexity theory is formulated from the theory of *recursive functions* and the definitions and applications of *recursively enumerable sets*. The terminology includes the following: A specific numeric case or class of cases having the same form is known as an *instance* (I). An instance of a problem is a particular case of the problem with all parameters quantitatively defined. The term *problem* (Π; also L, a reference to its basis in languages from *recursive enumeration*) refers to a class of instances and is defined as a general question to be answered that usually contains variables. It includes a broad description of all parameters and of what properties the answer (*solution*) must satisfy. A problem is then a general description of the properties of an optimal solution. Therefore, an instance can be thought of as a specific case of the problem. For example, TSP is a problem while an actual list of cities and weights is an instance. An *algorithm* is a general, step-by-step process for solving the problem. An algorithm can solve a problem if it can produce a solution for any instance of that problem. In order to solve a problem, an algorithm must always find the same, optimum (minimum or maximum) solution for a given instance (Garey and Johnson, 1979). [This strict definition of an algorithm from the field of mathematics as always providing the same result and one that is always the optimal result—techniques that include exhaustive search, the simplex algorithm for LP, and dynamic programming for PARTITION—is given here in the interest of accuracy but is not used throughout the remainder of this book in favor of the more prevalent definition from the field of computer science of an algorithm as being simply a step-by-step process for obtaining some solution to a problem (see Sec. 10.2 and Rosen, 1999); i.e., not necessarily an optimal solution and not necessarily a deterministic process always giving the same solution to an instance.] It is generally of interest to find an algorithm that gives a solution in the most efficient way, where the "most efficient" algorithm is generally considered to be the fastest. This is so because the time requirements are most often the measure of whether or not a given algorithm is of any practical use, since the time is reflective of the problem size. For every input size, the *time complexity function* for an algorithm gives the largest amount of time needed by that algorithm to solve an instance of that size. *Time complexity* is defined to be the same size as the worst-case runtime, though in this book time complexity is generically used to refer to any time study. A *polynomial-time algorithm* is one whose time complexity function is $O(p(x))$ for some polynomial function p with input of size x; this definition includes constant time, that is, $O(1)$. If (excluding certain nonpolynomial-time complexity functions) it cannot be bounded this way, it is called an *exponential-time algorithm*. As an example of the time differences between the two types of algorithms, a computer that can perform 1 million operations per second would require 13 minutes to process an instance of size $x = 60$ for a polynomial-time algorithm operating at $O(x^5)$, while taking 366 centuries to process the same size instance but using an exponential-time algorithm operating at $O(2^x)$ (Garey and Johnson, 1979). For all but the smallest problems, polynomial-time algorithms are much more desirable than exponential-time algorithms which (it is believed but not yet proven) render a problem *intractable* (where a problem cannot be solved by any known polynomial-time algorithm). Exceptions include algorithms that are worst-case exponential-time but tend to run much faster in average case, such as the simplex algorithm for the LINEAR PROGRAMMING problem and branch-and-bound for the KNAPSACK problem. There are two causes for intractability: problems so hard they require an exponential amount of time to obtain a solution, and problems whose solutions are exponential in size (this is typically indicative of an unrealistically defined problem since the solution cannot practically be used due to its size). These definitions assume a *reasonable machine*. Reasonable machines include personal computers, workstations, and supercomputers. They also include Turing machines [a one-tape deterministic Turing machine makes use of a list of n-tuples, entering at the first nonblank position on the tape using the information in an (the) initial state n-tuple that also corresponds to the 0, 1, or blank in the initial position on the tape, updates the machine's state, writes a 0, 1, or a blank, and moves right or left on the tape—all per that n-tuple's data—one position, then repeats until reaching a point where no n-tuple corresponds to the state required and the 0, 1, or blank currently in the position on the tape at which point the machine halts and the newly written sequence on the tape is the final tape output]. Machines that are not reasonable include a machine that can do an infinite number of things at once, a parallel computer with an unlimited number of nodes, and quantum computers (ultimate laptop, black hole computer). Subsequent definitions also require *reasonable encodings*. Reasonable encodings are concise; they do not contain extra information. An example of an encoding that is not reasonable would include an input to TSP that was a listing of all cycles in the graph and their costs. Each cycle could then be checked to see if it is Hamiltonian and then the one with the smallest cost could be selected. The number of cycles is the size of a TSP instance. Therefore, an algorithm can be found that solves TSP in polynomial time with respect to the number of cycles, so TSP would be tractable. However, this encoding is unreasonable—it is exponentially large with respect to the number of vertices. To summarize, polynomial-time complexity assumes the following: input size x is large enough, constant factors are ignored, instance encoding is reasonable, and the machine is reasonable.

Easy problems include LP, MINIMUM COST NETWORK FLOW, MATCHING, MINIMUM SPANNING TREE, and sorting. Hard problems include INTEGER PROGRAMMING, TSP, and JOB-SHOP SCHEDULING (Tovey, 2002). P denotes the class of easy problems; NP-hard is a larger class than NP-complete and includes NP-complete. Being NP-hard or NP-complete effectively means that it takes a long time to exactly solve all case of sufficiently large size (Tovey, 2002). The formality of NP-completeness is of special interest since, as put by Garey and Johnson (1979), there is a certain theoretical satisfaction in precisely locating the complexity of a problem via a "completeness" result that cannot be obtained with a "hardness" result.

Adding a threshold value to an optimization problem converts it to "yes-no" (*decision*) form, which is the standard format used in complexity analysis and NP-completeness proofs. This is due to complexity theory having its basis in recursive enumeration, which includes a function g that only takes two values (which are assumed to be 0 and 1). The yes-no version of a problem is analyzed to classify its optimization version. Therefore, hard problems should stay hard and easy problems should stay easy when converted to yes-no form. That is, the yes-no version should be hard if and only if the optimization version is hard. If the problem has correctly been converted to yes-no form, it will never be more difficult to solve than the original version. In the class NP, if lucky guesswork is allowed, then the problems can be solved in polynomial time. Most reasonable and practical problems are in NP.

6.3 NP-Completeness

Cook provided the foundations for NP-completeness in 1971 (Cook, 1971) by showing that a *polynomial time reduction* ($\leq_p$) from one problem to another ensures any polynomial-time algorithm used on one can be used to solve the other. In addition, Cook emphasized the class of NP-decision problems that can be solved on a nondeterministic computer (a fictitious computer that operates in a manner that is not predictable; i.e., a computer that guesses solutions). The final significant contribution of that paper was the proof that every problem in NP can be reduced to one particular problem in NP (the SATISFIABILITY problem, or SAT) and the suggestion that other problems in NP may share SAT's property of being the hardest problem in NP. This group of the hardest problems in NP is the class of *NP-complete problems*. [The term "complete" in reference to NP-complete problems was first used by Karp (1972) and comes from recursive enumerability and the notion of a language being "complete" for a class with respect to a given type of reducibility if it belongs to the class and every other member reduces to it; Karp defines the term to mean "equivalent" and used it to describe problems that could equivalently play the role of SAT in Cook's theorem.] Though it is suspected that NP-complete problems are intractable, this has never been proven or disproved. Once determined and proven, the answer to the question "is P = NP?" will effect every problem that is NP-complete. If and when answered, it will be known that no NP-complete problem will ever be solved in polynomial time or that all NP-complete problems can be optimally solved.

The theory of NP-completeness is applied only to *decision problems*. A decision problem Π consists of a set D of decision instances and a subset $Y \subseteq D$ of yes-instances along with any additional problem structure. Specifying a problem consists of two parts: describing a generic instance ("Instance") and stating the yes-no question ("Question") asked in terms of the generic instance. While the optimization problem asks for a structure of a certain type that has a minimum (or maximum) cost among all such structures, associating a numerical threshold as an additional parameter and asking whether there exists a structure of the required type having cost no more (or less) than that threshold creates the desired decision problem. As long as the cost function is relatively easy to evaluate, the decision problem can be no harder than the corresponding optimization problem. Therefore, even though the theory of NP-completeness is restricted to decision problems, it is known that the optimization version is at least as hard and hence, we can extend the implications of the theory. (Note that many decision problems—including TSP—can also be shown to be no easier than their corresponding optimization problems.) The reason that this theory is restricted to decision problems is due to their formal mathematical correspondence to languages (and ultimately, recursive enumeration) and use of encoding schemes.

Originally formalized with application to combinatorial problems in Karp (1972), the language framework is now defined. An *alphabet* Σ is a finite set of symbols (e.g., $\Sigma = \{0, 1\}$). A *string* is a sequence made up of the symbols in the alphabet (e.g., 01101). A *language* L over an alphabet Σ is any set of strings over Σ (e.g., $L = \{1, 10, 100, 010, 1001, \ldots\}$). The *empty string* is written ε while the set containing all possible strings from Σ is written Σ^*. It is assumed that any problem instance can be encoded as one of the strings in Σ^*, that is, as a binary string. Languages provide a way to easily list all the instances of a problem whose answer is "yes." Languages and problems are closely related. A language is equivalent to a problem; a string is equivalent to an instance; 1s and 0s can be used to encode real numbers where a 1 is equivalent to the answer "yes" and a 0 is equivalent to the answer "no." An encoding scheme is what is used when specifying problem instances. Karp (1972) provided a mathematical definition where an encoding $e : D \rightarrow \Sigma^*$ of a set $D' \subseteq D$ is recognizable in polynomial time if $e(D') \in$ P. Languages allow for the decision problem to be represented to and solved by a deterministic one-tape Turing machine (and ultimately a nondeterministic one-tape Turing machine). The class NP is intended to isolate decision problems in polynomial time. Given a problem instance I, a nondeterministic algorithm has two stages: a guessing stage where some structure S is proposed, and a checking stage where I and S are used to terminate with "yes," terminate with "no," or compute without halting (the latter two are effectively indistinguishable). The class of NP is then the class of all decision problems that can be solved (under reasonable encoding schemes) by polynomial-time nondeterministic algorithms (with the understanding that "solve" refers to verification of a "yes" or "no" answer). As a result, note that P $\subseteq$ NP though it is generally accepted (but never proven) that P $\neq$ NP. Applied from the mathematics of languages, a *polynomial transformation* ($\leq_p$; i.e., a polynomial time reduction) implies a function exists that can map the solution space of one problem to another. Karp (1972) in one of his definitions explains how the reduction from Π' to Π ($\Pi' \leq_p \Pi$) is actually the process of converting Π to Π' through the use of some function. Formally, $\Pi' \leq_p \Pi$ is a function $g : D' \rightarrow D$ where g is computable by a polynomial-time algorithm and $\forall I \in D', I \in Y' \Leftrightarrow g(I) \in Y$. $\Pi' \leq_p \Pi$ can be interpreted as meaning that Π is at least as hard as Π'. Also, if Π can be solved by a polynomial-time algorithm, so can Π'; if Π' is intractable, so is Π. $\leq_p$ is an equivalence relation and poses a partial order on the resulting classes of languages (i.e., decision problems). As a result, the class P forms the lowest equivalence class in this partial ordering with the decision problems viewed as the easiest problems computationally. The class of NP-complete languages (problems) forms another class, the one having the hardest languages (problems) in NP. A language L is NP-complete if (a) $L \in$ NP and (b) for all other languages L' such that $L' \in$ NP, $L' \leq_p L$. This prompts the identification of the NP-complete problems as the hardest problems in NP [it is property (b) by itself that causes problems to be in the class NP-hard]. If any NP-complete problem can be solved in polynomial time, then every problem in NP can be solved in polynomial time; if any problem in NP is intractable, then every problem in NP-complete is intractable. (Note that if P $\neq$ NP then there must be problems in NP that are neither solvable in polynomial time nor are they NP-complete, but these are generally not of interest.)

The process of devising an NP-completeness proof for a decision problem Π consists of the following four steps:

1. Show that $\Pi \in$ NP.
2. Select a known NP-complete problem Π'.
3. Construct a transform g from Π' to Π.
4. Prove that g is a polynomial transformation.

The main idea in an NP-completeness proof is to model a known NP-complete problem as the problem under consideration (not the reverse, which is what might be done to solve a problem). This is performed using a transformation—essentially a computer program—that has an *input* (an instance of Π', a known NP-complete problem) and an *output* (an instance of Π, the problem under consideration). The transformation must map "yes" instances of Π' to "yes" instances of Π, map "no" instances of Π' to "no" instances of Π, and it must be fast (run in polynomial time in the length of the input instance).

Cook provided the first NP-complete problem (SAT; Cook, 1971). The six other basic NP-complete problems include 3-SATISFIABILITY, 3-DIMENSIONAL MATCHING, VERTEX COVER, CLIQUE, HAMILTONIAN CIRCUIT, and PARTITION (Garey and Johnson, 1979). These six problems are commonly referred to as the "known NP-complete problems." Also, they are the most commonly used problems for NP-completeness proofs and serve as a core of known NP-complete problems. In addition, they all appeared in the original list of 21 NP-complete problems in Karp (1972).

Some techniques for proving NP-completeness include restriction, local replacement, and component design (Garey and Johnson, 1979).

An NP-completeness proof by *restriction* for a given problem $\Pi \in$ NP consists simply of showing that Π contains a known NP-complete problem Π' as a special case. It requires that additional restrictions be placed on the instances of Π so that the resulting restricted problem will be identical to Π' (not necessarily exact duplicates of one another but rather an obvious one-to-one correspondence between their instances that preserves "yes" and "no" answers). That is, considering a problem $\Pi = (D, Y)$ (the set D of all instances and the set Y of all "yes" instances), $\Pi' = (D', Y')$ is a subproblem of Π if

- $D' \subseteq D$
- $Y' = Y \cap D'$

Instead of transforming a known NP-complete problem to the target problem, an attempt is made to restrict inessential aspects of the target problem until a known NP-complete problem appears (many problems that arise in practice are simply more elaborate versions of known NP-complete problems).

With *local replacement*, some aspect of the structure of the known NP-complete problem instance is used to make up a collection of basic units. The corresponding instance of the target problem is obtained by uniformly replacing each basic unit with a different structure, each replacement constituting only a local modification of that structure. This is occasionally augmented by a limited amount of additional structure, known as an *enforcer*, which imposes certain additional restrictions on the ways in which a "yes" answer to the target instance can be obtained. The enforcer acts to limit the target problems instances to those that reflect the known NP-complete problem instance.

Component design proofs are any in which the constructed instance can be viewed as a collection of components, each performing some function in terms of the given instance. The basic idea is to use the constituents of the target problem instance to design certain components that can be combined to realize instances of the known NP-complete problem. Components are joined together in a target instance in such a way that the choices are communicated to *property testers* and the property testers then check whether the choices made satisfy the required constraints. Interactions between components occur both through direct connections and through global constraints.

6.4 NP-Completeness in the Strong Sense

NP-completeness provides strong evidence that a problem cannot be solved with a polynomial-time algorithm (the theory of NP-completeness is applied only to decision problems). Many problems that are polynomial solvable differ only slightly from other problems that are NP-complete. 3-SATISFIABILITY and 3-DIMENSIONAL MATCHING are NP-complete, while the related 2-SATISFIABILITY and 2-DIMENSIONAL MATCHING problems can be solved in polynomial time (polynomial-time algorithm design is detailed in Aho et al., 1974 and Reingold et al., 1977). If a problem is NP-complete, some subproblem (created by additional restrictions being placed on the allowed instances) may be solvable in polynomial time. Certain encoding schemes and mathematical techniques can be used on some size-limited cases of NP-complete problems (an example is a dynamic programming approach to PARTITION that results in a search space size equal to the size of the instance multiplied by the logarithm of the sum of each number in the instance) enabling solution by what is know as a "pseudopolynomial-time algorithm." In the PARTITION case this results in a polynomial-time solution as long as extremely large input numbers are not found in the instance. Problems for which no pseudopolynomial-time algorithm exists (assuming $P \neq NP$) are known as "NP-complete in the strong sense" (also, *unary* NP-complete—referring to allowance for a nonconcise or "unary" encoding scheme notation where a string of n 1s represents the number n—with an alternative being *binary* NP-complete). Typical limiting factors for NP-complete problems to have a pseudopolynomial-time algorithm include the requirement that the problem be a number problem (versus a graph problem, logic problem, etc.) and be size constrained, typically in the number of instance elements (n in DLBP) and the size of the instance elements (PRT_k in DLBP). Formally, an algorithm that solves a problem Π will be called a *pseudopolynomial-time algorithm* for Π if its time complexity function is bounded above as a function of the instance I by a polynomial function of the two variables Max[I] (problem size) and Length[I] (data length, the number of digits required to write each number—note that this can be enormous). A general NP-completeness result simply implies that the problem cannot be solved in polynomial time in all the chosen parameters.

KNAPSACK has been shown to be solvable by a pseudopolynomial-time algorithm in a dynamic-programming fashion similar to PARTITION by Dantzig (1957). MULTIPROCESSOR SCHEDULING and SEQUENCING WITHIN INTERVALS have both been shown to be NP-complete in the strong sense. The same is true for 3-PARTITION, which is similar to PARTITION but has m (instead of two) disjoint sets and each of the disjoint sets has exactly three elements. By considering subproblems created by placing restrictions on one or more of the natural problem parameters, useful information about what types of algorithms are possible for use with the general problem can be gleaned.

6.5 NP-Hardness

The techniques used for proving NP-completeness can also be used for proving problems outside of NP are hard. Any decision problem Π, whether a member of NP or not, to which an NP-complete problem can be transformed will have the property that it cannot be solved in polynomial time unless P = NP. Such a problem Π is NP-hard, since it is at least as hard as the NP-complete problems. Similarly, a search (optimization) problem Π, whether a member of NP or not, is NP-hard if there exists some NP-complete problem Π′ that reduces to Π. Furthermore, by the previous association of languages with string relations and decision problems with search problems, it can be said then that all NP-complete languages (i.e., all NP-complete problems) are NP-hard. So NP-complete $\subseteq$ NP and NP-complete $\subseteq$ NP-hard; that is, NP-complete = NP-hard $\cap$ NP. Whenever it is shown that a polynomial-time algorithm for the optimization problem can be used to solve the corresponding decision problem in polynomial time, a reduction between them is being made, and hence an NP-completeness result for the decision problem can be translated into an NP-hardness result for the optimization problem. That is, showing that an NP-complete decision problem is no harder than its optimization problem, along with the fact that the decision problem is NP-complete, constitutes a proof that the optimization problem is NP-hard. NP-hard can be interpreted as meaning "as hard as the hardest problem in NP" though the term has differing meanings in some of the literature (the definition of NP-complete, however, is generally standardized and universally understood).

Things that make a problem hard include dividing work or resources up evenly, making sequencing decisions that depend on current and past positions, splitting a set of objects into subsets where each must satisfy a constraint, finding a largest substructure that satisfies some property, maximizing or minimizing intersections or unions of sets, and interactions among three or more objects or classes of constraints (Tovey, 2002) (the BIN-PACKING problem is NP-hard, though there are fast, approximate solutions such as first-fit, first-fit-decreasing, and best-fit; Hu and Shing, 2002; Garey and Johnson, 1979). Problems that are not susceptible to solution by dynamic programming are called unary NP-hard or NP-hard in the strong sense (defined similarly to strong NP-completeness). For example, dynamic programming works well for 2-MACHINE-LINE BALANCING; it works but is slower by a factor of n for 3-MACHINE-LINE BALANCING; it is impractical for 5-MACHINE-LINE BALANCING. Theoretically, if m is allowed to vary and get large, the m-machine problem is like 3-PARTITION or BIN PACKING and is NP-hard in the strong sense; although the problem is not NP-hard in the strong sense for any fixed value of m, $m = 5$ is large, practically speaking.

6.6 Overview of NP-Complete Problems

To summarize the previously described classes informally, a problem is in the class NP if a guessed answer can be checked against the instance's threshold(s) in polynomial time. A problem is in the class NP-hard if a known NP-complete problem can be transformed to it in polynomial time. A problem is in the class NP-complete if both these requirements are met.

6.7 Solving NP-Complete Problems

There are normally considered to be two general categories of techniques for addressing NP-complete problems (Garey and Johnson, 1979).

The first category includes approaches that attempt to improve upon exhaustive search as much as possible. If realistic cases that are modeled using integer programming are not very large, commercial math-programming software may be effective. Among the most widely used approaches to reducing the search effort of problems with exponential time complexity are those based on branch-and-bound or implicit enumeration techniques (see, e.g., Garfinkel and Nemhauser, 1972). These generate "partial solutions" within a tree-structured search format and utilize powerful bounding methods to recognize partial solutions that cannot possibly be extended to actual solutions, thereby eliminating entire branches of the search in a single step. Other approaches include dynamic programming, cutting plane methods (see Hu, 1969; Garfinkel and Nemhauser, 1972), and Lagrangian techniques (Geoffrion, 1974; Held and Karp, 1971). Dynamic programming can solve KNAPSACK, PARTITION, and similar problems as long as the instance is not too large. However, other NP-hard problems including 3-PARTITION and BIN PACKING are not susceptible to quick solution by dynamic programming. Dynamic programming of a KNAPSACK problem that is made up of integers and having a knapsack size of a uses a table that is x by $a + 1$ in size, requiring $O(ax)$ work. This algorithm is pseudopolynomial time because it is only fast when a is small. As discussed previously, if the dynamic-programming algorithm is polynomial in the parameters of problem size and data length (note, however, that if a were an 80-digit number, the required table could not fit into the combined disk memory of all Internet-linked computers; Tovey, 2002), the algorithm is pseudopolynomial time. In practice, dynamic-programming algorithms are apt to run out of memory before they bog down with CPU usage (Tovey, 2002).

The second category allows for the attainment of suboptimal (though ideally, near-optimal or optimal) solutions—to optimization problems only—in a reasonable amount of time through the use of heuristics. These are especially effective when one is unsure of the model (e.g., when the objective can only be approximately quantified), when the data is not accurate, when an exact solution is not required, when the problem is transient or unstable and robustness in the solution method becomes important, when the instances are enormous or the problem is very difficult in practice (e.g., JOB-SHOP SCHEDULING or the QUADRATIC ASSIGNMENT PROBLEM), when solution speed is critical, or when the necessary expertise is not available. Heuristic approaches are frequently based on intuitive rules-of-thumb. These methods tend to be problem specific. The most widely applied technique is that of a *neighborhood search* where a preselected set of local operations is used to repeatedly improve an initial solution, continuing until no further local improvements can be made and a "locally optimum" solution has been obtained. Heuristic methods designed via this and other approaches have often proved successful in practice, although a considerable amount of problem-specific fine-tuning is usually required in order to achieve satisfactory performance. As a consequence, it is rarely possible to predict how well such methods will perform by formally analyzing them beforehand. Instead, these heuristics are usually evaluated and compared through a combination of empirical studies and commonsense arguments.

Papadimitriou and Steiglitz (1998) further expand on these two categories to allow for six alternatives to address NP-complete problems. In order to solve an exponential-time problem, they propose one of the following approaches be used:

Approximation: Application of an algorithm that quickly finds a suboptimal solution that is within a certain (known) range or percentage of the optimal solution.

Probabilistic: Application of an algorithm that provably yields good average runtime behavior for a distribution of the problem instances.

Special cases: Application of an algorithm that is provably fast if the problem instances belong to a certain special case.

Exponential: Strictly speaking, pseudopolynomial-time algorithms are exponential; also, many effective search techniques (e.g., branch-and-bound, simplex) have exponential worst-case complexity.

Local search: Recognized to be one of the most effective search techniques for hard combinatorial optimization problems, local (or *neighborhood*) search is the discrete analog of "hill climbing."

Heuristic: Application of an algorithm that works reasonably well on many cases, but cannot be proven to always be fast or optimal.

In Part II of this book, several of these search techniques are demonstrated using some of the established DLBP instances.

6.8 Other Classes of Problems

Tovey (2002) provides a detailed explanation of other classes of problems that are widely believed to be harder than both the NP-complete and the co-NP-complete (i.e., believed to not be in NP; e.g., co-TSP: is there *no* Hamiltonian cycle with cost $> c$, i.e., do all of the Hamiltonian cycles cost $> c$?—requires exponential time to check) classes. The most important of these is the class of PSPACE-complete time. Usually if a problem is PSPACE-complete, it will be harder to solve in practice than an NP-complete problem; it probably cannot even be verified in polynomial time. A problem of this type is typically found in combinatorial games such as those involving an alternating sequence of difficult decisions between two opposing sides. Many familiar games, such as chess and checkers, are in this category. Several important operations research and management science problems are as well, including stochastic scheduling, periodic scheduling, Markov decision problems, queueing networks, and systems of differential equations. Adding stochasticity or uncertainty generally makes hard problems substantially harder. Finally, computational complexity theory provides little information about continuous nonlinear problems. In practice, convexity is usually the dividing line between easy and hard optimization (nonsmoothness can also make things difficult). The DLBP's objective function is nonlinear.

6.9 Heuristic Performance Analysis

Of interest in the assessment of heuristics is their "worst-case" and "average-case" performance (Tovey, 2002). Average-case analysis often involves simply devising a benchmark (a set of supposedly typical instances), running the heuristic and its competitors on them, and comparing the results. Some experiments that have been done tend to confirm the suspicion that average behavior is generally much better than worst-case behavior (Garey and Johnson, 1979). It should be noted that designing a benchmark is considered to be difficult and it can affect the results. However, in order to compute average-case performance, some probability distribution on the instances must first be assumed and it is often not clear what distribution to choose (one way to avoid this is to use algorithms that do their own randomizing; Rabin, 1976). The distributions under which heuristics can be analyzed with currently available techniques can be quite different from the actual distributions that occur in practice, where instances tend to be highly structured (including biases that can be difficult to capture mathematically) and distributions that can change in unpredictable ways as time passes. Moreover, average-case results do not reveal anything about how heuristics will perform for particular instances, whereas worst-case results guarantee at least a bound on this performance. Therefore, it is preferential to analyze them in as many ways as possible, including both worst-case and average-case perspectives. These are included as part of the analysis in this book.

6.10 Hardware and Software Considerations

When implementing any chosen algorithms—and especially when comparing multiple algorithms—in software, there are a variety of items to consider. When a single algorithm is applied, taking advantage of available commercial, academic, or open-source software can save significant time and can be expected to minimize the chance of erroneous results. However, if multiple algorithms are being compared, it may be desirable that all of the search algorithm software used be expressly designed, engineered, written, refined, and tested by the researcher; that is, not make use of any off-the-shelf programs. This provides consistency in software architecture, data structures, and, ultimately, the time complexity. In addition, any sections of software code that could be used by multiple programs should be reused (*software reusability* can readily be performed using, e.g., header files in languages with a structure similar to that of the American National Standards Institute's C++). After the software engineering is completed, each subroutine and each entire program should then be investigated on a variety of test cases. The language and programming environment used (e.g., *Visual C++*) and the computer system used (e.g., 1.6 GHz PM x86-family workstation with 512 megabytes of random access memory) should also be documented as part of the final results.

During data collection using case-study instances, no programs other than the executable under evaluation and those required by the operating system should be either running or open. The computer software implementation of any search algorithm should be run at least three times to obtain an average of the computation time. Solution data for use in efficacy analysis should be collected a minimum of five times when methodologies possess a probabilistic component (e.g., genetic algorithm and ant colony optimization) with the results averaged to avoid reporting unusually favorable or unusually poor results. The data collection should ideally consist of more runs, however, noting that statistically, less than 30 is usually considered a "small" sample while a "large" sample is typically not recognized until 100 (note that these values correspond to Student-*t* and normal distribution table sizes). It is especially desirable that in however many runs are used, no wide variation in either runtime or solution efficacy is seen. A small number of runs may be more practical if it is desired that the total time any software is running be minimized in order to reduce the total runtime (which may be on the order of hours or even days for some search techniques and their associated parameters) in order to ensure memory leaks, garbage collection, and other operating system self-maintenance does not corrupt or skew the results.

PART II

Disassembly-Line Balancing

CHAPTER 7
Disassembly-Line Balancing Overview

CHAPTER 8
Description of the Disassembly Line and the Mathematical Model

CHAPTER 9
Computational Complexity of DLBP

CHAPTER 10
Combinatorial Optimization Searches

CHAPTER 11
Experimental Instances

CHAPTER 12
Analytical Methodologies

CHAPTER 13
Exhaustive Search

CHAPTER 14
Genetic Algorithm

CHAPTER 15
Ant Colony Optimization

CHAPTER 16
Greedy Algorihm

CHAPTER 17
Greedy/Adjacent Element Hill Climbing Hybrid

CHAPTER 18
Greedy/2-Opt Hybrid

CHAPTER 19
H-K Heuristic

CHAPTER 20
Quantitative and Qualitative Comparative Analysis

CHAPTER 21
Other Disassembly-Line Balancing Research

CHAPTER 7

Disassembly-Line Balancing Overview

7.1 Introduction

This chapter discusses the specific problem studied in Part II of this book in Sec. 7.2. Section 7.3 addresses considerations and assumptions about the DISASSEMBLY LINE BALANCING PROBLEM (DLBP) model and Sec. 7.4 provides the balancing objectives.

7.2 The Disassembly-Line Balancing Problem

In a general product recovery system, end-of-life products are taken back from their last owners, in some cases to fulfill a certain demand for their items (components) and materials. After retrieval, the end-of-life products are sent to a product recovery facility where they are sorted, cleaned, and prepared for further processing. The products are then sent to a disassembly facility where they are disassembled for their items and materials. Items that are demanded for reuse and those to be stored are disassembled using *nondestructive disassembly.* The rest of the items, which are demanded for recycling or those subject to disposal, are potentially disassembled by using *destructive disassembly*.

Disassembly is one of the most important elements of product recovery operations. In product recovery operations the main options are reuse, remanufacturing, recycling, storage, and proper disposal. Often, the end-of-life product has to be disassembled prior to choosing any product recovery option. With the expense involved in the labor-intensive disassembly process, an efficient means of disassembly is desirable; this is generally considered to be a line. However, paced lines can hold inefficiencies as well, so the line must be designed to require the minimum number of workers (workstations) and ensure similar work times at each station (station time).

A disassembly line has other concerns, unique to product recovery and not found in assembly. One disassembly-unique objective of a product recovery system is the need to determine the optimal number of end-of-life products to be ordered from the last owners to meet the demand of items to be disassembled for reuse, recycling, storage, and disposal. Also, disassembly operations are often associated with high levels of uncertainty, which adds to the unpredictability of the system. This is because most end-of-life products are exposed to varying conditions during their useful lives. For example, some items in the product structure may have been broken and/or replaced with other items, or the joint types might have been changed obstructing the planned disassembly operation. Other factors include potential environmental damage, the type of motions required to orient a part for removal, types of part removal tools required, selection of complete or *incomplete (partial) disassembly*, product part additions/deletions/modifications, destructive or nondestructive disassembly, and others as can be found in Güngör and Gupta (2002).

Disassembly sequencing provides the order in which the components of the end-of-life products are disassembled on the line. The main challenge in generating sequences is the computational complexity of the associated optimization problem. As with most sequencing and scheduling problems, generating an optimal combinatoric ordering is expected to be an NP-complete problem; as the number of items in the product structure grows, the time required to solve the problem increases exponentially.

One of the ways of dealing with increasing complexity is to employ heuristic methodologies to solve the problem. Many researchers have utilized the heuristics used in NP-complete scheduling and sequencing problems due to their ability to reduce high calculation times. Even though these methods do not always guarantee optimal solutions they often provide acceptable or near-optimal solutions.

In essence, Part II of this book addresses the unique characteristics of the paced disassembly line while providing efficient methodologies for balancing the line and providing a feasible disassembly sequence. Since single-objective models fall short in handling these problems, multicriteria decision-making models are considered for all cases. Precedence relationships and other constraints are also accounted for in each methodology.

7.3 Model Considerations

The disassembly line can be formulated as a scheduling problem. Scheduling problems are described in a triplet

$$\alpha \mid \beta \mid \gamma$$

where the single-entry α-field describes the machine environment, the β-field lists processing characteristics and constraints (and may contain one, none, or multiple entries), and the single-entry γ-field lists the objective to be minimized (Pinedo, 2002).

A basic disassembly line can be modeled as a first-in first-out (FIFO) flow shop (Fm). An FIFO flow shop is referenced as a permutation flow shop with $\alpha = Fm$ and β = prmu. The disassembly line also has precedence constraints that add the term "prec" to the β-field. In its most basic form, the objective can be thought of as minimizing the makespan C_{max} [equal to $\max(C_1, C_2, \ldots, C_n)$ and defined as the completion time of the last task to be completed in the flow shop—effectively seeking to minimize the total number of workstations and/or the cycle times at each workstation, though not necessarily for the last one]. This would make $\gamma = C_{max}$ with the resulting disassembly-line description formed as

$$Fm \mid \text{prec, prmu} \mid C_{max}$$

The DLBP investigated in this book is concerned with the paced disassembly line for a single model of product that undergoes complete disassembly. Part removal times are treated as being known and discrete; hence they do not posses any probabilistic component. A summary of model assumptions is listed at the end of this section.

The desired solution to a DLBP instance consists of an ordered sequence (i.e., n-tuple) of work elements (also referred to as tasks, components, or parts). For example, if a solution consisted of the eight-tuple ⟨5, 2, 8, 1, 4, 7, 6, 3⟩, then component 5 would be removed first, followed by component 2, then component 8, and so on.

While different authors use a variety of definitions for the term "balanced" in reference to assembly (Elsayed and Boucher, 1994) and disassembly lines, because many of these uses never actually address equalizing idle times (i.e., the balancing of the line) the following definition (McGovern et al., 2003; McGovern and Gupta, 2006c) is used consistently throughout this book.

Definition 7.1

A disassembly line is optimally *balanced* when the fewest possible number of workstations is needed and the variation in idle times between all workstations is minimized, while observing all constraints. This is mathematically described by

$$\text{Minimize NWS}$$

then

$$\text{Minimize } [\max(\text{ST}_x) - \min(\text{ST}_y)] \ \forall\, x, y \in \{1, 2, \ldots, \text{NWS}\}.$$

Line balancing can be visualized as in Fig. 7.1. The five large boxes represent workstations with the total height of these boxes indicating cycle time CT (the maximum time available at each workstation). The smaller numbered boxes represent each part (1 through

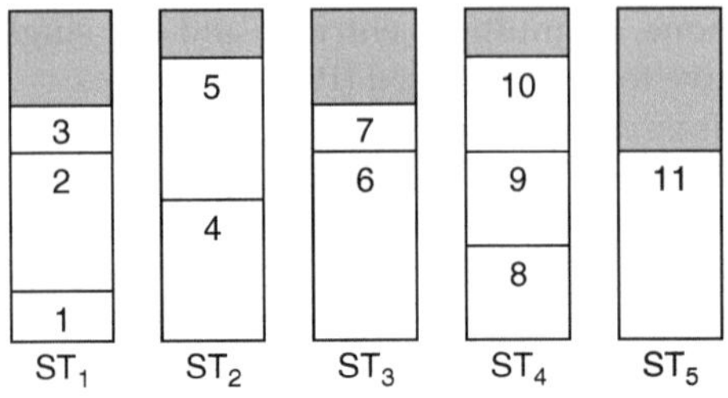

Figure 7.1 Line-balancing depiction.

11 here) with each being proportionate in height to its corresponding part removal time (and the sum of these in each workstation j totaling that workstation's station time ST_j). The gray area is indicative of the idle time I_j at each workstation j.

Except for some minor variation in the greedy approach, all of the combinatorial optimization techniques demonstrated here use a similar methodology to address the multicriteria aspects of the DLBP. Since measure of balance is the primary consideration in this book, additional objectives are only considered subsequently; that is, the methodologies first seek to select the best performing measure of balance solution; equal balance solutions are then evaluated for hazardous part removal positions; equal balance and hazard measure solutions are evaluated for high-demand part removal positions; and equal balance, hazard measure, and high-demand part removal position solutions are evaluated for the number of direction changes. This priority ranking approach is selected over a weighting scheme for its simplicity, ease in reranking the priorities, ease in expanding or reducing the number of priorities, due to the fact that other weighting methods can be readily addressed at a later time, and primarily to enable unencumbered *efficacy* (a method's effectiveness in finding good solutions) analysis of the combinatorial optimization methodologies and problem data instances under consideration (Sec. 12.3 provides a more detailed explanation of the selection and formulation of the multicriteria approach).

A mathematical model has been formulated to quantitatively address the DLBP (McGovern et al., 2003; McGovern and Gupta, 2003a, 2007a). Model assumptions include the following:

- The line is paced.
- Part removal times are deterministic, constant, and discrete (or able to be converted to integer form).
- Part removal times may refer to virtual parts (i.e., tasks not associated with a unique part's immediate removal).
- There is no preemption (i.e., tasks are not interrupted, once a task starts, it is completed).

- Each product undergoes complete disassembly (even if demands are zero).
- All products contain all parts with no additions, deletions, modifications, or physical defects.
- Each task is assigned to one and only one workstation.
- The sum of the part removal times of all the tasks assigned to a workstation must not exceed the cycle time.
- The precedence relationships among the parts must not be violated.

7.4 Balancing Objectives

The overall purpose of this book is to detail the development of environmentally benign and economically sound product recovery systems. All the demonstrated methodologies aim at specific objectives while fulfilling this primary goal.

In attaining this, Part II of this book establishes a first goal of accurately describing and then mathematically modeling the multicriteria DISASSEMBLY LINE BALANCING PROBLEM. The second goal is to establish exactly how difficult a problem disassembly-line balancing is by using the field of complexity theory. The next goal is the determination of data sets and evaluation criteria for use in the analysis of the problem as well as in analysis of the solution-generating techniques. Based on the complexity theory results, appropriate methodologies are then employed and compared in solving the DISASSEMBLY LINE BALANCING PROBLEM (i.e., generating feasible disassembly sequences that balance the disassembly line and address other product recovery concerns).

These goals are achieved by making use of several fields of study. These include set theory, logic, probability, counting techniques (generating functions), statistics, computer science, software engineering, complexity theory, combinatorial optimization, multicriteria decision-making, algorithm design, data structures, heuristics/metaheuristics/hybrids, scheduling and sequencing, production analysis, graph theory, combinatorics, simulation, manufacturing, discrete mathematics, and operations research.

CHAPTER 8

Description of the Disassembly Line and the Mathematical Model

8.1 Introduction

This chapter provides the mathematical foundations for the study of the DISASSEMBLY LINE BALANCING PROBLEM (DLBP). The multiple objectives for the DLBP are first described, and then each of these is generated using mathematical formulae. These formulae are essential in enabling a thorough efficacy analysis of a given solution sequence as performed in later chapters and as may be expected to be conducted in future studies.

The chapter is organized as follows: Section 8.2 explains the five objectives in solving the DISASSEMBLY LINE BALANCING PROBLEM, provides a generalized description of the problem, and generates the first descriptive formulae. Section 8.3 presents the balance measure, the lower bound on the number of workstations and its proof, the upper bound on the number of workstations, and upper and lower bounds of the balance measure. Section 8.4 presents the hazardous part measure and provides its lower and upper theoretical bounds. Section 8.5 presents the high-demand parts measure and provides its lower and upper theoretical bounds. Section 8.6 presents the part removal direction measure and provides its lower and upper theoretical bounds. Section 8.7 explains extensions to these formulae, while Sec. 8.8 details the use of matrices in formatting the precedence relationships, primarily for use by computer software.

Note that the upper and lower bounds on the balance, hazard, demand, and direction formulae and their theoretical bounds are dependent upon favorable precedence constraints that will allow for generation of these measures. In addition, these bounds are only applicable to the individual measures and may not be attainable when all measures are considered simultaneously.

8.2 Problem Overview

The particular problem investigated in Part II seeks to fulfill the following five objectives:

1. Minimize the number of disassembly workstations and hence, minimize the total idle time.
2. Ensure the idle times at each workstation are similar.
3. Remove hazardous components early in the disassembly sequence.
4. Remove high-demand components before low-demand components.
5. Minimize the number of direction changes required for disassembly.

A major constraint is the requirement to provide a feasible (i.e., precedence preserving) disassembly sequence for the product being investigated. The result is modeled as an integer, deterministic, n-dimensional, multicriteria decision-making problem with an exponentially growing search space (where n represents the number of parts for removal). Testing a given solution against the precedence constraints fulfills the major constraint of precedence preservation. Minimizing the sum of the workstation idle times I, which will also minimize the total number of workstations NWS, attains objective 1 and is described by

$$I = (\mathrm{NWS} \cdot \mathrm{CT}) - \sum_{k=1}^{n} \mathrm{PRT}_k \tag{8.1}$$

or

$$I = \sum_{j=1}^{\mathrm{NWS}} (\mathrm{CT} - \mathrm{ST}_j) \tag{8.2}$$

This objective is represented as

$$\text{Minimize } Z_1 = \sum_{j=1}^{\mathrm{NWS}} (\mathrm{CT} - \mathrm{ST}_j) \tag{8.3}$$

where each station time ST_j is given by

$$\mathrm{ST}_j = \sum_{k=1}^{n} \mathrm{PRT}_k \cdot X_{k,j} \; \forall j \tag{8.4}$$

and

$$X_{k,j} = \begin{cases} 1, \text{ if part } k \text{ is assigned to workstation } j \\ 0, \text{ otherwise} \end{cases} \tag{8.5}$$

8.3 Balance Measure and Theoretical Bounds Formulation

Line balancing seeks to achieve *perfect balance* (all idle times equal to zero). When this is not achievable, either *line efficiency* or the *smoothness index* is often used as a performance evaluation tool (Elsayed and Boucher, 1994).

The smoothness index rewards similar idle times at each workstation, but at the expense of allowing for a large (suboptimal) number of workstations. This is because the smoothness index compares workstation-elapsed times to the largest ST_j instead of to CT. [This index is very similar in format to the *sample standard deviation* from the field of statistics, but using $\max(ST_j) \mid j \in \{1, 2, \ldots, NWS\}$ rather than the mean of the station times.] Line efficiency rewards the minimum number of workstations but allows unlimited *variance* in idle times between workstations because no comparison is made between ST_j's. This text makes use of a measure of balance that combines the two and is easier to calculate. The balancing method developed by McGovern et al. (2003; McGovern and Gupta, 2003b) seeks to simultaneously minimize the number of workstations while ensuring that idle times at each workstation are similar, though at the expense of the generation of a nonlinear objective function. The method is computed based on the minimum number of workstations required as well as the sum of the square of the idle times for each of the workstations. This penalizes solutions where, even though the number of workstations may be minimized, one or more have an exorbitant amount of idle time when compared to the other workstations. It also provides for leveling the workload between different workstations on the disassembly line. Therefore, a resulting minimum numerical performance value is the more desirable solution, indicating both a minimum number of workstations and similar idle times across all workstations. The measure of balance F is represented as

$$F = \sum_{j=1}^{NWS} (CT - ST_j)^2 \tag{8.6}$$

with the DLBP balancing objective represented as

$$\text{Minimize } Z_2 = \sum_{j=1}^{NWS} (CT - ST_j)^2 \tag{8.7}$$

Perfect balance is indicated by

$$Z_2 = 0 \tag{8.8}$$

Note that mathematically, Eq. (8.7) effectively makes Eq. (8.3) redundant due to the fact that it concurrently minimizes the number of workstations.

Theorem 8.1 Let PRT_k be the part removal time for the kth of n parts where CT is the maximum amount of time available to complete all tasks assigned to each workstation. Then for the most efficient distribution of tasks, the lower bound on the number of workstations $\text{NWS}_{\text{lower}}$ satisfies

$$\text{NWS}_{\text{lower}} = \left\lceil \frac{\sum_{k=1}^{n} \text{PRT}_k}{\text{CT}} \right\rceil \tag{8.9}$$

Proof: If the above equality is not satisfied, then there must be at least one workstation completing tasks requiring more than CT of time, which is a contradiction.

Subsequent bounds are shown to be true in a similar fashion and are presented throughout the chapter without proof.

The optimal number of workstations NWS^* cannot be any better than the lower bound, therefore,

$$\text{NWS}^* \geq \text{NWS}_{\text{lower}} \tag{8.10}$$

The upper bound for the number of workstations $\text{NWS}_{\text{upper}}$ is given by

$$\text{NWS}_{\text{upper}} = n \tag{8.11}$$

therefore,

$$\left\lceil \frac{\sum_{k=1}^{n} \text{PRT}_k}{\text{CT}} \right\rceil \leq \text{NWS} \leq n \tag{8.12}$$

The lower bound F_{lower} on F is seen to be

$$F_{\text{lower}} = \left(\frac{I}{\text{NWS}_{\text{lower}}} \right)^2 \cdot \text{NWS}_{\text{lower}}$$

which simplifies to

$$F_{\text{lower}} = \frac{I^2}{\text{NWS}_{\text{lower}}} \tag{8.13}$$

and is related to the optimal balance F^* by

$$F^* \geq F_{\text{lower}} \tag{8.14}$$

while the upper bound is described by the worst-case balance F_{upper} as

$$F_{\text{upper}} = \sum_{k=1}^{n} (\text{CT} - \text{PRT}_k)^2 \tag{8.15}$$

therefore,

$$\frac{I^2}{\text{NWS}_{\text{lower}}} \leq F \leq \sum_{k=1}^{n} (\text{CT} - \text{PRT}_k)^2 \tag{8.16}$$

The cycle time is bounded by

$$\max(\text{PRT}_k) \leq \text{CT} \leq \sum_{k=1}^{n} \text{PRT}_k \qquad \text{CT} \in \mathbf{N}, \forall k \in P \tag{8.17}$$

where **N** represents set of natural numbers; that is, {0, 1, 2, …} and P is the set of n part removal tasks.

8.4 Hazard Measure and Theoretical Bounds Formulation

A hazard measure H quantifies each solution sequence's performance in removing hazardous parts, with a lower calculated value being more desirable (McGovern and Gupta, 2003a, 2007a). This measure is based on binary variables that indicate whether a part is considered to contain hazardous material (set equal to 1 if the part is hazardous, else 0) and its position in the sequence. A given solution sequence hazard measure is defined as the sum of hazard binary flags multiplied by their position in the solution sequence, thereby rewarding the removal of hazardous parts early in the part removal sequence. This measure is represented as

$$H = \sum_{k=1}^{n} (k \cdot h_{\text{PS}_k}) \qquad h_{\text{PS}_k} = \begin{cases} 1, & \text{hazardous} \\ 0, & \text{otherwise} \end{cases} \tag{8.18}$$

where PS_k identifies the kth part in the solution sequence PS; for example, for solution ⟨3, 1, 2⟩, $\text{PS}_2 = 1$. The DLBP hazardous part objective is represented as

$$\text{Minimize } Z_3 = \sum_{k=1}^{n} (k \cdot h_{\text{PS}_k}) \tag{8.19}$$

The lower bound H_{lower} on H is given by

$$H_{\text{lower}} = \sum_{p=1}^{|\text{HP}|} p \tag{8.20}$$

where the set of hazardous parts HP is defined as

$$\mathrm{HP} = \{k : h_k \neq 0 \ \forall \ k \in P\} \tag{8.21}$$

Its cardinality can be calculated with

$$|\mathrm{HP}| = \sum_{k=1}^{n} h_k \tag{8.22}$$

The lower bound is related to the optimal hazard measure H^* by

$$H^* \geq H_{\text{lower}} \tag{8.23}$$

For example, three hazardous parts would give a best-case H_{lower} value of $1 + 2 + 3 = 6$. The upper bound on the hazardous part measure is given by

$$H_{\text{upper}} = \sum_{p=n-|\mathrm{HP}|+1}^{n} p \tag{8.24}$$

For example, three hazardous parts in a total of twenty would give a worst-case H_{upper} value of $18 + 19 + 20 = 57$. Equations (8.20) and (8.24) are combined to give

$$\sum_{p=1}^{|\mathrm{HP}|} p \leq H \leq \sum_{p=n-|\mathrm{HP}|+1}^{n} p \tag{8.25}$$

8.5 Demand Measure and Theoretical Bounds Formulation

A demand measure D quantifies each solution sequence's performance in removing high-demand parts, with a lower calculated value being more desirable (McGovern and Gupta, 2003a, 2007a). This measure is based on positive integer values that indicate the quantity required of a given part after it is removed—or 0 if it is not desired—and its position in the sequence. A solution sequence demand measure is then defined as the sum of the demand value multiplied by that part's position in the sequence, rewarding the removal of high-demand parts early in the part removal sequence. This measure is represented as

$$D = \sum_{k=1}^{n} (k \cdot d_{\mathrm{PS}_k}) \qquad d_{\mathrm{PS}_k} \in \mathbf{N} \ \forall \ \mathrm{PS}_k. \tag{8.26}$$

The DLBP part-demand objective is represented as

$$\text{Minimize } Z_4 = \sum_{k=1}^{n} (k \cdot d_{\mathrm{PS}_k}) \tag{8.27}$$

The lower bound on the demand measure D_{lower} is given by Eq. (8.26) where

$$d_{\text{PS}_1} \geq d_{\text{PS}_2} \geq \cdots \geq d_{\text{PS}_n} \tag{8.28}$$

and is related to the optimal demand measure D^* by

$$D^* \geq D_{\text{lower}} \tag{8.29}$$

For example, three parts with demands of 4, 5, and 6 respectively would give a best-case value of $(1 \cdot 6) + (2 \cdot 5) + (3 \cdot 4) = 28$. The upper bound on the demand measure (D_{upper}) is given by Eq. (8.26) where

$$d_{\text{PS}_1} \leq d_{\text{PS}_2} \leq \cdots \leq d_{\text{PS}_n} \tag{8.30}$$

For example, three parts with demands of 4, 5, and 6 respectively would give a worst-case value of $(1 \cdot 4) + (2 \cdot 5) + (3 \cdot 6) = 32$.

8.6 Direction Measure and Theoretical Bounds Formulation

Finally, a direction measure R quantifies each solution sequence's performance in removing parts with the same orientation together, with a lower calculated value indicating minimal direction changes and hence a more desirable solution (McGovern and Gupta, 2003a, 2007a). This measure is based on a count of the direction changes. Integer values represent each possible direction (typically $r = \{+x, -x, +y, -y, +z, -z\}$; in this case $|r| = 6$). These directions are expressed as

$$r_{\text{PS}_k} = \begin{cases} +1, & \text{direction } +x \\ -1, & \text{direction } -x \\ +2, & \text{direction } +y \\ -2, & \text{direction } -y \\ +3, & \text{direction } +z \\ -3, & \text{direction } -z \end{cases} \tag{8.31}$$

and are easily expanded to other or different directions in a similar manner. The direction measure is represented as

$$R = \sum_{k=1}^{n-1} R_k \qquad R_k = \begin{cases} 1, & r_{\text{PS}_k} \neq r_{\text{PS}_{k+1}} \\ 0, & \text{otherwise} \end{cases} \tag{8.32}$$

with the DLBP part-direction objective represented as

$$\text{Minimize } Z_5 = \sum_{k=1}^{n-1} R_k \tag{8.33}$$

The lower bound R_{lower} on the direction measure R is given by

$$R_{\text{lower}} = |r| - 1 \tag{8.34}$$

and is related to the optimal direction measure R^* by

$$R^* \geq R_{\text{lower}} \tag{8.35}$$

For example, for a given product containing six parts that are installed/removed in directions $(-y, +x, -y, -y, +x, +x)$, the resulting best-case value would be $2 - 1 = 1$ (e.g., one possible R^* solution containing the optimal, single-change of product direction would be: $\langle -y, -y, -y, +x, +x, +x\rangle$). In the specific case where the number of unique direction changes is one less than the total number of parts n, the upper bound on the direction measure would be given by

$$R_{\text{upper}} = |r| \qquad \text{where } |r| = n - 1 \tag{8.36}$$

Otherwise, the measure varies depending on the number of parts having a given removal direction and the total number of removal directions. It is bounded by

$$|r| \leq R_{\text{upper}} \leq n - 1 \qquad \text{where } |r| < n - 1 \tag{8.37}$$

For example, six parts installed/removed in directions $(+x, +x, +x, -y, +x, +x)$ would give an R_{upper} value of 2 as given by the lower bound of Eq. (8.37) with a possible solution sequence of, for instance, $\langle +x, +x, -y, +x, +x, +x\rangle$. Six parts installed/removed in directions $(-y, +x, -y, -y, +x, +x)$ would give an R_{upper} value of $6 - 1 = 5$ as given by the upper bound of Eq. (8.37) with a solution sequence, for example, of $\langle -y, +x, -y, +x, -y, +x\rangle$.

In the special case where each part has a unique removal direction, the measures for R_{lower} and R_{upper} are equal and given by

$$R_{\text{lower}} = R_{\text{upper}} = n - 1 \qquad \text{where } |r| = n \tag{8.38}$$

8.7 Models and Measures as Prototypes

It is important to note that the H, D, and R metrics are also intended as forming the three basic prototypes of any additional disassembly-line evaluation criteria. These three different models are then the basis for developing differing or additional objectives.

The H metric is used as the prototype for any binary criteria; for example, a part could be listed according to the two categories "valuable" and "not valuable."

The D metric is used as the prototype for any known value (integer or real) criteria; for example, a part can be assigned a D-type metric which contains the part's actual dollar value. It could also be used to model a hazard metric where different numbers are assigned to hazardous parts with the size of the assigned value in proportion to the level of hazard (e.g., 0 would represent a part that was of no hazard, a value of 1 would indicate a slight hazard, 2 slightly higher, etc.).

The *R* metric is used as the prototype for any adjacency or grouping criteria; for example, a part could be categorized as "glass," "metal," or "plastic" if it were desirable to remove parts together in this form of grouping. These groupings also have a hazardous material application as well. That is, certain hazardous materials may be desired to be all removed at the same or at nearby workstations. Alternatively, certain materials might be incompatible and could be hazardous or dangerous if combined. In these cases, the *R* metric would be the appropriate choice.

It is again emphasized that attainment of the previously defined hazard, demand, and direction formulae upper and lower bounds is dependent upon favorable precedence constraints that will allow for generation of these optimal or nominal measures.

8.8 Matrices for Precedence Representation

Processing a model's representation on a computer requires that the information about the product be presented in a way that is complete, concise, and lends itself to manipulation in a format suitable for and understandable by a machine. For many researchers, the mathematical format provided by the use of matrices best meets all of these criteria.

One example of presenting the data needed for disassembly calculations in a matrix format can be seen in Li et al. (2010). This format is made up of two matrices: the contacting matrix and the disassembly precedence matrix.

The *contacting matrix* is square matrix of order n. It is a binary matrix [or logical matrix or (0, 1)-matrix] where 0s indicate there is no contact between the part in the column and the part in the row and 1s indicate there is some type of physical contact (not necessarily a bond; e.g., nut and bolt, adhesive, screw, or weld requiring breaking—mechanical, chemical, or otherwise—between the two). As such, this matrix is a symmetric matrix with zeros on the main diagonal. In the example seen in Fig. 8.1, part 5 (i.e., row 5) is in contact with parts 1, 2, 3, and 6 (columns 1, 2, 3, and 6) and vice versa.

The *disassembly precedence matrix* is also a square matrix of order n. It is again a binary matrix with zeros on the main diagonal, however it is not symmetric. This matrix indicates blocking rather than contact. That is, the data in the matrix identifies parts that must be removed (listed in the column header) before a given part can be removed (listed in the row header). Note that the parts that must be removed first do not necessarily need to be in contact with the part they obstruct (see Figs. 8.1 and 8.2, which are both generated from the same notional product); for

FIGURE 8.1 Sample contacting matrix.

	1	2	3	4	5	6
1	0	0	0	0	1	1
2	0	0	0	0	1	1
3	0	0	0	1	1	0
4	0	0	1	0	0	0
5	1	1	1	0	0	1
6	1	1	0	0	1	0

FIGURE 8.2 Sample disassembly precedence matrix.

	1	2	3	4	5	6
1	0	1	1	1	0	1
2	0	0	1	1	0	0
3	0	0	0	1	0	0
4	0	0	0	0	0	0
5	0	0	0	0	0	0
6	0	1	1	1	0	0

example, part 1 and part 5 are in contact, but neither obstructs the other's removal. In the disassembly precedence matrix, 0s indicate there is no blocking between the part in the column and the part in the row, and 1s indicate there is some type of restriction that must be addressed. Since the matrix is not symmetric, the most intuitive way to read the data may be from the column to the row. In the example seen in Fig. 8.2, part 4 (column 4) must be removed before parts 1, 2, 3, or 6 (rows 1, 2, 3, and 6) could possibly be removed. Also, reading from row to column, it can be seen that, for example, part 1 (row 1) cannot be removed until parts 2, 3, 4, and 6 (columns 2, 3, 4, and 6) are removed.

To summarize, the contacting matrix indicates all pairs of parts that are in contact with each other (but not necessarily blocking the removal of one or the other), while the disassembly precedence matrix indicates parts that block the removal of other parts (though the parts are not necessarily in contact).

CHAPTER 9
Computational Complexity of DLBP

9.1 Introduction

While Chap. 8 mathematically defined the DISASSEMBLY LINE BALANCING PROBLEM (DLBP), this chapter proves that it belongs to the class of NP-complete problems, necessitating specialized solution techniques.

In order to attempt to solve the DLBP, its structure and complexity first must be established and understood. In this chapter, Sec. 9.2 proves that the decision version of DLBP is NP-complete. Section 9.3 proves that the decision version of DLBP is NP-complete in the strong sense. Section 9.4 proves that DLBP is NP-hard.

This chapter uses the compact proof format used by Garey and Johnson (1979). Although this format is slightly more detailed than that used by Karp (1972), additional information is still provided in some of the proofs in this chapter including the listing of analogous elements in DLBP and MULTIPROCESSOR SCHEDULING prior to Theorem 9.2 and the discussion showing DLBP $\in$ NP in Theorem 9.2.

9.2 DLBP NP-Completeness

The decision version of DLBP is NP-complete (McGovern and Gupta, 2004a, 2006b, 2007a). This can be shown through a proof by restriction to one of the known NP-complete problems (PARTITION). PARTITION is described by the following.

Instance: A finite set A of tasks, a size $s(a) \in \mathbf{Z}^+$ for each $a \in A$.

Question: Is there a partition $A' \subseteq A$ into two disjoint sets such that $\Sigma_{a \in A'} s(a) = \Sigma_{a \in A-A'} s(a)$?

Theorem 9.1 The decision version of DLBP is NP-complete (verification is omitted for brevity; the decision version of DLBP is easily seen to be solvable in polynomial time by a nondeterministic algorithm; see Theorem 9.2).

Instance: A finite set P of tasks, partial order $\prec$ on P, task time $\mathrm{PRT}_k \in \mathbf{Z}^+$, hazardous part binary value $h_k \in \{0, 1\}$, part demand $d_k \in \mathbf{N}$, and part removal direction $r_k \in \mathbf{Z}$ for each $k \in P$, workstation capacity $\mathrm{CT} \in \mathbf{Z}^+$, number $\mathrm{NWS} \in \mathbf{Z}^+$ of workstations, difference between largest and smallest idle time $V \in \mathbf{N}$, hazard measure $H \in \mathbf{N}$, demand measure $D \in \mathbf{N}$, and direction change measure $R \in \mathbf{N}$.

Question: Is there a partition of P into disjoint sets $P_A, P_B, \ldots, P_{\mathrm{NWS}}$ such that the sum of the sizes of the tasks in each P_X is CT or less, the difference between largest and smallest idle times is V or less, the sum of the hazardous part binary values multiplied by their sequence position is H or less, the sum of the demanded part values multiplied by their sequence position is D or less, the sum of the number of part removal direction changes is R or less, and it obeys the precedence constraints?

Proof: Clearly the decision version of DLBP is in NP (see Theorem 9.2 for details). PARTITION $\leq_P$ DLBP. Restrict to PARTITION by allowing only instances in which $\mathrm{CT} = (\Sigma_{k=1}^{n} \mathrm{PRT}_k)/2 \in \mathbf{Z}^+$, $V = 0$, $\prec$ is empty, and $h_x = h_y$, $d_x = d_y$, $r_x = r_y$ $\forall\, x, y \in P$. Therefore, the decision version of DLBP is NP-complete.

Informally, PARTITION is a subset of DLBP; that is, some DLBP instances look like PARTITION instances. It can also be viewed that DLBP contains PARTITION within it. Other problems that are formulated similarly to DLBP substantiate the suspicion that this problem is NP-complete. In addition, these similar problems can be easily used to prove DLBP belongs to the class NP-complete. They include problems from the areas of storage and retrieval (BIN PACKING), mathematical programming (KNAPSACK), sets and partitions (3-PARTITION), and sequencing and scheduling (MULTIPROCESSOR SCHEDULING, PRECEDENCE CONSTRAINED SCHEDULING).

Note that the addition of precedence constraints (i.e., $\prec$ not empty) can both reduce the search space and may provide promise for an effective pseudopolynomial-time algorithm. This observation may be primarily academic since, unless it is a process similar to branch-and-bound or some type of preprocessing is performed on the instance, many search techniques (including the ones in this book) first consider a solution, and then determine if it is feasible. Also, it is the requirement to check every possibility that causes many problems having partial orders (e.g., PARTIALLY ORDERED KNAPSACK) to be NP-complete. In addition, precedence constraints invariably add a layer of complexity to the search, with problems possessing precedence constraints still being found to be NP-complete (primarily sequencing and scheduling problems such as the previously mentioned MULTIPROCESSOR SCHEDULING and PRECEDENCE CONSTRAINED SCHEDULING problems). Note that just determining if a feasible solution to DLBP exists can be seen to be the same problem as DIRECTED HAMILTONIAN CIRCUIT, an NP-complete problem [e.g., transform HAMILTONIAN CIRCUIT to DIRECTED HAMILTONIAN CIRCUIT by replacing each edge $[p, q]$ in the undirected graph with the two arcs (p, q) and (q, p)]. Finally, with DLBP able to be modeled both as a flow shop and as a multicriteria version of an assembly line, it should come as no surprise that DLBP is NP-complete since it is obviously in NP and the simple assembly-line balancing type I (SALB-I) problem is regularly described as NP-hard; also, the FLOW-SHOP SCHEDULING problem has been proven to be NP-complete (Garey and Johnson, 1979).

9.3 DLBP NP-Completeness in the Strong Sense

The decision version of DLBP is NP-complete in the strong sense (McGovern and Gupta, 2005c, 2006b, 2007c). This can be shown through a proof by restriction to MULTIPROCESSOR SCHEDULING; note that some authors (e.g., Papadimitriou and Steiglitz, 1998) define the problem differently (e.g., they include precedence constraints and require task lengths to be equal). Garey and Johnson (1979) describe MULTIPROCESSOR SCHEDULING as explained next.

Instance: A finite set A of tasks, a length $l(a) \in \mathbf{Z}^+$ for each $a \in A$, a number $m \in \mathbf{Z}^+$ of processors, and a deadline $B \in Z^+$.

Question: Is there a partition $A = A_1 \cup A_2 \cup \cdots \cup A_m$ of A into m disjoint sets such that $\max[\Sigma_{a \in A_i} l(a) : 1 \leq i \leq m] \leq B$?

In DLBP, NWS is equivalent to m in MULTIPROCESSOR SCHEDULING, P is equivalent to A, while CT is equivalent to B.

The next theorem provides the proof.

Theorem 9.2 The decision version of DLBP is NP-complete in the strong sense.

Instance: A finite set P of tasks, partial order $\prec$ on P, task time $\text{PRT}_k \in Z^+$, hazardous part binary value $h_k \in \{0, 1\}$, part demand $d_k \in \mathbf{N}$, and part removal direction $r_k \in \mathbf{Z}$ for each $k \in P$, workstation capacity $\text{CT} \in \mathbf{Z}^+$, number $\text{NWS} \in \mathbf{Z}^+$ of workstations, difference between largest and smallest idle time $V \in \mathbf{N}$, hazard measure $H \in \mathbf{N}$, demand measure $D \in \mathbf{N}$, and direction change measure $R \in \mathbf{N}$.

Question: Is there a partition of P into disjoint sets $P_A, P_B, \ldots, P_{\text{NWS}}$ such that the sum of the sizes of the tasks in each P_X is CT or less, the difference between largest and smallest idle times is V or less, the sum of the hazardous part binary values multiplied by their sequence position is H or less, the sum of the demanded part values multiplied by their sequence position is D or less, the sum of the number of part removal direction changes is R or less, and it obeys the precedence constraints?

Proof: DLBP $\in$ NP. Given an instance, it can be verified in polynomial time if the answer is "yes" by counting the number of disjoint sets and showing that they are NWS or less, summing the sizes of the tasks in each disjoint set and showing that they are CT or less, examining each of the disjoint sets and noting the largest and the smallest idle times then subtracting the smallest from the largest and showing that this value is V or less, summing the hazardous part binary values multiplied by their sequence position and showing this is H or less, summing the demanded part values multiplied by their sequence position and showing that this is D or less, summing the number of changes in part removal direction and showing that this is R or less, and checking that each task has no predecessors listed after it in the sequence.

MULTIPROCESSOR SCHEDULING $\leq_p$ DLBP. Restrict to MULTIPROCESSOR SCHEDULING by allowing only instances in which V = CT, $\prec$ is empty, and $h_x = h_y$, $d_x = d_y$, $r_x = r_y$ $\forall$ $x, y \in P$. Therefore, the decision version of DLBP is NP-complete in the strong sense.

It is easy to see that the BIN-PACKING problem is also similar to DLBP. The BIN-PACKING problem is NP-complete in the strong sense (it includes 3-PARTITION as a special case) which indicates that there is little hope in finding even a pseudopolynomial time optimization algorithm for it.

9.4 DLBP NP-Hardness

DLBP is NP-hard (McGovern and Gupta, 2005b, 2006b). This can be shown in two steps (Garey and Johnson, 1979) by showing

1. The NP-complete decision problem is no harder than its optimization problem.
2. The decision problem is NP-complete.

This constitutes a proof that the optimization problem is NP-hard. Note that SALB-I is claimed to be NP-hard (Elsayed and Boucher, 1994).

Theorem 9.3 DLBP is NP-hard.

Proof: Shown in two steps:

1. The decision version of DLBP is no harder than its optimization version. Both versions require preservation of the precedence constraints. The optimization version of DLBP asks for a sequence that has the minimum difference between largest and smallest idle times among all disjoint sets, the minimum sum of hazardous part binary values multiplied by their sequence position, the minimum sum of demanded part values multiplied by their sequence position, and the minimum number of part removal direction changes. The decision version includes numerical bounds V, H, D, and R as additional parameters and asks whether there exists NWS disjoint sets with a difference between largest and smallest idle times no more than V, a sum of hazardous part binary values multiplied by their sequence position no more than H, a sum of demanded part values multiplied by their sequence position no more than D, and a number of part removal direction changes no more than R. So long as the difference between largest and smallest idle times, the sum of hazardous part binary values multiplied by their sequence position, the sum of demanded part values multiplied by their sequence position, and the number of part removal direction changes is relatively easy to evaluate, the decision problem can be no harder than the corresponding optimization

problem. If a minimum difference between the largest and smallest idle times, a minimum sum of hazardous part binary values multiplied by their sequence position, a minimum sum of demanded part values multiplied by their sequence position, and a minimum number of part removal direction changes could be found for the DLBP optimization problem in polynomial time, then the associated decision problem could be solved in polynomial time. This would require finding the minimum idle time and the maximum idle time from all NWS subsets, compute the difference, and compare that difference between largest and smallest idle times to the given bound V; sum all the hazardous part binary values multiplied by their sequence position and compare that sum to the given bound H; sum all the demanded part values multiplied by their sequence position and compare that sum to the given bound D; and sum all of the part removal direction changes and compare that sum to the given bound R. Therefore, the decision version of DLBP is no harder than its optimization version.

2. From Theorems 9.1 and 9.2, the decision version of DLBP is NP-complete.

Therefore, from (1) and (2), DLBP is NP-hard.

CHAPTER 10

Combinatorial Optimization Searches

10.1 Introduction

Due to the complexity of the DISASSEMBLY LINE BALANCING PROBLEM (DLBP) there is currently no known way to optimally solve even moderately sized instances of the problem; therefore, alternatives must be considered. In fully conducting a combinatorial optimization treatment of this problem, several combinatorial optimization techniques are selected in order to solve the DLBP. While each was selected based upon its unique approach to obtaining a solution, because none can be expected to consistently generate the optimal solution, these methodologies are also analyzed both by studying their solutions to multiple and varied problem instances and then by comparing the solutions generated by each to one another.

This chapter focuses on providing an introduction to the diverse group of seven searches used here. Section 10.2 introduces combinatorial optimization and then lists the four areas from which the techniques used are selected from. Section 10.3 introduces exhaustive search. Section 10.4 reviews the genetic algorithm (GA) and then Sec. 10.5 provides an overview of ant colony optimization (ACO). Section 10.6 reviews the greedy algorithm, while Sec. 10.7 introduces the purpose-built adjacent element hill-climbing heuristic. The *k*-optimal (*k*-opt; note that "*k*-opt" is the search's commonly used name but it has no relation to our use of the variable *k* in the DLBP) heuristic is seen in Sec. 10.8. Finally, the deterministic, uniformed search technique of the H-K general purpose heuristic is detailed in Sec. 10.9.

10.2 Combinatorial Optimization Methodologies

This book demonstrates techniques from combinatorial optimization, a field that combines a variety of solution methodologies to solve *combinatoric* (i.e., the study of discrete objects, often specifically referring to permutations) problems, of which the DLBP is one. Combinatorial optimization is a branch of optimization in applied mathematics and computer science that is related to operations research, algorithm theory, and computational complexity theory. Sometimes referred to as "discrete optimization," the domain of combinatorial optimization is optimization problems where the set of feasible solutions is discrete (or can be reduced to a discrete one) with the goal of finding the best possible solution. Solution methods are generally suboptimal and include heuristics (such as tabu search and simulated annealing) and metaheuristics (such as swarm intelligence and genetic algorithms).

While exhaustive search consistently provides the problem's optimal solution, its time complexity quickly limits its practicality. Combinatorial optimization techniques allow for the generation of a solution, though the solution may often be suboptimal. Here, seven techniques are used for study and comparison. They include two probabilistic distributed intelligent agent metaheuristics, two basic deterministic searches, and two hybrid techniques, as well an optimal exhaustive search to serve as a benchmark.

The probabilistic *distributed intelligent agent* metaheuristics include the genetic algorithm and ant colony optimization; the purely deterministic searches include the greedy algorithm and the H-K general-purpose heuristic; the hybrid techniques include the greedy/hill climbing hybrid heuristic and the greedy/2-optimal (2-opt) heuristic. These processes are then applied to four instances for analysis and evaluation.

From the name of the nineth century Persian mathematician Abu Abdullah Muhammad bin Musa al-Khwarizmi, the word "algorism" originally referred only to the rules of performing arithmetic using Arabic numerals then became "algorithm" by the eighteenth century. The word has now evolved to include all formalized procedures for solving problems or performing tasks. In the field of computer science a *search algorithm* takes a problem and its instance as input and returns a solution, usually after evaluating a number of possible solutions. The set of all possible solutions to a problem instance is the *search space* (the set of all permutations of n for the DLBP; note that this is different from the *feasible search space*, generally a smaller set that consists of all permutations of n that are also feasible solution sequences).

Brute-force search and *British Museum search* are both terms used to refer to exhaustive search, which requires the checking of every possible solution in order to determine the optima, typically by performing a *depth-first search* or *breath-first search* throughout the search space. The drawback is that most search spaces are extremely large and an exhaustive search of a combinatoric problem instance will only take a reasonable amount of time for small examples. Exhaustive search techniques (e.g., pure depth-first or pure breadth-first) will fail to find a solution to any but the smallest instances within any practical length of time.

Blind search, weak search, naïve search, and *uninformed search* are all terms used to refer to algorithms that use the simplest, most intuitive method of searching through the search space, whereas *informed search* algorithms use heuristics to apply knowledge about the structure of the search space. An uninformed search algorithm is one that does not take into account the specific nature of the problem. This allows uninformed searches to be implemented in general, with the same implementation able to be used in a wide range of problems. In this book, the uninformed searches include Exhaustive Search and H-K. While the Exhaustive Search used here is a traditional depth-first search modified for the DISASSEMBLY LINE BALANCING PROBLEM, H-K seeks to take advantage of the benefits of uninformed search while addressing its drawbacks of runtime growth with instance size.

A *heuristic* is any problem-specific step-by-step set of procedures or rules. It is commonly understood to refer to a "rule-of-thumb" and comes from the Greek verb for "to find or discover." In combinatorial optimization it is a general class that refers to any approach that does not provide a formal guarantee of optimality (Papadimitriou and Steiglitz, 1998). *Local search* is a type of heuristic often used in solving combinatorial optimization problems. Given a search space, local search starts from an initial solution (often randomly generated) and then iteratively moves from one possible (*candidate*) solution to another in search of a solution that is both feasible and better performing than the best found so far. This continues until some stopping condition is met, usually when there are no better performing candidates in the neighborhood. Examples of local search include *k*-opt and hill-climbing algorithms. *Hill-climbing* and *k-opt* algorithms stop when a node is reached where all the node's children have lower performance measures. It performs well if the performance measure is locally consistent, that is, if the nodes with the best performance eventually lead to the goal. Limitations of hill climbing include a tendency to get trapped in *local optima,* resulting in an inability to leave and find the desired *global optimum*. This book makes use of hill-climbing and 2-opt local search heuristics. Note that Papadimitriou and Steiglitz (1998) have observed that many metaheuristics are simply clever variants of local search.

Other heuristic searches include greedy search. *Greedy algorithms* adhere to the problem-solving heuristic of making the locally optimum choice at each decision point in search of the global optima. For example, a greedy algorithm for the TRAVELING SALESPERSON PROBLEM (TSP) would be, at each decision point visit the nearest city that has not yet been visited. While greedy algorithms do not consistently find the globally optimal solution (due to committing-decisions being made early in the search process and due to a lack of a global view of the problem), the algorithms are very easy to conceptualize and implement and they have the ability to give at least good approximations to the optimal solution, while taking minimal execution time. Here, a greedy algorithm is first used alone and then as a first stage in two hybrid search processes.

A *hybrid* makes use of two or more solution-generating techniques in sequence, or repeatedly, during each iteration. An example of a hybrid is the *memetic algorithm* that combines local search heuristics with genetic algorithm crossover operators. This book makes use of hybrids that include a two-phase greedy algorithm and hill-climbing heuristic, as well as a two-phase greedy algorithm and 2-opt local search heuristic.

A *metaheuristic* is a general-purpose heuristic that provides a top-level strategy to guide other heuristics in finding feasible solutions within the search space. Metaheuristics have also been defined as "an experimental heuristic method for solving a general class of computational problems by combining user procedures in the hope of obtaining a more efficient or robust procedure." In practice, they are often seen having a stochastic component to allow for perturbation or mutation. "Metaheuristic" adds the Greek prefix "meta" meaning "beyond, in an upper level." Metaheuristics include genetic algorithms and swarm intelligence.

Genetic algorithms originated from the studies of cellular automata conducted by John Holland at the University of Michigan (Holland, 1975). A genetic algorithm is a search technique used to find approximate solutions to combinatorial optimization problems. Genetic algorithms are a particular class of *evolutionary algorithms* that use techniques inspired by evolutionary biology such as inheritance, mutation, natural selection, and recombination (or crossover). They are implemented as a computer simulation in which a population of abstract representations of candidate solutions to an optimization problem evolves toward better solutions. The evolution starts from a population of either completely random or *hot-started* (i.e., preselected) initial solutions and continues for multiple generations. In each generation the fitness of each solution is evaluated, a fixed number of solutions are stochastically selected from the current population (based on their fitness) and modified (mutated or recombined) to form a new population. The process continues until some stopping criterion is reached. This research remained largely theoretical until the mid-1980s and *The First International Conference on Genetic Algorithms* held at the University of Illinois. As academic interest grew, the increase in desktop computational power allowed for practical application of the new technique. In the late 1980s, the first commercially available desktop genetic algorithm *Evolver* became available. Genetic algorithms are now regularly used to solve difficult scheduling, data fitting, trend spotting, budgeting, and other types of applied combinatorial optimization problems.

First published in 1992, by Marco Dorigo as his Ph.D. dissertation in Italy, *ant colony optimization* is a multiagent-based global search for difficult combinatorial optimization problems like the TSP and the QUADRATIC ASSIGNMENT PROBLEM. Inspired by colonies of ants, the ant colony optimization metaheuristic makes use of computer agents known as "ants" that initiate search randomly and lay pheromone to mark trails. The pheromone evaporates over time; however, good trails are reinforced by other ants and their pheromone in subsequent cycles, while poor trails are eventually abandoned. Novel aspects include its constructive methodology of building solutions as well as the concept of information being shared by colony members.

This book makes use of metaheuristics that include a genetic algorithm and an ant colony optimization model.

McGovern et al. (2003) first proposed combinatorial optimization techniques for the DLBP. The seven methodologies investigated here and applied to the DLBP include an optimal algorithm, four heuristics, and two metaheuristics. The optimal algorithm is an uninformed, depth-first exhaustive search. The first metaheuristic considered is a genetic algorithm, which involves randomly generated and hot-started initial populations with crossover and mutation performed over a fixed number of generations. The ant colony optimization metaheuristic considered for the DLBP is an ant system algorithm known as the ant-cycle model. The first of the four heuristics, the greedy algorithm is based on first-fit-decreasing (FFD) rules (developed for the BIN-PACKING problem and effectively used in computer processor scheduling). A hill-climbing algorithm is then used on the greedy solution to ensure that the idle times at each workstation are similar (the hill-climbing heuristic only compares tasks assigned in adjacent workstations; this is done both to conserve search time and to only investigate swapping tasks that will most likely result in a feasible sequence). In addition, a version of a 2-opt algorithm is applied in a hybrid fashion to the greedy solution. Finally, a DLBP implementation of the H-K general-purpose optimization heuristic—a deterministic uninformed search technique based on an enhanced exhaustive search—is then considered. All techniques are structured to seek to preserve precedence relationships and address multicriteria objectives.

10.3 Exhaustive Search

An exhaustive search algorithm is presented for obtaining the optimal solution to small instances of the DLBP. Exhaustive search provides the optimal solution to many problems, including NP-complete problems. However, its exponential time complexity quickly reduces its practicality necessitating the use of combinatorial optimization techniques, which are instrumental in obtaining optimal or near-optimal solutions to problems with intractably large solution spaces.

Exhaustive search is a deterministic method and is one of three processes used here that are not iterative in nature; that is, they are not designed to look for incremental improvements on each new solution found. Both of these observations are trivial when considering exhaustive search since it looks at all answers, with the final result being optimal (therefore, it gives the same answer every time it is run and there is no room for improvement in the quality of the solution).

Chapter 13 reviews the design and structure of a DLBP exhaustive search algorithm and then considers its performance in order to provide a baseline for later comparison with the six other combinatorial optimization searches.

10.4 Genetic Algorithm

A genetic algorithm (a parallel neighborhood, stochastic-directed search technique) provides an environment where solutions continuously crossbreed, mutate, and compete with each other until they evolve into an optimal or near-optimal solution. Due to its structure and search method, a GA is often able to find a global solution, unlike many other heuristics that use hill climbing to find a best solution nearby resulting in only a local optimum. In addition, a GA does not need specific details about a problem nor is the problem's structure relevant; a function can be linear, nonlinear, stochastic, combinatorial, noisy, and so on.

GA has a solution structure defined as a *chromosome,* which is made up of *genes* and generated by two *parent* chromosomes from the *pool* of solutions, each having its own measure of *fitness*. New solutions are generated from old using the techniques of *crossover* (sever parents genes and swap severed sections) R_x and *mutation* (randomly vary genes within a chromosome) R_m. Typically, the main challenge with any genetic algorithm implementation is determining a chromosome representation that remains valid after each generation.

10.5 Ant Colony Optimization

The DISASSEMBLY LINE BALANCING PROBLEM is addressed using an ant colony optimization metaheuristic. The application of an ACO algorithm is instrumental in obtaining optimal or near-optimal solutions to the DLBP's intractably large solution space. The ACO considered here is an *ant system algorithm* known as the *ant-cycle model* (Dorigo et al., 1999), which is then enhanced for the DLBP.

Ant colony optimization is a probabilistic evolutionary algorithm based on a *distributed autocatalytic* (i.e., positive feedback;

Dorigo et al., 1996) *process* (i.e., a process that reinforces itself in a way that causes very rapid convergence; Dorigo et al., 1996) that makes use of *agents* called ants (due to these agents' similar attributes to the insects). Just as a colony of ants can find a short distance to a food source, these ACO agents work cooperatively toward an optimal solution. Multiple agents are placed at multiple starting nodes, such as cities for the TRAVELING SALESPERSON PROBLEM or parts for the DLBP. Each of the *m* ants is allowed to visit all remaining edges. Each ant's possible subsequent steps are evaluated for desirability and each is assigned a proportionate probability. Based on these probabilities, the next step in the tour is randomly selected for each ant. After completing an entire tour, all feasible ants are given the equivalent of additional pheromone, which is added to each step it has taken. All paths are then decreased in their pheromone strength according to a measure of evaporation and the entire process is repeated.

10.6 Greedy Algorithm

Originally referred to by Garey and Johnson (1979) as *nearest neighbor* and attributed to a 1965 paper by Gavett, this search is commonly known as a *greedy algorithm,* possibly due to a book chapter (Lawler et al., 1985) reference by Johnson and Papadimitriou where they describe the process as "being 'greedy'." A greedy strategy always makes the choice that looks the best at the moment. That is, it makes a locally optimal choice in the hope that this choice will lead to a globally optimal solution. Greedy algorithms do not always yield optimal solutions but for many problems, they do (Cormen et al., 2001).

The greedy algorithm considered here is based on first-fit-decreasing rules. The first-fit-decreasing algorithm was developed for the BIN-PACKING problem (Garey and Johnson, 1979) and effectively used in computer processor scheduling. The greedy algorithm is the second of the five deterministic methods (including exhaustive search) used in Part II and is the first of the four methodologies that are considered as heuristics here. It is one of three processes in Part II that are not *iterative* in nature; that is, not designed to look for incremental improvements on each new solution found. Also, the greedy algorithm here is one of the two problem-specific heuristics demonstrated here that are especially designed for the DISASSEMBLY LINE BALANCING PROBLEM and may have little or no application outside of this problem and ones like it (e.g., the BIN-PACKING problem). While greedy heuristics are universally referred to as "greedy algorithms," they do not fit the mathematical definition of an algorithm as given in Sec. 6.2 and this is part of the impetus for using the more general definition of "algorithm" in this book (per Sec. 10.2).

10.7 Adjacent Element Hill-Climbing Heuristic

While the greedy process is very fast, it has no ability to balance the workstations. A hill-climbing heuristic can be used to balance a part removal solution sequence. Hill climbing is an iterated improvement algorithm, basically a gradient descent/ascent. It makes use of an iterative greedy strategy, which is to move in the direction of increasing value. A hill-climbing algorithm evaluates the successor states and keeps only the best one (Hopgood, 1993).

The adjacent element hill-climbing (AEHC) heuristic only compares tasks assigned in adjacent workstations. This is done both to conserve search time (by not investigating all tasks in all workstations) and to only investigate swapping tasks that will most likely result in a feasible sequence (since the farther apart the positional changes, the less likely that precedence will be preserved for both of the tasks exchanged and for all of the tasks between them).

The hill-climbing heuristic is the third of the five deterministic methods used in Part II and is the second of four methodologies that will be classified here as purely heuristic (i.e., unable to be classified as a metaheuristic). As with all hill-climbing searches AEHC is iterative in nature; it is reapplied to each new solution found in search of incremental improvements. Also, the hill-climbing heuristic is the second of two problem-specific heuristics in this research that are especially designed for the DISASSEMBLY LINE BALANCING PROBLEM and may have little or no application outside of this problem and ones like it. While applicable to any BIN-PACKING-type problem, especially those having precedence constraints, AEHC is tailored to the DLBP to take advantage of knowledge about the problem's format and constraints in order to quickly provide a solution that is better balanced.

10.8 *k*-Opt Heuristic

An alternative to AEHC is a *tour improvement procedure*. The best-known tour improvement procedures are the edge exchange procedures. In an *k*-opt algorithm, all exchanges of *k* edges are tested until there is no feasible exchange that improves the current solution; this solution is said to be *k*-optimal. Since the number of operations increases rapidly with increases in k, $k = 2$ and $k = 3$ are most commonly used.

In this text, a modified 2-opt local search algorithm is applied to the DLBP. The 2-opt algorithm is the second (and final) hybrid methodology and the fourth of the five deterministic models used in the Part II demonstrations. It is the third of the four methodologies here listed purely as a heuristic. As with all *k*-opt neighborhood searches, 2-opt is iterative in nature. The algorithm is reapplied to each new solution n-tuple generated in search of neighborhood improvements.

This methodology considers many more part swaps than AEHC, enabling a potential reduction in the number of workstations as well as better overall solutions, though at a significant time complexity cost over AEHC.

10.9 H-K General-Purpose Heuristic

10.9.1 Introduction

Influenced by the "hunter-killer" search tactics of military helicopters, the H-K heuristic is an uninformed search; it methodically moves through the search space regardless of prior, current, or forecast solution results. The H-K general-purpose heuristic is the last of the five deterministic methods used in this study and is the last of the four methodologies referred to in this book as heuristics with no metaheuristic (e.g., GA and ACO) or optimal (e.g., Exhaustive Search) component. While the original H-K concept includes an iterative component with more refined searches taking place on the best solutions found or its use as a first phase in place of random initial solution-generating methods as commonly used by other searches (including GA), the version of H-K used here is not iterative in nature. It is applied once to forward and reverse listings of the data with the single best answer being saved; it is not reapplied to each new solution found in search of incremental improvements.

10.9.2 Heuristic Background and Motivation

Exhaustive search is optimal because it looks at every possible answer. While an optimal solution can be found, this technique is impractical for all but the simplest combinatorial problems due to the explosive growth in search time. In many physical search applications (e.g., antisubmarine warfare, search and rescue) exhaustive search is not possible due to time or sensor limitations. In these cases, it becomes practical to sample the search space and operate under the assumption that, for example, the highest point of land found during the conduct of a limited search either is the highest point in a given search area or is reasonably near the highest point. The search technique (an early version of the H-K heuristic was demonstrated by McGovern and Gupta, 2004b) in this book works by sampling the exhaustive solution set; that is, search the solution space in a method similar to an exhaustive search but in a pattern that skips solutions (conceptually similar to the STEP functionality in a FOR loop as found in computer programming) to significantly minimize the search space (Fig. 10.1; the shading indicates solutions visited, the border represents the search space).

This pattern is analogous to the radar acquisition search pattern known as "spiral scan," the search and rescue pattern of the "expanding square," or the antisubmarine warfare aircraft "magnetic anomaly

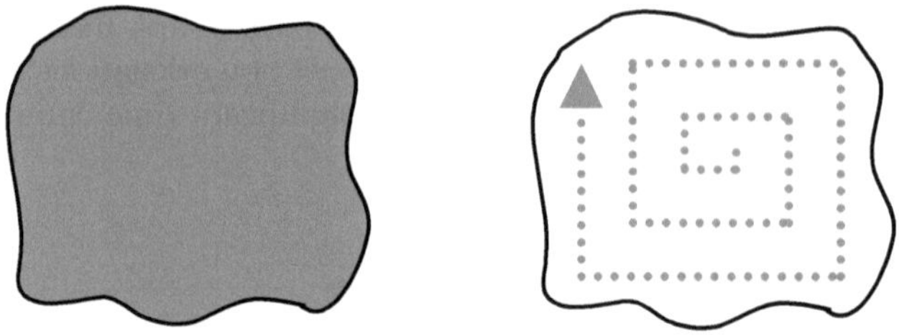

FIGURE 10.1 Exhaustive search space and the H-K search space and methodology.

detector hunting circle." Once the solution is generated, the space can be further searched with additional applications of the H-K heuristic (with modifications from the previous H-K) or the best-to-date solution can be further refined by performing subsequent local searches (such as 2-opt or smaller, localized H-K searches). Depending on the application, H-K can be run once; multiple times on subsequent solutions; multiple times from the same starting point using different skip measure (potentially as a multiprocessor or cluster computing application using parallel algorithms or as a parallel or distributed computing application; e.g., grid computing—a form of network-distributed parallel processing and defined as a model for solving massive computational problems using large numbers of computers arranged as clusters embedded in a distributed telecommunications infrastructure—or cloud computing); multiple times from a different starting point using the same skip measure (again, potentially as a multiprocessor or distributed computing application); or followed up with an H-K or another, differing local search on the best or several of the best suboptimal solutions generated. While termination normally takes place after all sequences are generated for a given skip size, termination can also be effected based on time elapsed or once finding a solution that is within a predetermined bound. H-K can also be used as the first phase of a hybrid algorithm or to hot start another methodology (e.g., to provide the initial population in a GA). One interesting use for H-K is application to the unusual problem where quantifying a small improvement (i.e., a greedy decision, such as would be found in ant colony optimization where the ant agents build a solution incrementally and, therefore, need to know which of the available solution elements reflects an improvement) is not possible or is not understood, or where the incremental greedy improvements may not lead to a global optima. Finally, H-K would also be useful in quickly gathering a sampling of the solution space to allow for a statistical or other study of the data (e.g., H-K could enable the determination of the approximate worst-case and best-case solutions as well as solution efficacy indices mean, median, and mode).

The solutions considered by H-K as shown here consist of an ordered sequence (i.e., n-tuple) of elements. The *skip size* ψ, or more generally ψ_k (the kth element's skip measure; i.e., for the solution's third element, visit every second possible task for $\psi_3 = 2$), is an integer that can be as small as $\psi = 1$ or as large as $\psi = n$ (where n is the number of parts). Since $\psi = 1$ is equivalent to exhaustive search and $\psi = n$ generates a trivial solution (it returns only one solution, that being the data in the same sequence as it is given to H-K, that is $PS_k = \langle 1, 2, 3, \ldots, n \rangle$ where PS_k identifies the kth element in a solution sequence PS; e.g., for solution $\langle 3, 1, 2 \rangle$, $PS_2 = 1$; also, in the single-phase H-K this solution is already considered by any value of ψ), in general all skip values can be further constrained as

$$2 \le \psi_k \le n - 1 : \psi_k \quad n \in \mathbf{Z}^+ \tag{10.1}$$

where $\mathbf{Z}^+$ represents set of positive integers; that is, $\{1, 2, \ldots\}$.

10.9.3 Comparison to Other Methodologies

The H-K general-purpose heuristic shows a variety of similarities to many currently accepted combinatorial optimization methodologies. For example, like ACO and GA—but unlike tabu search—H-K generates the best solutions when suboptimal solutions bear a resemblance to the optimal solution. In three dimensions, this would appear as a set having topography with relatively shallow gradients (i.e., no sharp spikes); the more shallow the gradient, the better the chance that the found solution is near the optimal solution. These data sets are effective in ACO applications since the ant agents work to reinforce good solution sections by adding trail (similar to pheromones in actual ants). They are effective in GA applications since those algorithms break apart good solutions and recombine them with other good solution sections, again, reinforcing preferred tours. The drawback to these is that if suboptimal solutions do not bear a resemblance to the optimal solution, the optimal solution may not be found. However, in many applications it is not typical for the data to perform in this manner. Also, it is not a significant drawback since the general H-K takes a deterministic, nonweighted path; therefore, an isolated solution is as likely to be visited as the optimal shallow gradient solution. Isolated or steep gradient solutions are addressed in ACO and GA with probabilistically selected tours, allowing for potential excursions to seemingly poor-performing areas of the search space.

Like simulated annealing—and unlike ACO, GA, and tabu—the multiphase version of H-K looks at a large area initially, then smaller and smaller areas in more detail.

Like tabu search (and unlike ACO and GA), the single-phase version of H-K does not revisit solutions found previously during the same phase of search.

Unlike many applications of ACO, tabu, and simulated annealing, (but like GA) H-K does not generate solutions similar to a branch-and-bound type of method where the best solution is grown by making decisions on which branch to take as the sequence moves from beginning to end. Rather, an entire solution sequence is generated prior to evaluation. In addition, H-K does not contain any greedy component since the movement through the search space is uninformed and no branching decisions are required, which are typically made based upon the best short-term (greedy) move.

Like branch-and-bound, in the version of H-K that considers partial orders (such as precedence constraints for the DLBP), if a vertex is visited that would result in an infeasible solution, no further investigation of any of the subsequent children of that vertex is ever made, effectively removing that branch from the digraph.

As with each of these heuristics, H-K can be applied to a wide range of combinatorial optimization problems. Like greedy and *k*-opt heuristics (Lawler et al., 1985) and traditional mathematical programming techniques (such as the simplex algorithm for linear programming), H-K selects solutions for consideration based in a deterministic manner, while metaheuristics such as ACO and GA have a strong probabilistic component. As a result, H-K is repeatable, providing the same solution every time it is run. However, probabilistic components can be added in the form of randomized starting point(s), skip size(s), and/or skip type(s).

Unlike traditional mathematical programming techniques such as simplex, H-K is not limited to linear problem descriptions and, as shown in Part II using the DLBP, readily lends itself to multicriteria decision making as well as nonlinear problem formats. However, like most heuristics, it cannot consistently provide the optimal solution, while most mathematical programming methods generally do. Also common to heuristics, there is some degree of tuning and design required by the user; for example, selection of number of generations, mutation rate, crossover rate, and crossover method in GA, or pheromone evaporation rate, number of ants, number of cycles, and visibility description for ACO. H-K decisions include

- **Solution structure:** Combination, permutation, differing element value ranges, number of solution elements (if other than n), other
- **Number of H-K processes:** One, multiple
- **Starting point/initial solution:** Constant, deterministic varying (i.e., different but known starting points for use with multiple H-K processes), random
- **Data presentation order:** Forward (e.g., starting at $\langle 1, 2, 3, \ldots, n \rangle$), reverse (e.g., starting at $\langle n, \ldots, 3, 2, 1 \rangle$), random, other
- **Skip type (type of skipping between solution elements):** Constant type, deterministic varying type, random type

- **Skip size (size of data skipped in each solution element):** Constant size, deterministic varying size, random size
- **Stopping criteria:** Full run, time limited, end on solution found within a given bound, or end on solution found within a given percentage
- **Follow-on solution refinement:** None ("single phase"), different H-K (different starting point, skip type, skip size, etc.), local H-K (H-K search about a previous solution), other (such as 2-opt or GA; effectively a hybrid)

Depending on these structural decisions, H-K can take on a variety of forms, from a classical optimization algorithm in its most basic form, to a general evolutionary algorithm with the use of multiple H-K processes, to a *biological* or *natural process algorithm* by electing random functionality.

10.9.4 The H-K Process

As far as the H-K process itself, since it is a modified exhaustive search that makes use of solution-space sampling, it searches for solutions similar to depth-first search, iteratively seeking the next solution—allowing for skips in the sequence—in lexicographic order.

In the basic H-K searching a permutation and with, for example, $\psi = 2$ (for all values of k is implied when no subscript is displayed with the skip parameter), $n = 4$, and forward ordering only, the first element in the first solution would be 1, the next element position would first consider 1, but since 1 is already in the solution (in element position one), element position two would be incremented by one and 2 would be considered and be acceptable. This is repeated for all of the elements until the first solution (i.e., $PS_k = \langle 1, 2, 3, \ldots, n \rangle$) is generated.

For the next solution visited, the rightmost element that is able to be incremented by ψ (ψ is equal to 2 in this example) while not exceeding n would be incremented by ψ; if the element's new value is equal to some preceding element's value, the current element's value is incremented by one (again, as long as this does not result in exceeding n). Any remaining element-positions to the right would be filled lexicographically (smallest to largest, left to right) with the remaining values.

This process of generating solutions is continued to the left until the first element position ($k = 1$) is reached. The part under consideration would then be $PS_1 = 1$ which would be incremented by $\psi = 2$ and, therefore, 3 would be considered and inserted as the first element position value. Since part 1 is not yet in the sequence, it would be placed in the second element position, part 2 in the third, and so on (i.e., $PS_k = \langle 3, 1, 2, \ldots, n \rangle$), and the process continues.

For example, with $n = 4$ giving the set of the solution space elements as $P = \{1, 2, 3, 4\}$ and no precedence constraints, instead of

$PS_k = \langle 1, 2, 3, 4, 5 \rangle$	$PS_k = \langle 3, 4, 1, 2, 5 \rangle$
$PS_k = \langle 1, 2, 5, 3, 4 \rangle$	$PS_k = \langle 3, 4, 1, 5, 2 \rangle$
$PS_k = \langle 1, 4, 2, 3, 5 \rangle$	$PS_k = \langle 3, 4, 5, 1, 2 \rangle$
$PS_k = \langle 1, 4, 2, 5, 3 \rangle$	$PS_k = \langle 5, 1, 2, 3, 4 \rangle$
$PS_k = \langle 1, 4, 5, 2, 3 \rangle$	$PS_k = \langle 5, 1, 4, 2, 3 \rangle$
$PS_k = \langle 3, 1, 2, 4, 5 \rangle$	$PS_k = \langle 5, 3, 1, 2, 4 \rangle$
$PS_k = \langle 3, 1, 4, 2, 5 \rangle$	$PS_k = \langle 5, 3, 1, 4, 2 \rangle$
$PS_k = \langle 3, 1, 4, 5, 2 \rangle$	$PS_k = \langle 5, 3, 4, 1, 2 \rangle$

TABLE 10.1 H-K Results at $n = 5$ and $\psi = 2$

considering the 4! = 24 possible permutations, only five are visited by the single-phase H-K with $\psi = 2$ and using forward-only data: $PS_k = \langle 1, 2, 3, 4 \rangle$, $PS_k = \langle 1, 4, 2, 3 \rangle$, $PS_k = \langle 3, 1, 2, 4 \rangle$, $PS_k = \langle 3, 1, 4, 2 \rangle$, and then $PS_k = \langle 3, 4, 1, 2 \rangle$. With $n = 5$, $P = \{1, 2, 3, 4, 5\}$, and no precedence constraints, instead of considering the 5! = 120 possible permutations, only 16 are considered by the single-phase H-K with $\psi = 2$ and using forward-only data as demonstrated in Table 10.1.

Each generated solution is compared to the best solution found up to that point in the search (or alternatively, sorted with some subgroup of the best or all solutions found up to that point) using one or more predetermined metrics, typically defined by mathematical formulae. At the conclusion of the search, the best solution(s) found are then known.

Skip size affects various measures including solution performance metrics and time complexity. The general form of the skip-size to problem-size relationship is formulated as

$$\psi_k = n - \Delta\psi_k : \psi_k \in \mathbf{N} \tag{10.2}$$

where $\mathbf{N}$ is the set of natural numbers (i.e., {0, 1, 2, ...}) and $\Delta\psi_k$ represents the kth solution-element's *delta-skip measure*—the difference between problem size n and skip size ψ_k—which is used to determine the minimum skip size (e.g., for $\Delta\psi = 10$ and $n = 80$, $\psi = 70$) in order to ensure reasonable search times. That is, settling on fixed $\Delta\psi_k$s regardless of instance size, and then using those $\Delta\psi_k$s to calculate each ψ_k is a mechanism to establish the skip size for any size problem instance. The result is that larger instances do not take exponentially more search time than do smaller instances (thought at the expense of a lower percentage of the solution space being visited).

Since any values of ψ that are larger than the chosen skip value for a given H-K instance take significantly less processing time, considering all larger skip values should also be considered in order to increase the search space at the expense of a minimal increase in search time. In other words, H-K can be run repeatedly on a given instance using all skip values from a smallest ψ_k (user selected based upon time complexity considerations) to the largest [i.e., $n - 1$ per Eq. (10.1)] without a significant time penalty. In this case, any ψ_k would be constrained using Eqs. (10.1) and (10.2) as

$$n - \Delta\psi_k \leq \psi_k \leq n - 1 \qquad \text{where } 1 \leq \Delta\psi_k \leq n - 2 \tag{10.3}$$

If this technique is used (as it is in this text), it should also be noted that multiples of ψ visit the same solutions; for example, for $n = 12$ and $2 \leq \psi \leq 10$, the four solutions considered by $\psi = 10$ are also visited by $\psi = 2$ and $\psi = 5$.

10.9.5 Other H-K Formulations

While demonstrated in this book for use in searching the solution space described by a permutation and having a solution space size of $n!$ (i.e., the number of permutations of n distinct objects), the heuristic is readily modified to search a solution space described by a more general permutation as given by the familiar

$${}_nP_k = \frac{n!}{(n-k)!} \tag{10.4}$$

(i.e., the number of permutations of n distinct objects taken k at a time) as well as other types of permutations. It may also be applied to different combinations; for example, the number of combinations of n distinct objects taken k at a time with a search space commonly described by

$$\binom{n}{k} = \frac{n!}{k!(n-k)!} \tag{10.5}$$

Another common search space is defined by the generalized multiplication rule covering k operations. Using the generalized multiplication rule covering k operations enables a visualization of H-K search mechanics via a two-dimensional (2-D) representation of all of the solutions visited by H-K and of the entire search space. Applying the H-K heuristic to two elements, each containing one of a possible 10 objects (i.e., $P = \{1, 2, 3, \ldots, 10\}$) gives a search space of 100 solutions; H-K with $n = 10$, $k = 2$, and $\psi = 2$ (i.e., $\psi_1 = \psi_2$) and using forward-only data gives 25 solutions to be visited as demonstrated in Table 10.2.

Using this example of a discrete two-element solution consisting of all ordered pairs of values from 1 to 10, Fig. 10.2 shows the limited, but systematic and diverse search provided by H-K.

While starting at a value of one in this book, solutions may also allow for values of zero. In addition, continuous solutions can be

$PS_k = \langle 1, 2 \rangle$	$PS_k = \langle 3, 1 \rangle$	$PS_k = \langle 5, 1 \rangle$	$PS_k = \langle 7, 1 \rangle$	$PS_k = \langle 9, 1 \rangle$
$PS_k = \langle 1, 4 \rangle$	$PS_k = \langle 3, 4 \rangle$	$PS_k = \langle 5, 3 \rangle$	$PS_k = \langle 7, 3 \rangle$	$PS_k = \langle 9, 3 \rangle$
$PS_k = \langle 1, 6 \rangle$	$PS_k = \langle 3, 6 \rangle$	$PS_k = \langle 5, 6 \rangle$	$PS_k = \langle 7, 5 \rangle$	$PS_k = \langle 9, 5 \rangle$
$PS_k = \langle 1, 8 \rangle$	$PS_k = \langle 3, 8 \rangle$	$PS_k = \langle 5, 8 \rangle$	$PS_k = \langle 7, 8 \rangle$	$PS_k = \langle 9, 7 \rangle$
$PS_k = \langle 1, 10 \rangle$	$PS_k = \langle 3, 10 \rangle$	$PS_k = \langle 5, 10 \rangle$	$PS_k = \langle 7, 10 \rangle$	$PS_k = \langle 9, 10 \rangle$

TABLE 10.2 Generalized Multiplication Rule H-K Results at $n = 10$, $k = 2$, and $\psi = 2$

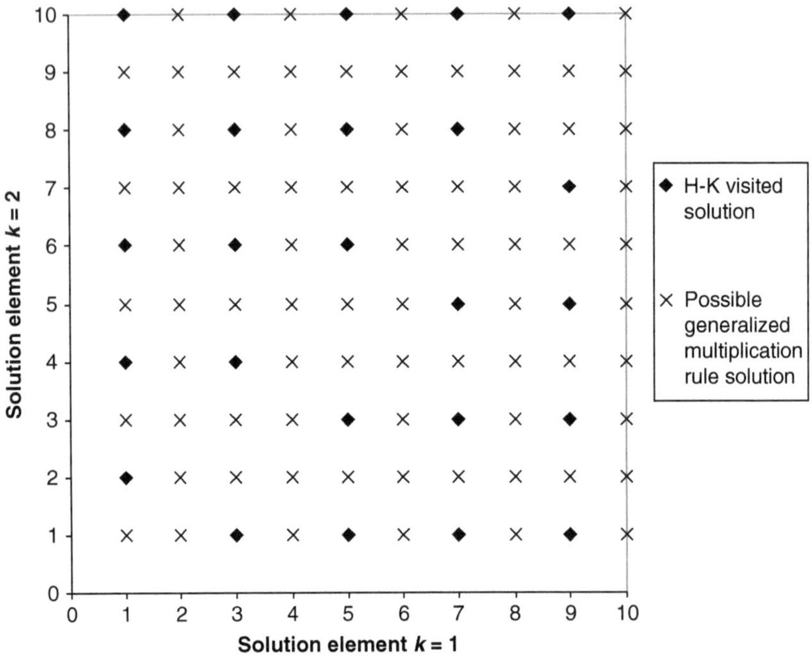

FIGURE 10.2 Two-dimensional multiplication rule H-K results at $n = 10$, $k = 2$, and $\psi = 2$.

modeled by shifting the decimal point and treating significant digits as integers (e.g., a single significant digit would result in an additional nine solution points in the search space for each original solution point) or by modeling the data in binary and using negative exponents to obtain values to the right of the decimal point (e.g., $2^{-3} = 0.125$).

CHAPTER 11

Experimental Instances

11.1 Introduction

Since the DISASSEMBLY LINE BALANCING PROBLEM (DLBP) is a recent problem, few problem instances exist to study the performance of different heuristic solutions. Four that are used here include the personal computer (PC) instance, the 10-part instance, the cellular telephone instance, and a variable-size known-solution benchmark set of instances. Each of these instances is presented with a list of parts and their associated part removal times, hazardous content, demand, and removal direction, as well as their precedence diagram.

As a point of reference, an automobile assembly plant has a rather short cycle time of CT = 60 seconds while at the other extreme is a general aviation aircraft manufacturer that provides 3.5 hours of cycle time (Boatman, 2004). The number of components can run into the thousands with even an efficient contemporary automobile design running over 4000 individual parts and taking a total of 20 hours to build (e.g., 2005 Ford Mustang).

This chapter focuses on providing a thorough description of each of the four data sets as well as any modifications or enhancements made to the original case-study data and any derivations performed. Section 11.2 describes the modified personal computer instance, an eight-part data set developed from research on an actual product. Section 11.3 reviews the enhanced 10-part instance, a case study from the disassembly sequencing literature. Section 11.4 presents the cellular telephone instance, a 25-part data set based on a study performed on a popular electronics product. Section 11.5 considers a variable-size, known-solution data set that enables efficacy and time complexity studies to be performed on a range of DLBP solution techniques.

11.2 Personal Computer Instance

Güngör and Gupta (2002) developed the personal computer instance based on the conduct of an actual disassembly study. With more than 20 million new personal computers purchased in the United States each year and the National Safety Council estimating 10,000 televisions and personal computers taken out of service each day (Gualtieri, 2005), this provides a practical and relevant example from the literature. The instance consists of the data for the disassembly of a personal computer as shown in Table 11.1 where the objective is to completely disassemble a PC consisting of eight subassemblies on a paced disassembly line operating at a speed which allows 40 seconds for each workstation to perform its required disassembly tasks [Fig. 11.1; note the dotted lines as described in Sec. 5.5 that represent the OR relationship in the product; e.g., remove part (2 OR 3) prior to part 6].

The data in Table 11.1 was modified (McGovern and Gupta, 2005a) from the original by reducing the number of part removal directions from between one and four, uniformly to one. Specifically, the part removal directions for the product were changed from ($-x$, $+x$, $-x$, $+x/-x/+y/-y$, $+y$, $+z$, $+x/-x/+y$, $+z$) to ($-x$, $+x$, $-x$, $-x$, $+y$, $+z$, $-x$, $+z$); that is, parts with multiple removal directions having $-x$ in common were reduced to being removed only in the direction $-x$. This was done both due to the fact that the other directions did not enhance disassembly (the optimal solution makes use of direction $-x$ when given all of the part removal direction options for parts 4 and 7) and to allow for more streamlined software (eliminating the need for a processor-time-intensive part removal option selection procedure which could skew or pad a solution technique's actual time complexity, especially in light of the fact—of the four instances here—that only the PC instance would require this functionality). Also, changing the direction

Task	Part Removal Description	Time	Hazardous	Demand	Direction	Predecessors
1	Cover	14	No	360	$-x$	n/a
2	Media drive	10	No	500	$+x$	1
3	Hard drive	12	No	620	$-x$	1
4	Back plane	18	No	480	$-x$	7
5	Adaptor cards	23	No	540	$+y$	1
6	Memory modules (2)	16	No	750	$+z$	2 OR 3
7	Power supply	20	Yes	295	$-x$	8
8	Motherboard	36	No	720	$+z$	2, 3, 5, 6

TABLE 11.1 Knowledge Base of the Personal Computer Instance

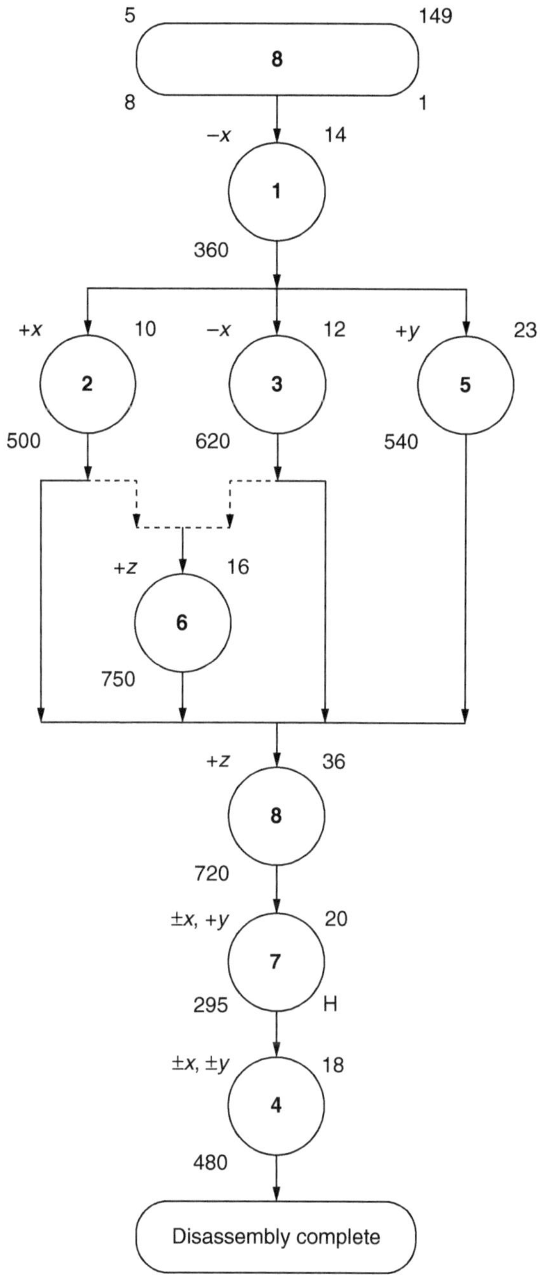

FIGURE 11.1 Personal computer precedence relationships.

representation {$+x$, $-x$, $+y$, $-y$, $+z$, $-z$} from {+1, −1, +2, −2, +3, −3} as portrayed in Eq. (8.31) to {0, 1, 2, 3, 4, 5} can simplify software design in some cases. This change can be seen in all of tables depicting solution sequences; however, it does not affect selection of any given solution nor does it impact any solution's calculated part removal measure and is made only in the interest of software engineering.

The search space is 8! or 40,320 (however, it can be seen from Fig. 11.1 that due to precedence constraints, there are only $6 + 6 + 4 = 16$ feasible solutions). With the typical computer containing 3 to 8 lb of lead, the cost of properly disposing of a PC estimated to be $30 (Gualtieri, 2005), computer monitor cathode-ray tubes accounting for 40 percent of all the lead in the American waste stream, and the Environmental Protection Agency able to fine businesses $15,000 for every piece of equipment found that ends up in a landfill and $25,000 a day until the machines are cleaned up, this data set provides a timely instance for the DLBP, while a count of 315 million obsolete computers adds a sense of urgency (Stape, 2004).

11.3 The 10-Part Instance

Kongar and Gupta (2002a) provided the basis for the 10-part DLBP instance. McGovern and Gupta (2003a) then modified this instance from its original use in disassembly sequencing and augmented it with disassembly-line specific attributes. Here the objective is to completely disassemble a notional product (Table 11.2) consisting of $n = 10$ components and several precedence relationships (e.g., parts 5 AND 6 need to be removed prior to part 7). The problem and its data were modified for use on a paced disassembly line operating at a speed which allows CT = 40 seconds for each workstation to perform its required disassembly tasks. This example consists of the data for the disassembly of a product as shown in Fig. 11.2.

Task	Time	Hazardous	Demand	Direction
1	14	No	No	$+y$
2	10	No	500	$+x$
3	12	No	No	$+x$
4	17	No	No	$+y$
5	23	No	No	$-z$
6	14	No	750	$-z$
7	19	Yes	295	$+y$
8	36	No	No	$-x$
9	14	No	360	$-z$
10	10	No	No	$-y$

Table 11.2 Knowledge Base of the 10-Part Instance

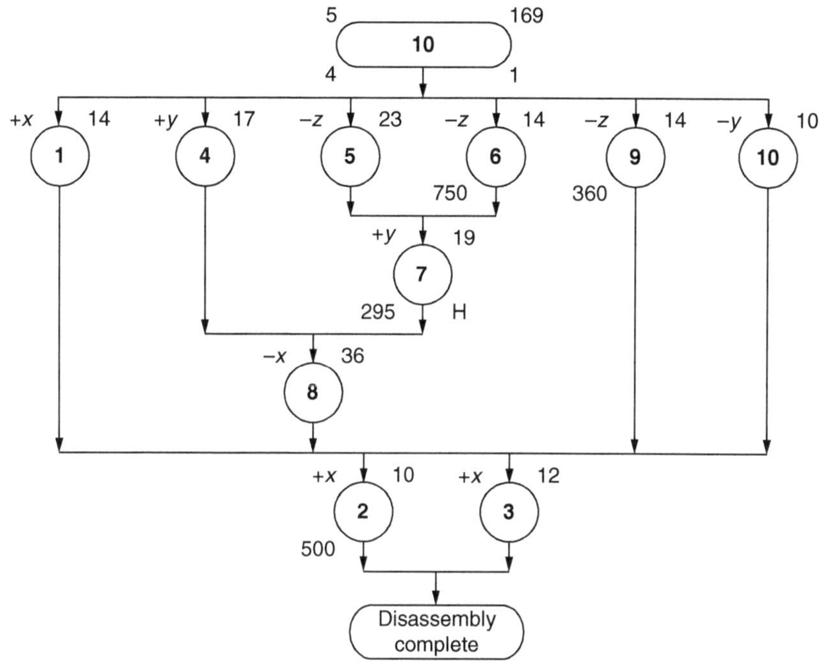

Figure 11.2 The 10-part product precedence relationships.

As with all the data sets, the direction representation {+*x*, −*x*, +*y*, −*y*, +*z*, −*z*} can be changed from {+1, −1, +2, −2, +3, −3} as portrayed in Eq. (8.31) to {0, 1, 2, 3, 4, 5}. The search space of the 10-part instance is 10! or 3,628,800.

11.4 Cellular Telephone Instance

Gupta et al. (2004) provided the basis for the cellular telephone DLBP instance. The growth of cellular telephone use and rapid changes in technology and features has prompted the entry of new models on a regular basis while, according to Collective Good International, 100 million cellular telephones are discarded each year (Gualtieri, 2005). Unwanted cell phones typically end up in landfills and usually contain numerous hazardous parts that may contain mercury, cadmium, arsenic, zinc, nickel, lead, gallium arsenide, and beryllium, any of which can pose a threat to the environment. Gupta et al. (2004) selected a 2001 model year Samsung SCH-3500 cell phone for disassembly analysis. The result is an appropriate, real-world instance consisting of $n = 25$ components having several precedence relationships. The data set includes a paced disassembly line operating at a speed which allows CT = 18 seconds per workstation. Collected data on the SCH-3500 is listed in Table 11.3. Demand is estimated based on part value and/or

Task	Part Removal Description	Time	Hazardous	Demand	Direction
1	Antenna	3	Yes	4	$+y$
2	Battery	2	Yes	7	$-y$
3	Antenna guide	3	No	1	$-z$
4	Bolt (type 1) a	10	No	1	$-z$
5	Bolt (type 1) b	10	No	1	$-z$
6	Bolt (type 2) 1	15	No	1	$-z$
7	Bolt (type 2) 2	15	No	1	$-z$
8	Bolt (type 2) 3	15	No	1	$-z$
9	Bolt (type 2) 4	15	No	1	$-z$
10	Clip	2	No	2	$+z$
11	Rubber seal	2	No	1	$+z$
12	Speaker	2	Yes	4	$+z$
13	White cable	2	No	1	$-z$
14	Red/blue cable	2	No	1	$+y$
15	Orange cable	2	No	1	$+x$
16	Metal top	2	No	1	$+y$
17	Front cover	2	No	2	$+z$
18	Back cover	3	No	2	$-z$
19	Circuit board	18	Yes	8	$-z$
20	Plastic screen	5	No	1	$+z$
21	Keyboard	1	No	4	$+z$
22	Liquid crystal display	5	No	6	$+z$
23	Sub-keyboard	15	Yes	7	$+z$
24	Internal circuit	2	No	1	$+z$
25	Microphone	2	Yes	4	$+z$

TABLE 11.3 Knowledge Base of the Cellular Telephone Instance

recycling value. Part removal times and precedence relationships (Fig. 11.3) were determined experimentally; part removal times were repeatedly collected until a consistent part removal performance was attained. The search space is 25! or 1.55112×10^{25}.

Note that in 2005, companies that refurbished cellular telephones typically paid consumers $2 to $20 per phone for the more popular Motorola and Nokia cellular telephones while cellular telephone recyclers (that may only extract precious metals such as gold from the circuit boards) paid between $1 and $6 per phone (Metz, 2005).

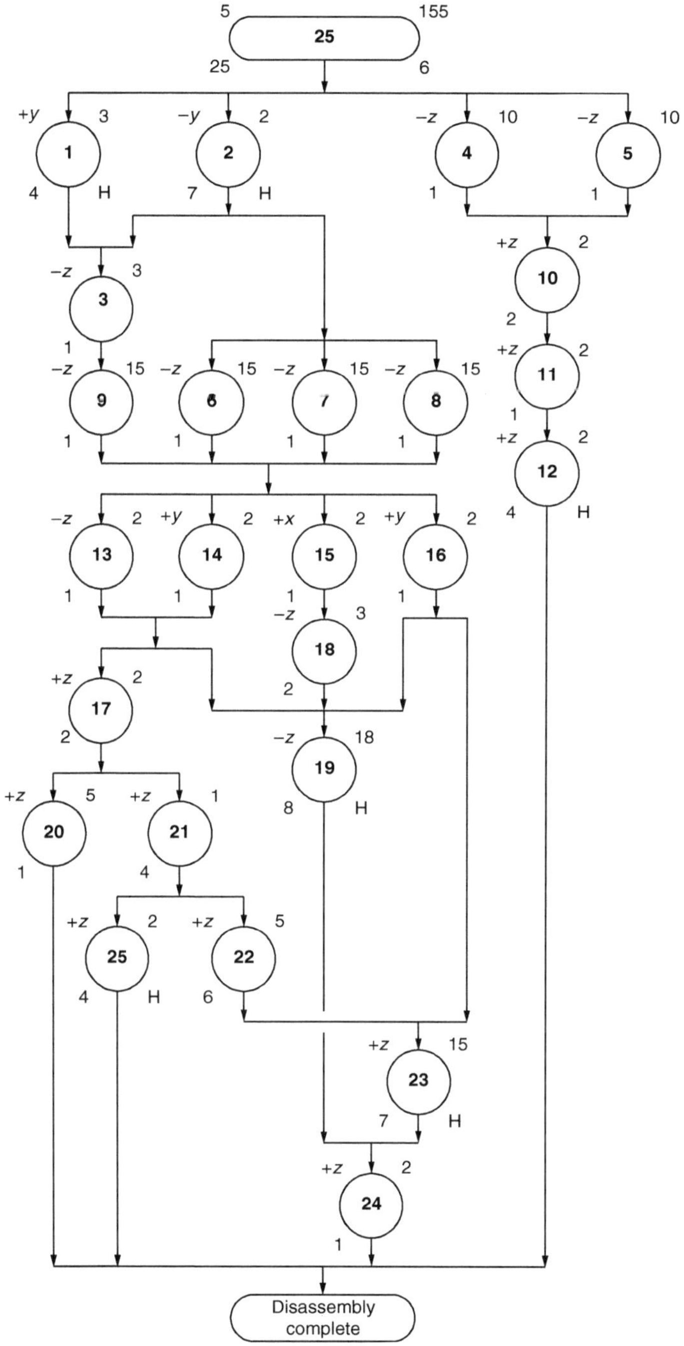

Figure 11.3 Cellular telephone precedence relationships.

11.5 DLBP A Priori Optimal Solution Benchmark Instances

11.5.1 Background

Benchmark data sets are valuable for use in evaluating new search methodologies and in thoroughly testing their developed software. Once testing is completed, large, established benchmarks can be used to demonstrate a given solution-generating methodology's performance as well as to identify its limitations.

Benchmark data sets are common for many NP-complete problems, such as *Oliver30* and *RY48P* for application to the TRAVELING SALESPERSON PROBLEM and *Nugent15/20/30*, *Elshafei19*, and *Krarup30* for the QUADRATIC ASSIGNMENT PROBLEM. Unfortunately, because of their size and their design, most of these existing data sets have no known optimal answer and new solutions are not compared to the optimal solution, but rather the best solution to-date. In the case of the DLBP, since it is a recently defined problem, no appropriate benchmark data sets exist. For these reasons, a DLBP benchmark data set is of value. This section describes a known-optimal solution benchmark line-balance data set which is useful in evaluating DLBP heuristics.

11.5.2 Mathematical Formulation

This size-independent A Priori benchmark data set was generated (McGovern and Gupta, 2007b) based on the following. Since solutions to larger and larger instances cannot be verified as optimal (due to the time complexity of exhaustive search), instances can be generated in such a way as to always provide a known solution. This is done by using part removal times consisting exclusively of prime numbers, further selected to ensure that no combinations of these part removal times allow for any equal summations (in order to reduce the number of possible optimal solutions). For example, part removal times of 1, 3, 5, and 7, and CT = 16 would have minimum idle time solutions of not only one 1, one 3, one 5, and one 7 at each workstation, but various additional combinations of these as well since 1 + 7 = 3 + 5 = ½ CT. Subsequently, the chosen instances are made up of parts with removal times of 3, 5, 7, and 11, and CT = 26. As a result, the optimal balance for all subsequent instances consists of a perfect balance of precedence-preserving permutations of 3, 5, 7, and 11 at each workstation with idle times of zero. The size of this A Priori data set is then constrained by

$$n = x \cdot |\mathrm{PRT}| : x \in \mathbf{Z}^{+} \tag{11.1}$$

(Note that $|\mathrm{PRT}| \le n$ since PRT_k is *onto* mapped to PRT, though not necessarily *one-to-one*, since multiple parts may have equal part removal times; i.e., PRT_k is a *surjection* and may or may not be a *bijection* to PRT.)

To further complicate the data (i.e., provide a large, feasible search space), only one part is listed as hazardous and this was one of the parts with the largest part removal time (specifically, the last one listed in the initial data). In addition, one part (the last listed of the second-largest part-removal-time component) is listed as being demanded. This is done so that only the hazardous and the demand sequencing is demonstrated while providing a slight solution sequence disadvantage to any purely greedy methodology (since two parts with part removal times of 3 and 5 are needed along with the parts with the larger part removal times to reach F^*, assigning hazard and high-demand attributes to those parts with smaller part removal times may prevent some methodologies from artificially obtaining an F^* sequence). From each part removal time size, the first listed part is selected to have a removal direction differing from the other parts with the same part removal time. This is done to demonstrate direction selection while requiring any solution-generating methodology to move these first parts of each part removal time size encountered to the end of the sequence (i.e., into the last workstation) in order to obtain the optimal direction value of $R^* = 1$ (assuming the solution technique being evaluated is able to successfully place the hazardous and demanded parts toward the front of the sequence). Also, there are no precedence constraints placed on the sequence, a deletion that further challenges any method's ability to attain an optimal solution (by maximizing the feasible search space). This has the added benefit of more precisely modeling the restricted version of the decision version of the DLBP seen in Theorem 9.1 and Theorem 9.2.

Known optimal results include $F^* = 0$, $H^* = 1$, $D^* = 2$, $R^* = 1$. While this book makes use of data with $|\text{PRT}| = 4$ unique part removal times, in general for any n parts consisting of this type of data, the following can be calculated as

$$\text{NWS}^* = \text{NWS}_{\text{lower}} = \frac{n}{|\text{PRT}|} \tag{11.2}$$

$$\text{NWS}_{\text{upper}} = n \tag{11.3}$$

$$I^* = I_{\text{lower}} = 0 \tag{11.4}$$

$$I_{\text{upper}} = \frac{n \cdot \text{CT} \cdot (|\text{PRT}| - 1)}{|\text{PRT}|} \tag{11.5}$$

$$F^* = F_{\text{lower}} = 0 \tag{11.6}$$

with F_{upper} given by Eq. (8.15).

Formulae have been developed to generate all data parameters—as well as for calculating optimal and nominal measures—for any size instance [as constrained by Eq. (11.1)]. Hazard measures and values are given by

$$h_k = \begin{cases} 1, & k = n \\ 0, & \text{otherwise} \end{cases} \tag{11.7}$$

$$H^* = H_{\text{lower}} = 1 \tag{11.8}$$

$$H_{\text{upper}} = n \tag{11.9}$$

with demand measures and values given by

$$d_k = \begin{cases} 1, & k = \dfrac{n \cdot (|\text{PRT}| - 1)}{|\text{PRT}|} \\ 0, & \text{otherwise} \end{cases} \tag{11.10}$$

$$D^* = D_{\text{lower}} = \begin{cases} 2, & H = 1 \\ 1, & \text{otherwise} \end{cases} \tag{11.11}$$

$$D_{\text{upper}} = \begin{cases} n-1, & H = n \\ n, & \text{otherwise} \end{cases} \tag{11.12}$$

and part removal direction measures and values given by

$$r_k = \begin{cases} 1, & k = 1, \dfrac{n}{|\text{PRT}|}+1, \dfrac{2n}{|\text{PRT}|}+1, \ldots, \dfrac{(|\text{PRT}|-1)\cdot n}{|\text{PRT}|}+1 \\ 0, & \text{otherwise} \end{cases} \tag{11.13}$$

$$R^* = R_{\text{lower}} = 1 \tag{11.14}$$

$$R_{\text{upper}} = \begin{cases} 0, & n = |\text{PRT}| \\ 2 \cdot |\text{PRT}| - 1, & n = 2 \cdot |\text{PRT}| \\ 2 \cdot |\text{PRT}|, & \text{otherwise} \end{cases} \tag{11.15}$$

Since $|\text{PRT}| = 4$ in this part of the text, each part removal time is generated by

$$\text{PRT}_k = \begin{cases} 3, & 0 < k \le \dfrac{n}{4} \\ 5, & \dfrac{n}{4} < k \le \dfrac{n}{2} \\ 7, & \dfrac{n}{2} < k \le \dfrac{3n}{4} \\ 11, & \dfrac{3n}{4} < k \le n \end{cases} \tag{11.16}$$

While the demand values as generated by Eq. (11.10) are the preferred representation (due to the fact that the resulting small numerical values make it easy to interpret demand efficacy since $D = k$, i.e., the sequence position of the sole demanded part), algorithms that allow incomplete disassembly may terminate after placing the single demanded part in the solution sequence. In this case, Eqs. (11.10) through (11.12) may be modified to give

$$d_k = \begin{cases} 2, & k = \dfrac{n \cdot (|\mathrm{PRT}| - 1)}{|\mathrm{PRT}|} \\ 1, & \text{otherwise} \end{cases} \tag{11.17}$$

$$D^* = D_{\text{lower}} = \begin{cases} 2 + \sum_{p=1}^{n} p, & H = 1 \\ 1 + \sum_{p=1}^{n} p, & \text{otherwise} \end{cases} \tag{11.18}$$

$$D_{\text{upper}} = \begin{cases} n - 1 + \sum_{p=1}^{n} p, & H = n \\ n + \sum_{p=1}^{n} p, & \text{otherwise} \end{cases} \tag{11.19}$$

11.5.3 Probabilistic Analysis of the Benchmark Data Set

It is noted that a data set containing parts with equal part removal times and no precedence constraints will result in *multiple optimum extreme points*. To properly gauge the performance of any solution-generating technique on the DLBP A Priori data, the size of the optimal solution set needs to be quantified.

From probability theory (specifically, counting sample points using the *generalized multiplication rule* covering n operations; Miller and Freund, 1985) we know that, for example, with $n = 12$ and $|\mathrm{PRT}| = 4$, the size of the set of optimally balanced solutions $|F^*|$ when using the DLBP A Priori data can be calculated as $(12 \cdot 9 \cdot 6 \cdot 3) \cdot (8 \cdot 6 \cdot 4 \cdot 2) \cdot (4 \cdot 3 \cdot 2 \cdot 1) = 17{,}915{,}904$ using Table 11.4.

k	1	2	3	4	5	6	7	8	9	10	11	12
Count	12	9	6	3	8	6	4	2	4	3	2	1

Table 11.4 Number of Possible Entries in Each Element Position Resulting in Perfect Balance Using the DLBP A Priori Data with $n = 12$ and $|\mathrm{PRT}| = 4$

Grouping these counts by workstation and reversing their ordering enables one to more easily recognize a pattern:

(1	2	3	4)
(2	4	6	8)
(3	6	9	12)

It can be seen that the first row can be generalized as $(1 \cdot 2 \cdot 3 \cdot \cdots \cdot |\mathrm{PRT}|)$, the second as $[2 \cdot 4 \cdot 6 \cdot \cdots \cdot (2 \cdot |\mathrm{PRT}|)]$, and the third as $[3 \cdot 6 \cdot 9 \cdot \cdots \cdot (3 \cdot |\mathrm{PRT}|)]$. Expanding in this way, the number of optimally balanced solutions can be written as

$$|F^*| = [1 \cdot 2 \cdot 3 \cdot \cdots \cdot (1 \cdot |\mathrm{PRT}|)] \cdot [2 \cdot 4 \cdot 6 \cdot \cdots \cdot (2 \cdot |\mathrm{PRT}|)] \cdot \cdots \cdot \left(\frac{n}{|\mathrm{PRT}|} \cdot \frac{2n}{|\mathrm{PRT}|} \cdot \frac{3n}{|\mathrm{PRT}|} \cdot \cdots \cdot n \right)$$

This can be written as

$$|F^*| = \prod_{x=1}^{|\mathrm{PRT}|} x \cdot \prod_{x=1}^{|\mathrm{PRT}|} 2x \cdot \prod_{x=1}^{|\mathrm{PRT}|} 3x \cdot \cdots \cdot \prod_{x=1}^{|\mathrm{PRT}|} \frac{n}{|\mathrm{PRT}|} \cdot x$$

and finally as

$$|F^*| = \prod_{x=1}^{|\mathrm{PRT}|} x^{\frac{n}{|\mathrm{PRT}|}} \cdot \prod_{y=1}^{\frac{n}{|\mathrm{PRT}|}} y$$

or

$$|F^*| = \prod_{x=1}^{|\mathrm{PRT}|} \prod_{y=1}^{\frac{n}{|\mathrm{PRT}|}} x^{\frac{n}{|\mathrm{PRT}|}} \cdot y \tag{11.20}$$

Since $|\mathrm{PRT}| = 4$ in this chapter, Eq. (11.20) becomes

$$|F^*| = \prod_{x=1}^{4} \prod_{y=1}^{\frac{n}{4}} x^{\frac{n}{4}} \cdot y \tag{11.21}$$

In our example with $n = 12$ and $|\mathrm{PRT}| = 4$, Eq. (11.21) is solved as

$$|F^*| = \prod_{x=1}^{4} \prod_{y=1}^{\frac{12}{4}} x^{\frac{12}{4}} \cdot y = \prod_{x=1}^{4} \prod_{y=1}^{3} x^3 \cdot y = \prod_{x=1}^{4} x^3 \cdot (1 \cdot 2 \cdot 3)$$

or

$$|F^*| = 1 \cdot 1 \cdot 1 \cdot (1 \cdot 2 \cdot 3) \cdot 2 \cdot 2 \cdot 2 \cdot (1 \cdot 2 \cdot 3) \cdot 3 \cdot 3 \cdot 3 \cdot (1 \cdot 2 \cdot 3) \cdot 4 \cdot 4 \cdot 4 \cdot (1 \cdot 2 \cdot 3)$$

k	1	2	3	4	5	6	7	8	9	10	11	12
Count	1	1	6	3	4	3	2	1	4	3	2	1

Table 11.5 Number of Possible Entries in Each Element Position Resulting in Optimal in *F, H, D,* and *R* Using A Priori Data with $n = 12$ and $|\text{PRT}| = 4$

which, when rearranged, can be written as the more familiar

$$|F^*| = (12 \cdot 9 \cdot 6 \cdot 3) \cdot (8 \cdot 6 \cdot 4 \cdot 2) \cdot (4 \cdot 3 \cdot 2 \cdot 1) = 17{,}915{,}904$$

Even when all objectives are considered, there still exist multiple optimal solutions, again due to the use of a data set containing parts with equal part removal times and no precedence constraints. Using probability theory and the example having $n = 12$ and $|\text{PRT}| = 4$, it is known that the size of the set of solutions optimal in $F, H, D,$ and R, that is, $|F^* \cap H^* \cap D^* \cap R^*|$, when using the DLBP A Priori data can be calculated as $(1 \cdot 1 \cdot 6 \cdot 3) \cdot (4 \cdot 3 \cdot 2 \cdot 1) \cdot (4 \cdot 3 \cdot 2 \cdot 1) = 10{,}368$ using Table 11.5.

Repeating the technique of grouping these counts by workstation and reversing their ordering again reveals a pattern:

$$\begin{matrix} (1 & 2 & 3 & 4) \\ (1 & 2 & 3 & 4) \\ (3 & 6 & 1 & 1) \end{matrix}$$

The middle row elements will always be the same as those given by Eq. (11.20) but with two fewer sets (due to different first and last workstation elements). The first row is always $(1 \cdot 2 \cdot 3 \cdot \cdots \cdot |\text{PRT}|)$ since the directional elements in the A Priori data should always be together at the end (the beginning in this case since we reversed the sequence for readability) of any optimal solution sequence. The last row (again, reversed) is always $((n/|\text{PRT}|) \cdot (2n/|\text{PRT}|) \cdot (3n/|\text{PRT}|) \cdot \cdots \cdot ((|\text{PRT}| - 2) \cdot n/|PRT|) \cdot 1 \cdot 1)$ since there is only one hazardous part (optimal element position $k = 1$) and only one demanded part (optimal element position $k = 2$). Combining these components, the number of fully optimal solutions can be written as

$$|F^* \cap H^* \cap D^* \cap R^*| = (1 \cdot 2 \cdot 3 \cdot \cdots \cdot |\text{PRT}|)$$
$$\left((1 \cdot 2 \cdot 3 \cdot \cdots \cdot (1 \cdot |\text{PRT}|)) \cdot (2 \cdot 4 \cdot 6 \cdot \cdots \cdot (2 \cdot |\text{PRT}|)) \cdot \cdots \cdot \right.$$
$$\left(1 \cdot \left(\frac{n}{|\text{PRT}|} - 2\right) \cdot 2 \cdot \left(\frac{n}{|\text{PRT}|} - 2\right) \cdot 3 \cdot \left(\frac{n}{|\text{PRT}|} - 2\right) \cdot \cdots \cdot \right.$$
$$\left.\left.|\text{PRT}| \cdot \left(\frac{n}{|\text{PRT}|} - 2\right)\right)\right)$$

$$\left(\left(\frac{n}{|\mathrm{PRT}|}\right)\cdot\left(\frac{2n}{|\mathrm{PRT}|}\right)\cdot\left(\frac{3n}{|\mathrm{PRT}|}\right)\cdots\cdot\left(\frac{(|\mathrm{PRT}|-2)\cdot n}{|\mathrm{PRT}|}\right)\cdot 1\cdot 1\right)$$

By replacing the second term with a modified version of Eq. (11.20) and simplifying the first and third terms, this can be written as

$$|F^*\cap H^*\cap D^*\cap R^*| = \left(\prod_{y=1}^{|\mathrm{PRT}|} y\right)\cdot\left(\prod_{y=1}^{|\mathrm{PRT}|}\prod_{z=1}^{\frac{n}{|\mathrm{PRT}|}-2} y^{\frac{n}{|\mathrm{PRT}|}-2}\cdot z\right)\cdot\left(\prod_{x=1}^{|\mathrm{PRT}|-2}\frac{n}{|\mathrm{PRT}|}\cdot x\right)$$

Expanding the second term gives

$$|F^*\cap H^*\cap D^*\cap R^*| = \left(\prod_{y=1}^{|\mathrm{PRT}|} y\right)\cdot\left(\prod_{y=1}^{|\mathrm{PRT}|} y^{\frac{n}{|\mathrm{PRT}|}-2}\left(\prod_{z=1}^{\frac{n}{|\mathrm{PRT}|}-2} z\right)\right)\cdot\left(\prod_{x=1}^{|\mathrm{PRT}|-2}\frac{n}{|\mathrm{PRT}|}\cdot x\right)$$

Combining the first and second terms results in

$$|F^*\cap H^*\cap D^*\cap R^*| = \left(\prod_{y=1}^{|\mathrm{PRT}|} y^{\frac{n}{|\mathrm{PRT}|}-1}\left(\prod_{z=1}^{\frac{n}{|\mathrm{PRT}|}-2} z\right)\right)\cdot\left(\prod_{x=1}^{|\mathrm{PRT}|-2}\frac{n}{|\mathrm{PRT}|}\cdot x\right)$$

or

$$|F^*\cap H^*\cap D^*\cap R^*| = \prod_{x=1}^{|\mathrm{PRT}|-2}\frac{nx}{|\mathrm{PRT}|}\cdot\prod_{y=1}^{|\mathrm{PRT}|}\prod_{z=1}^{\frac{n}{|\mathrm{PRT}|}-2} y^{\frac{n}{|\mathrm{PRT}|}-1}\cdot z$$

with a constraint that

$$\frac{n}{|\mathrm{PRT}|} > 2$$

Alternatively, this can be written as

$$|F^*\cap H^*\cap D^*\cap R^*| = \prod_{x=1}^{|\mathrm{PRT}|-2}\frac{nx}{|\mathrm{PRT}|}\cdot\prod_{y=1}^{|\mathrm{PRT}|}\prod_{z=1}^{a} y^{\frac{n}{|\mathrm{PRT}|}-1}\cdot z \qquad (11.22)$$

where

$$a = \begin{cases}\frac{n}{|\mathrm{PRT}|}-2, & \text{if } \frac{n}{|\mathrm{PRT}|} > 2\\ 1, & \text{otherwise}\end{cases}$$

Since |PRT| = 4 is being used in these examples, Eq. (11.22) becomes

$$|F^* \cap H^* \cap D^* \cap R^*| = \prod_{x=1}^{2} \frac{nx}{4} \cdot \prod_{y=1}^{4} \prod_{z=1}^{\frac{n}{4}-2} y^{\frac{n}{4}-1} \cdot z \qquad (11.23)$$

In our example with $n = 12$ and |PRT| = 4, Eq. (11.23) is solved as

$$|F^* \cap H^* \cap D^* \cap R^*| = \prod_{x=1}^{2} 3x \cdot \prod_{y=1}^{4} \prod_{z=1}^{1} y^2 \cdot z$$

or

$$|F^* \cap H^* \cap D^* \cap R^*| = \prod_{x=1}^{2} 3x \cdot \prod_{y=1}^{4} y^2$$

giving

$$|F^* \cap H^* \cap D^* \cap R^*| = [(3 \cdot 1) \cdot (3 \cdot 2)] \cdot [(1 \cdot 1) \cdot (2 \cdot 2) \cdot (3 \cdot 3) \cdot (4 \cdot 4)]$$

which, when rearranged, can be written as the more familiar

$$|F^* \cap H^* \cap D^* \cap R^*| = (1 \cdot 1 \cdot 6 \cdot 3) \cdot (4 \cdot 3 \cdot 2 \cdot 1) \cdot (4 \cdot 3 \cdot 2 \cdot 1) = 10{,}368$$

Although the sizes of both DLBP A Priori optimal solution sets is quite large in this example, they are also significantly smaller than the search space of n! = 479,001,600. As shown in Table 11.6, the number of solutions that are optimal in balance alone goes from 100 percent of n at $n = 4$, to 22.9 percent at $n = 8$, and to less than 1 percent at $n = 16$; as n grows, this percentage gets closer and closer to 0. The number of solutions optimal in all objectives goes from less than 8.3 percent of n at $n = 4$, to 0.12 percent at $n = 8$, dropping to effectively 0 percent at $n = 16$; again, as n grows, the percentage of optimal solutions gets closer and closer to zero.

n	*n*!	**Number Optimal in Balance**	**Number Optimal in All**	**Percentage Optimal in Balance**	**Percentage Optimal in All**
4	24	24	2	100.00%	8.33%
8	40,320	9,216	48	22.86%	0.12%
12	479,001,600	17,915,904	10,368	3.74%	0.00%
16	2.09228E+13	1.10075E+11	7,077,888	0.53%	0.00%

Table 11.6 Comparison of Possible Solutions to Optimal Solutions for a Given *n* Using the DLBP A Priori Data

The final configuration of the benchmark as used in the Part II analysis is 19 instances with instance size evenly distributed from $n = 8$ to $n = 80$ in steps of $|PRT| = 4$. This provides numerous instances of predetermined, calculable solutions with the largest instance ten times larger than the smallest instance. The size and range of the instances provides for the testing of small *n*s—which decreases the NWS value and tends to exaggerate less than optimal performance—as well as large, which demonstrates time complexity growth and efficacy changes with n.

To summarize, the DLBP A Priori test data as used here will consist of $n = \{8, 12, 16, \ldots, 80\}$ parts with four unique part removal times giving PRT = {3, 5, 7, 11}. Only the last part having a part removal time of 11 is hazardous; only the last part having a part removal time of 7 is demanded. The first of each part with part removal times equal to 3, 5, 7, and 11 are removed in direction $+x$, while all others are in direction $-x$. The disassembly line is paced and operated at a speed that allows 26 seconds (CT = 26) for each workstation (McGovern and Gupta, 2004c, 2007b). Known optimal results include $F^* = 0$, $H^* = 1$, $D^* = 2$, and $R^* = 1$.

CHAPTER 12

Analytical Methodologies

12.1 Introduction

Although combinatorial optimization holds promise in solving the DISASSEMBLY LINE BALANCING PROBLEM (DLBP), one of the concerns when using heuristics is the idea that very little has been rigorously established in reference to their performance. Developing ways of explaining and predicting their performance is considered to be one of the most important challenges currently facing the fields of optimization and algorithms (Papadimitriou and Steiglitz, 1998). These challenges exist in the variety of evaluation criteria available (addressed in Secs. 8.3 to 8.6), a lack of data sets for testing (disassembly-specific instances are addressed in Chap. 10), and a lack of performance analysis tools. In this chapter, mathematical and graphical tools for quantitative and qualitative analysis are reviewed, focusing on the analytical methodologies used in evaluating combinatorial optimization searches. Section 12.2 discusses the graphical techniques used for qualitative analysis and seen in later chapters. Section 12.3 reviews the multicriteria approach used by all of the combinatorial optimization techniques. Section 12.4 introduces the mathematical formulae and statistics used for quantitative analysis. Finally, Sec. 12.5 provides an overview of simulation, an alternative analysis technique.

12.2 Graphical Analysis Tools

Charts and tables provide an intuitive view into the workings and performance of search methodologies. Both are used here to enhance the qualitative understanding of a methodology's execution and status of its terminal state as well as to allow for a comparison of relative performance with instance size and when compared to other methodologies.

The tables are used to observe the terminal solution of any single instance. The tables used here present a solution in the following format: the sequence n-tuple is listed in the first row, followed by the corresponding part removal times, then the workstation assignments, then the hazard values, followed by the demand values, and finally the direction values [keeping in mind the comment in Sec. 11.2 that the direction representation $\{+x, -x, +y, -y, +z, -z\}$ is changed from $\{+1, -1, +2, -2, +3, -3\}$ as portrayed in Eq. (8.31) to $\{0, 1, 2, 3, 4, 5\}$ for purposes of software engineering]. To improve readability, the columns are shaded corresponding to the workstation assignment using alternating shades of gray. Use of this format (i.e., tables) allows for study of the final solution state as well as potentially enabling improvements in algorithm performance due to insights gained by this type of observation.

The second graphical format demonstrated to allow for qualitative study of techniques and their solutions consists of a graphical comparison of known best- and worst-case results with the results/averaged results (deterministic techniques/stochastic techniques) of a solution technique under consideration. The charts are used to observe multiple, varying-size solutions of the DLBP A Priori instances. Multiple charts are used to display the various performance measures, which are typically demonstrated with the DLBP A Priori benchmark data sets of sizes $n = \{8, 12, 16, \ldots, 80\}$. The near-optimal solutions, coupled with the known optimal and nominal solutions for all problem sizes under study, provide a method for not only comparing the methodologies to the best and worst cases, but to other methodologies as well. Computational complexity is portrayed using time complexity (analysis of the time required to solve a particular size instance of a given problem) while *space complexity* (analysis of the computer memory required to solve a particular size instance of a given problem) is not considered (Rosen, 1999). All time complexities are provided in asymptotic notation (*big-oh*, *big-omega*, and *big-theta*) when commonly known or when calculated wherever possible in Chap. 13 through Chap. 19. "Time complexity" typically refers to worst-case runtimes, while in the "Numerical Results" sections of these chapters (Sec. 13.3, 14.6, 15.5, 16.3, 17.3, 18.3, and 19.4) the runtimes provide a qualitative description of the studied methodologies, so the experimentally determined time complexities are presented with the understanding that the information is the average-case time complexity of the particular software written for the problem used with a specific instance (see Sec. 6.2). In Part II the charts used include number of workstations with the optimal number of workstations, balance measure with the optimal balance measure, normalized (as defined in Sec. 12.4) balance measure with the optimal and nominal balance measures, hazard measure with the optimal and nominal hazard measures, demand measure with the optimal and nominal demand measures, direction measure with the optimal and nominal direction measures, average-case time complexity with overlaid first-, second-, or third-order polynomial trend line (linear or polynomial regression models), average-case time complexity with third-order and exhaustive growth curves, and average-case time complexity with a second-order curve. Note that "number of workstations" and "idle time" measures are analogous (e.g., one can be calculated from the other) so only "number of workstations" is calculated and displayed. Also, while "number of workstations" and "balance" are both calculated in various ways and displayed in separate graphs, they are strongly related as well. Both are presented to allow insight into the search processes and further quantify the efficacy of their solutions; however, it should be noted that, for example, a solution optimal in balance—using the definition from Sec. 7.3—must also obviously be optimal in the number of workstations.

The overlaid linear or polynomial fitted regression lines assist with determining the order of some of the very fast heuristics used here. With the software set up to measure timing down to 1/100th of a second and some of the heuristics running on that order (or in some cases even faster), these average-case time complexity curves may give the appearance of making dramatic steps up or down when this is actually more of an aberration of the order of the timing data that is collected. For that reason, showing the average-case time complexity with its regression line displays both the actual data and more importantly, the shape of the time growth in n.

Note that with the third-order and exhaustive growth curves and with the average-case time complexity and second-order curves, the actual average-case time complexity curve under consideration is often not even readily visible. Even so, average-case time complexity with the third-order and exhaustive growth curves helps to show how relatively fast all of these techniques are, while average-case time complexity with the second-order curve defines the methodology's speed and rate of growth in even more detail while relating it to a known quantity.

12.3 Multiple-Criteria Decision-Making Considerations

One of the ways in which the complexity of the DLBP manifests itself is with the multiple, often conflicting objectives as defined in Sec. 8.2. The field of *multiple-criteria decision making* (MCDM) provides a variety of means for addressing the selection of a solution where several objectives exist. The bulk of MCDM methods involve multicriteria versions of linear programming (LP) problems. Since the DLBP requires integers exclusively as its solution, it cannot be formulated as a linear programming model. Additionally, since the objective described by Eq. (8.7) is nonlinear, the DLBP is not linear either (a requirement of linear programming, though this can be remedied using a version of the descriptions in Definition 7.1 of Sec. 7.3). Also, many MCDM methods rely on *weighting*. These weights are in proportion to the

```
Procedure BETTER_SOLUTION (new_solution, best_solution) {
IF      (new_solution.F < best_solution.F
        ∨
        (new_solution.F ≤ best_solution.F ∧
        new_solution.H < best_solution.H)
        ∨
        (new_solution.F ≤ best_solution.F ∧
        new_solution.H ≤ best_solution.H ∧
        new_solution.D < best_solution.D)
        ∨
        (new_solution.F ≤ best_solution.F ∧
        new_solution.H ≤ best_solution.H ∧
        new_solution.D ≤ best_solution.D ∧
        new_solution.R < best_solution.R)){
            RETURN (TRUE)
        }
        RETURN (FALSE)
}
```

FIGURE 12.1 Multicriteria selection procedure.

importance of each objective. Weights are not desirable for the examples given since any results would be expected to be influenced by the weights selected. While this is appropriate for an application of these methodologies to an applied problem and using experts to select the appropriate weights, here weighting would only serve to add an additional level of complexity to the comprehension of the problem and the demonstrated solutions. In addition, since the examples in Part II are not applied to a particular, unique disassembly situation but rather to the DLBP in general, the assignment of weighting values would be completely arbitrary and hence add little, if any, value to the final analysis of any results. Finally, the use of weights may not adequately reflect the generalized performance of the combinatorial optimization methods being studied; nuances in the methods, the data, and the weights themselves may generate atypical, unforeseen, or nonrepeatable results. For these reasons, a simplified process is demonstrated to select the best solution (Fig. 12.1). Based on the priorities listed in Sec. 8.2, the balance is the primary objective used to search for an optimal solution [note the use of "less than" (<) and "less than or equal to" (≤) signs in Fig. 12.1 indicating the desire for the better solution to be on the left side of the inequality since we are seeking to minimize all measures]. Given multiple optimum extreme points in F, early removal of hazardous parts is then considered. Given multiple optimum extreme points in F and H, early removal of high-demand parts is considered next. Finally, given multiple optimum extreme points in F, H, and D, adjacent removal of parts with equivalent part removal directions is considered.

This process has its basis in two MCDM techniques.

The *feasible region* in a linear programming problem (and in the DLBP) is usually a multidimensional subset of $\mathbf{R}^z$ containing an infinite (finite in the DLBP) number of points. Because it is formed by the intersection of a finite number of *closed half-spaces* (defined by ≤ and ≥, i.e., the inequality constraints) and *hyperplanes* (equality constraints) it is *polyhedral*. Thus, the feasible region is *closed* and *convex* with a finite number of extreme points. The simplex method for solving linear programming problems exploits the polyhedral properties in the sense that the optimal solutions can be found without having to examine all of the points in the feasible region. Taking the steepest *gradient* following each point examined accomplishes this. Conventional LP algorithms and software terminate their search once the first optimal extreme point is found (in our example, once the first F^* is found). They fail to identify alternative optima if they exist. In general, an LP instance may have one or more optimal extreme points and one or more *unbounded edge(s)* (though the latter would not be expected of the DLBP since it should be contained within the *convex hull* of a *polytope*, i.e., a finite region of n-dimensional space enclosed by a finite number of hyperplanes). The optimal set is the *convex combination* (i.e., the set of all the points) of all optimal extreme points and points on unbounded edges. It is therefore desired to test all optimal extreme points. This can be done by *pivoting* and is performed using what is known as the *phase III bookkeeping system* (Steuer, 1989). Determining all alternative optima is enabled in a similar way using the routine shown in Fig. 12.1 (as long as the combinatorial optimization technique in question is able to visit those extreme points).

An additional MCDM technique that the process in Fig. 12.1 borrows from is *preemptive (lexicographic) goal programming* (GP). GP was initially conceived by Charnes et al. (1955) and Charnes and Cooper (1961) and conceptualizes objectives as *goals* then assigns priorities to the achievement of these goals. In preemptive GP, goals are grouped according to priorities. The goals at the highest priority level are considered to be infinitely more important than goals at the second priority level, and the goals at the second priority level are considered to be infinitely more important than goals at the third priority level, and so forth (note that a search can effectively terminate using GP if a high-priority goal has a unique solution; as a result, lower-order goals would not have the opportunity to influence the GP-generated solution). This process can be readily seen in Fig. 12.1 where "infinitely more important" is enforced using the "less than or equal to" (≤) symbol.

These methods require an introduction to the MCDM concept of *dominance* (see Steuer, 1989). One solution dominates another if at least one of the multiple objective values is greater than its alternative, while the remaining objective values are all greater than or equal to each of their alternatives (when seeking to maximize the objectives; multiply objective function(s) by −1 when seeking to minimize). One solution *strongly dominates* another if all objective values are greater than their alternatives.

Finally, this book makes use of the term "optimal" to describe the best solution. It should be noted that in the field of MCDM this term is changed to "efficient"—also, *nondominated, noninferior,* or *Pareto optimum* (for Italian economist Vilfredo Pareto who described a situation where it is not possible to improve the economic situation of some people without making others worse off)—where there is no unique solution that maximizes all objectives simultaneously. With this understanding, "optimal" will continue to be used to refer to the best answer possible for a given instance and meeting the criteria set in Fig. 12.1. Additional information on the two listed MCDM techniques as well as formal mathematical definitions for the various terms used in this section can be found in Steuer (1989).

12.4 Normalization and Efficacy Index Equations

In order to make the balance results comparable in magnitude to all other measures and to allow for more legible graphical comparisons with worst-case calculations in the charts, the effects of squaring portions of Eq. (8.6) can be compensated for by taking the square root of the resulting F, F_{lower}, or F_{upper}. This will subsequently be referred to in this text as *normalizing* (to reflect the concept of a reduction in the values to a common magnitude). While using Eq. (8.6) is desirable to emphasize the importance of a solution's balance as well as to drive stochastic search processes toward the optimal solution, normalization allows for a more intuitive observation of the relative merits of any two solutions. For example, solutions having an equal number of workstations (e.g., NWS = 3) but differing idle times at each workstation (I_j) resulting in differing balance such as $I_j = \langle 1, 1, 4 \rangle$ and $I_j = \langle 2, 2, 2 \rangle$ (optimal) would have balance values of 18 and 12 respectively, while the normalized values would stand at 4.24 and 3.46, still indicating better balance with the latter solution, but also giving a sense of the relative improvement that solution provides, which the measure generated by Eq. (8.6) lacks.

The primary mathematical tool used here for quantitative analysis is shown in Eq. (12.1) and referred to as the *efficacy index*. The efficacy index is the ratio of the difference between a calculated measure x and its worst-case measure x_{upper} to the measure's *sample range* [i.e., the difference between the best-case measure x_{lower} and the worst-case measure as given by: $\max(X_y) - \min(X_z) \mid y, z \in \{1, 2, \ldots, |X|\}$ from the area of statistical quality control] expressed as a percentage and described by [with the vertical lines in Eqs. (12.1) and (12.2) representing absolute value versus cardinality as is seen elsewhere in this book]

$$\text{EI}_x = \frac{100 \cdot |x_{\text{upper}} - x|}{|x_{\text{upper}} - x_{\text{lower}}|} \tag{12.1}$$

This generates a value between 0 and 100 percent, indicating the percentage of optimum for any given measure and any given solution-generating methodology being evaluated. For example, the efficacy index formula for balance would read

$$\mathrm{EI}_F = \frac{100 \cdot |F_{\text{upper}} - F|}{|F_{\text{upper}} - F_{\text{lower}}|}$$

For multiple data sets being simultaneously studied, such as those given in Sec. 11.5, probability theory presents us with the concept of a *sample mean*. The sample mean of a method's efficacy index is calculated using

$$\overline{\mathrm{EI}}_x = \left(\sum_{i=1}^{y} \frac{100 \cdot |x_{\text{upper}} - x_i|}{|x_{\text{upper}} - x_{\text{lower}}|} \right) \Bigg/ y \qquad (12.2)$$

where y is the sample size (the number of data sets). While Eq. (12.1) provides individual data set size efficacy indices—especially useful in demonstrating worst and best case as well as trends with instance size—Eq. (12.2) allows a single numerical value that provides a quantitative measure of the location of the data center in a sample. Though not formulated here, another analysis tool for use when considering multiple data sets is standard deviation, which can provide additional insights into the data as well as to the solution methodology.

While they are not necessary for the DLBP, note that the absolute value function allows the efficacy index to be used for a problem other than the DLBP and in the case where a large number represents a better result, rather than requiring the upper- and lower-bound references be exchanged in Eqs. (12.1) and (12.2).

An additional quantitative tool can be borrowed from the field of statistics. *Simple linear regression* and *correlation* (using the *sample coefficient of determination*), and *polynomial regression* and its associated *coefficient of multiple determination* can be used to quantify the accuracy of the curves discussed in Sec. 12.2 and to provide the regression equation.

Once the degree of the polynomial regression model is qualitatively determined, the regression equation can be automatically calculated by mathematical software (using, e.g., a common spreadsheet software application), as can the coefficient of determination. In this text, the charts containing the combinatorial optimization methodology with the third-order and exhaustive growth curves, and with the second-order curve are used not only to provide a qualitative, graphical comparison, but also (along with the detailed time complexity curves) to determine the *degree* (*order*) of the fitted polynomial regression curve. Once the order is observed by comparison, either a linear or a polynomial regression model is selected. All six of the heuristic methodologies demonstrated here are either first (linear), second, or third order; this corresponds to average-case time complexities of $O(n)$, $O(n^2)$, or $O(n^3)$.

The coefficient of determination is the final portion of the quantitative component of the study. This value represents the portion of the *variation* in the collected data explained by the fitted curve. The coefficient of determination is then multiplied by 100 to illustrate the adequacy of the fitted regression model, indicating the percentage of variation time that can be attributed to the size of the data set. The closer the value comes to 100 percent, the more likely it is an accurate model. While the coefficients of the polynomial regression model are of interest in presenting as accurate a growth curve model as possible, of greater value is the order of the model since the largest exponent is the only variable of interest in complexity theory.

12.5 Simulation

Though not demonstrated in this text, another methodology for analyzing complex problems such as the DLBP is simulation. The Defense Acquisition University (2009b) provides a detailed background in the concepts of modeling and simulation starting with several definitions. A *model* is described as a physical, mathematical, or otherwise logical representation of a system, entity, phenomenon, or process, while *modeling* is listed as an application of a standard, rigorous, structured methodology to create and validate a physical, mathematical, or otherwise logical representation of a system, entity, phenomenon, or process. *Simulation* is then a method of implementing a model over time or as a sequence of events. Also, it is a technique for testing, analysis, or training in which real-world systems are used, or where real-world and conceptual systems are reproduced by a model. A *simulator* is described as a device, computer program, or system that performs simulation. Finally, *modeling and simulation* is the use of models, including emulators, prototypes, simulators, and stimulators, either statically or over time, to develop data as a basis for making managerial or technical decisions; it is noted that the terms "modeling" and "simulation" are often used interchangeably.

There are a variety of definitions of models. One way to distinguish among models is by classifying them as static and dynamic models; structural and behavioral models; and physical and computerized models. A more common listing of model types includes mathematical, static, dynamic, process, and physical.

A *mathematical model* is defined as a symbolic model whose properties are expressed as mathematical symbols and relationships (in contrast with graphical, narrative, software, or tabular models); for example, a model of a nation's economy expressed as a set of equations. A *static model* is a model of a system in which there is no change (e.g., a scale model of a bridge, studied for its appearance rather than for its performance under varying loads). A *dynamic model* is a model of a system in which there is change, such as the occurrence of events over time or the movement of objects through space (e.g., a model of a bridge that is subjected to a moving load to determine characteristics of the bridge under changing stress). A *process model* is a model of the processes performed by a system (e.g., a model that represents the software development process as a sequence of phases). Finally, a *physical model* is a model whose physical characteristics resemble the physical characteristics of the system being modeled (e.g., a plastic or wooden replica of an airplane, a mock-up of the inside of a car, a scale model of a building). While each of these model types has its own benefits and weaknesses, in the type of demanufacturing analysis considered here, a mathematical model is the most applicable.

There are also various types of simulations. Generally, a simulation is the use of a model to understand the system that has been represented. Usually this is a study of the time evolution of systems and how to carry out realizations of the model to make predictions. There are three types of simulations: constructive, virtual, and live. The virtual simulation is most commonly used in the types of studies in this text. A *virtual simulation* is generally one that is a synthetic representation of environments patterned after the actual organization, operations, or equipment. A *constructive simulation* is one that involves real people making inputs into a simulation that carries out those inputs by simulated people operating simulated systems. This is also referred to as human-in-the-loop. A *live simulation* is one that involves real people operating real systems in real environments.

Modeling and simulation requires verification and validation (V&V). *Verification* determines accuracy, completeness, consistency, and traceability within and among developmental products. It typically consists of providing a wide range of inputs to a module of the software to ensure proper operation of an individual software component. *Validation* evaluates the product as a whole, comparing it with respect to the intended use. It determines whether or not the software program as a whole provides a correct output for given input, possibly necessitating (in the case of the DLBP and similar problems) varying-size data sets with known optimal results. (A possible follow-on to V&V is *accreditation*, which is an official certification that a model or simulation is acceptable for use for a specific purpose.) V&V helps to understand strengths and weaknesses in order to gain acceptance of the model and its results, understand the model's capabilities, increase confidence in the simulation output, and provide evidence to substantiate informed decisions.

While simulation is an acceptable methodology—and often the only practical one (e.g., for the analysis of complex just-in-time systems)—for addressing a production-related problem such as the DLBP, it differs significantly from optimal methods and heuristics in a significant way. Simulation provides a descriptive solution as opposed to a prescriptive solution. (The two primary types of modeling techniques are *prescriptive* models—where the model prescribes a solution; e.g., linear programming—and *descriptive* models—where the model describes a situation to allow for analysis but does not provide a solution; e.g., queuing theory). As such, simulation can show what a complex model performs using given parameters, but cannot give the optimal or near-optimal solution (there are creative exceptions to this, including cases where some type of sensitivity analysis can be used to quickly modify significant parameters, as well as problems where there are a limited number of discrete values that need to be simulated; e.g., finding the optimal number of supermarket checkout lines where one would line be a minimum and the store size would physically bound the maximum).

CHAPTER 13
Exhaustive Search

13.1 Introduction

An exhaustive search algorithm is presented for obtaining the optimal solution to small instances of the DISASSEMBLY LINE BALANCING PROBLEM (DLBP). Exhaustive search provides the optimal solution to many problems, including NP-complete problems. However, its exponential time complexity quickly reduces its practicality necessitating the use of combinatorial optimization techniques, which are instrumental in obtaining optimal or near-optimal solutions to problems with intractably large solution spaces.

Exhaustive search is a deterministic search method and is one of the three processes demonstrated here that are not iterative in nature; that is, they are not designed to look for incremental improvements on each new solution found. Both of these observations are trivial when considering exhaustive search since it looks at all answers with the result being optimal (therefore, it gives the same answer every time it is run and there is no room for improvement in the quality of the solution).

This chapter focuses on the design and structure of an exhaustive search algorithm and its performance, providing a baseline for later comparison with the six other combinatorial optimization searches. Section 13.2 provides background on the exhaustive search itself and considers its theoretical time complexity. Section 13.3 documents the exhaustive search results when run using the instances from Chap. 11. Finally, Sec. 13.4 summarizes the results of the analysis performed in this chapter. (Sec. 10.3 provided an introduction to exhaustive search.)

13.2 Model Description

An exhaustive search algorithm (McGovern and Gupta, 2007c) is used to confirm the exponential time growth of exhaustive search, to provide a time complexity benchmark for the subsequent heuristic studies, and to determine the optimal solutions for any data set up to a given size (based on a reasonable search time). The exhaustive algorithm used here is built around a recursive function (Fig. 13.1)

```
Procedure GENERATE (q){
    IF ((q = = 0) ∧ (PRECEDENCE_PASS)){

        CALC_PERFORMANCE

        IF (FIRST_PERMUTATION_GENERATED){
            SET best_solution : = new_solution
        }

        IF (BETTER_SOLUTION (new_solution, best_solution)) {
            SET best_solution: = new_solution
        }
    }

    ∀ k∈ P{
        EXCHANGE (k, q – 1)
        GENERATE (q – 1)
        EXCHANGE (k, q – 1)
    }
    RETURN
}

Procedure EXCHANGE (x, y){
        SET temp : = PS_x
        SET PS_x : = PS_y
        SET PS_y : = temp
        RETURN
}
```

Figure 13.1 Recursive backtracking algorithm to generate all permutations of q numbers and save the best sequence visited.

that generates all permutations of a sequence of n numbers (given the input $q = n$ when it is first called) with each number representing a disassembly task and the order of the numbers representing the disassembly sequence. The purpose of this exhaustive algorithm is to find every subset of the data set (every permutation) and, from all of those subsets, find the solution set that best satisfies the multicriteria performance requirements by checking against all other permutations. The solution performance was computed based on the subroutine described in Sec. 12.3 after F, H, D, and R measures are calculated using Eqs. (8.6), (8.18), (8.26), and (8.32).

This procedure effectively performs a search of all arcs connecting all vertices (the only exception being an arc which would cause the precedence to fail as determined by the PRECEDENCE_PASS routine) in a depth-first search ordering. The algorithm grows each solution sequence from the first vertex to the last, checking for feasibility then, if precedence is maintained to this point in the sequence, calculating F, H, D, and R (and other values including number of workstations), then continues to the next arc and vertex. Once a sequence is successfully generated, the next sequence is started and the process repeats. The first sequence generated is defaulted to be the best solution found thus far. This "best solution" is later replaced by any new sequence that performs better in accordance with the multicriteria subroutine of Chap. 12. The worst-case time complexity of the recursive depth-first search routine [big-oh of $O(n!)$] is also equal to the best-case [big-omega of $\Omega(n!)$] with the tight time complexity asymptotic bound then listed as having a big-theta of $\Theta(n!)$.

13.3 Numerical Results

DLBP Exhaustive Search is used to solve all of the instances of size $n \leq 12$. Any problems larger than $n = 12$ are not attempted due to the excessive calculation time involved. This is not a drawback for the DLBP A Priori instances since, by design, their optimal solution is already known. In addition, optimal solutions for the personal computer (PC) instance and the 10-part instance can be successfully generated. The only instance for which the optimal solution is not known is the cellular telephone instance. Solutions to this instance will be treated in a traditional NP-complete benchmark format as discussed in Sec. 11.5.1.

13.3.1 Personal Computer Instance

DLBP Exhaustive Search is able to rapidly find the optimal solution to the PC instance, taking less than 1/100th of a second. The solution, as seen in Table 13.1 requires four workstations and results in $F^* = 33$, $H^* = 7$, $D^* = 19{,}025$, and $R^* = 6$. This solution consists of the workstations having 2 to 4 seconds per workstation of idle time, providing idle times at each workstation of at least 5 percent but not more than 10 percent of the total disassembly time of the 40 seconds allocated.

The speed in obtaining the optimal solution is primarily attributed to the small instance size. While DLBP Exhaustive Search generates all permutations of the search space, once a solution is deemed infeasible due to the precedence constraints, the algorithm does not perform any of the related (and time-consuming) performance calculations.

Part ID	1	5	3	6	2	8	7	4
PRT	14	23	12	16	10	36	20	18
Workstation	1	1	2	2	2	3	4	4
Hazardous	0	0	0	0	0	0	1	0
Demand	360	540	620	750	500	720	295	480
Direction	1	2	1	4	0	4	1	1

TABLE 13.1 Exhaustive Solution Using the Personal Computer Instance

Part ID	10	5	6	7	9	4	8	1	2	3
PRT	10	23	14	19	14	17	36	14	10	12
Workstation	1	1	2	2	3	3	4	5	5	5
Hazardous	0	0	0	1	0	0	0	0	0	0
Demand	0	0	750	295	360	0	0	0	500	0
Direction	3	5	5	2	5	2	1	2	0	0

TABLE 13.2 Exhaustive Solution Using the 10-Part Instance

13.3.2 The 10-Part Instance

DLBP Exhaustive Search is also able to rapidly find the optimal solution to the 10-part instance, averaging 0.03 seconds. The solution, as seen in Table 13.2, requires five workstations and results in $F^* = 211$, $H^* = 4$, $D^* = 9730$, and $R^* = 7$.

Again, the speed in obtaining the optimal solution is attributed both to the small instance size and the fact that the precedence constraints enables DLBP Exhaustive Search to minimize time spent on many of the possible (but infeasible) sequence permutations.

13.3.3 Cellular Telephone Instance

DLBP Exhaustive Search has not been used to solve the cellular telephone instance. Using the DLBP A Priori instances (which run slightly slower due to a lack of precedence constraints), $n = 8$ averaged 0.10 seconds, while $n = 12$ averaged 1,476.44 seconds (just under 25 minutes). At this rate it could be conservatively estimated (assuming a 10-fold increase for each single increment of n; this approximates the observed growth but is actually slower than its true $n!$ growth) that $n = 13$ would take over 4 hours, $n = 14$ would take almost 2 days, $n = 15$ would run over 2 weeks, $n = 16$ almost 6 months, $n = 17$ over 4½ years, and so forth. Based on its size of $n = 25$, it is roughly estimated that the cellular telephone instance would take DLBP Exhaustive Search over 465,000,000 years on the search space of 25! or 1.551 × 1025.

13.3.4 DLBP A Priori Problem Instances

The DLBP Exhaustive Search technique can be seen below on the DLBP A Priori data with $n = 12$ (Table 13.3). The optimal solution

Part ID	12	9	2	5	3	6	8	11	1	4	7	10
PRT	11	7	3	5	3	5	7	11	3	5	7	11
Workstation	1	1	1	1	2	2	2	2	3	3	3	3
Hazardous	1	0	0	0	0	0	0	0	0	0	0	0
Demand	0	1	0	0	0	0	0	0	0	0	0	0
Direction	0	0	0	0	0	0	0	0	1	1	1	1

TABLE 13.3 Exhaustive Solution Using the A Priori Instance at $n = 12$

found agrees with the Sec. 11.5 measures of NWS* = 3 workstations, $F^* = 0$, $H^* = 1$, $D^* = 2$, and $R^* = 1$. The time to complete the search averaged 24.61 minutes. Per Table 11.6, note that this is only one of 10,368 optimum extreme points at $n = 12$ (and based on the multicriteria subroutine in Chap. 12, it is also known to be the very first optimal solution visited by the Exhaustive Search process).

As the first of the multiple optimum extreme points that is found, the sequence also demonstrates the *backtracking* methodology. Ignoring the positions of the hazardous and high-demand parts as well as the parts with part removal directions of 1, it can be seen then from position $k = 3$ to $k = 8$ that the parts placed earliest in the sequence are those with the lowest part identification values. This indicates that Exhaustive Search proceeds from candidate sequence $\langle 1, 2, \ldots, n\rangle$ through to $\langle n, \ldots, 2, 1\rangle$.

Figure 13.2 provides the plot of runtime versus instance size using the DLBP A Priori benchmark for the exhaustive algorithm and a third-order curve for growth comparison and to demonstrate how the exhaustive search runtime rapidly increases with instance size. As discussed in Sec. 13.3.3, a DLBP A Priori instance size of $n = 12$ tasks is the largest practical size for analysis with the next size up ($n = 16$) considered to be effectively intractable. Faster computers and programming tricks

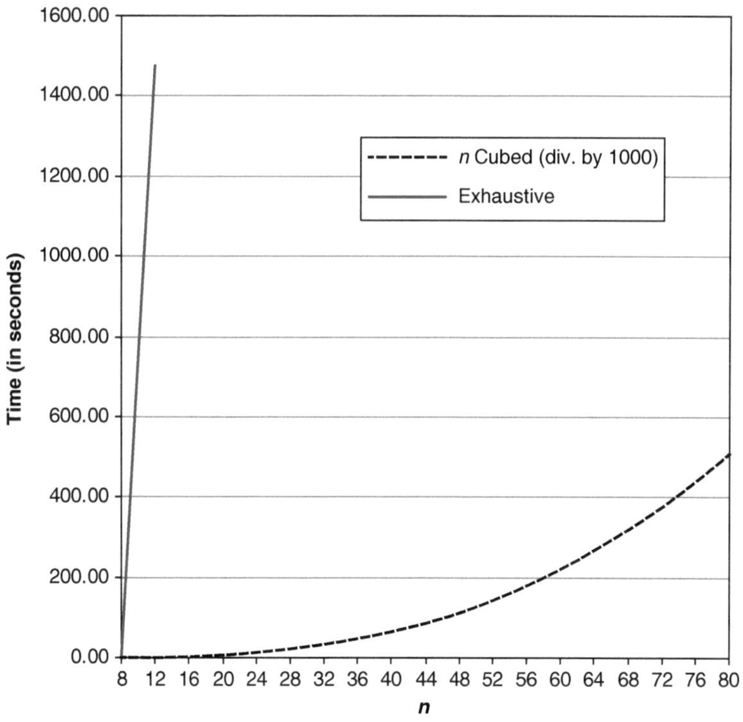

FIGURE 13.2 Exhaustive Search time complexity.

can significantly speed this up but the growth in exhaustive search is exponential nonetheless. Using the DLBP A Priori instances in the range of $n = \{8, 12, 16, \ldots, 80\}$, only the $n \leq 12$ sized instances can be solved in a reasonable amount of time using the recursive Exhaustive Search algorithm. Upon closer study, the algorithm's runtime $T(n)$ is seen to increase with instance size at just over $n!$, specifically $T(n) \propto 1.199(n!)$. This is in agreement with the time complexity of the recursive depth-first search (i.e., backtracking) as reported in Sec. 13.2. The time complexity of Exhaustive Search can then be described as $\Theta(n!)$ or factorial complexity (Rosen, 1999). While commonly considered to be exponential growth (since the big-oh estimate for $n! = 1 \cdot 2 \cdot 3 \cdot \cdots \cdot n \leq n \cdot n \cdot n \cdot \cdots \cdot n = n^n$), it is actually even slower than the definition of exponential growth, which is $O(b^n)$ where $b > 1$ (Rosen, 1999) and assuming b is constant. [Note that using the actual growth proportionate to $1.199(n!)$ gives an $n = 16$ runtime of 2.05 years and a cellular telephone instance runtime of 15,162,296,041 centuries; the universe is generally accepted to be 137,000,000 centuries old.]

While DLBP Exhaustive Search is limited in the size and number of instances it can solve, due to the nature of the DLBP A Priori benchmark instances, this does not pose any drawback since the optimal solution can always be calculated using the information and formulae in Sec. 11.5.2. This is the intent and part of the design of the benchmark instances. Searching for the optimal solution on smaller size instances, both confirm the functionality of the exhaustive search and provides for a baseline time complexity.

13.4 Conclusions

In this chapter an exhaustive search algorithm was presented for determining the optimal solution to several instances of the DISASSEMBLY LINE BALANCING PROBLEM as well as to provide a time complexity baseline for comparison with other search techniques. The exhaustive search shown is based on a depth-first, recursive algorithm that was modified for the DLBP and its multicriteria selection methodology. Though impractical for all but the smallest problems, exhaustive search can provide the optimal solution for all DLBP instances (and all NP-complete problems as well). In addition, at this time it is the only optimal methodology for all combinatoric problems. However, until the advent of computing machines that are not categorized as "reasonable," purely exhaustive search has little practical application.

CHAPTER 14
Genetic Algorithm

14.1 Introduction

A genetic algorithm (GA) is presented for obtaining solutions to the DISASSEMBLY LINE BALANCING PROBLEM (DLBP). The genetic algorithm considered here involves hot-started or randomly generated initial populations with crossover, mutation, and fitness competition conducted over many generations. The methodology is applied to the four experimental instances and the average results for each are then analyzed.

GA is the first of two stochastic methods used in Part II and is also one of two methodologies here classified as metaheuristics. This chapter introduces the genetic algorithm and the DLBP variant. Section 14.2 provides background on the genetic algorithm. This section also considers the theoretical time complexity of the GA. Section 14.3 introduces details of the DLBP implementation of GA, including the selection of a crossover operator. Section 14.4 provides the justification for modifications made to the architecture of a traditional GA in the interest of addressing algorithm actions that may be detrimental to the specific problem as presented by the DLBP. Section 14.5 reviews the numerical constants used in the DLBP GA, while Sec. 14.6 documents the GA search results when run using the instances from Chap. 10. Finally, Sec. 14.7 summarizes the results of this chapter. (Section 10.4 provides a background in genetic algorithms.)

14.2 Model Description

The chromosome (solution) consists of a sequence of genes (parts for the DLBP). A pool, or *population*, of size N is used. Only feasible disassembly sequences are allowed as members of the population or as offspring. The fitness is computed for each chromosome using the method for solution performance determination as described in Sec. 12.3.

The time complexity is a function of the number of generations, the population size N, and the chromosome size n. As such, the runtime is seen to be on the order of $n \cdot N \cdot$(number of generations). Since, in this chapter, both the population and the number of generations are considered to stay constant with instance size, the best-case time complexity of GA is seen to have an asymptotic lower bound of $\Omega(n)$. Because the worst-case runtime also requires no more processing time than $T(n) \propto n \cdot N \cdot$(number of generations), the worst-case time complexity of GA has the asymptotic upper bound $O(n)$, so GA therefore exhibits a tight time complexity bound of $\Theta(n)$.

14.3 DLBP-Specific Genetic Algorithm Architecture

The GA for the DLBP is constructed as follows (McGovern et al., 2003; McGovern and Gupta, 2007a). An initial, feasible population is randomly generated and the fitness of each chromosome in this generation is calculated. (The sole exception to this is taken for the cellular telephone instance where the primary technique of randomly generating the initial population is ineffective due to the instance's relatively large size and numerous precedence constraints, so four feasible and diverse solutions are manually generated to hot start DLBP GA.) An even integer of $R_x \cdot N$ parents is then randomly selected for crossover to produce $R_x \cdot N$ offspring [offspring make up $(R_x \cdot N \cdot 100)$ percent of each generation's population]. An elegant crossover, the precedence preservative crossover (PPX) developed by Bierwirth et al. (1996) is used here to create the offspring. As shown in Fig. 14.1, PPX first creates a mask (one for each child, every generation). The mask consists of random 1s

Figure 14.1 Precedence preservative crossover example.

Parent 1:	1 3 2 6 5 8 7 4
Parent 2:	1 2 3 5 6 8 7 4
Mask:	2 2 1 2 1 1 1 2
Child:	
Parent 1:	1 3 2 6 5 8 7 4
Parent 2:	**1 2** 3 5 6 8 7 4
Mask:	2 2 1 2 1 1 1 2
Child:	1 2
Parent 1:	x **3** x 6 5 8 7 4
Parent 2:	x x 3 5 6 8 7 4
Mask:	1 2 1 1 1 2
Child:	1 2 3
Parent 1:	x x x 6 5 8 7 4
Parent 2:	x x x **5** 6 8 7 4
Mask:	2 1 1 1 2
Child:	1 2 3 5
Parent 1:	x x x **6** x **8 7** 4
Parent 2:	x x x x 6 8 7 4
Mask:	1 1 1 2
Child:	1 2 3 5 6 8 7
Parent 1:	x x x x x x x 4
Parent 2:	x x x x x x x **4**
Mask:	2
Child:	1 2 3 5 6 8 7 4

and 2s indicating which parent part information should be taken from. If, for example, the mask for child 1 reads 22121112, the first two parts (i.e., from left to right) in parent 2 would make up the first two genes of child 1 (and these parts would be stricken from the parts available to take from both parent 1 and 2); the first available (i.e., not stricken) part in parent 1 would make up gene three of child 1; the next available part in parent 2 would make up gene four of child 1; the next three available parts in parent 1 would make up genes five, six, and seven of child 1; the last part in parent 2 would make up gene eight of child 1. This technique is then repeated using a new mask for child 2.

After crossover, mutation is randomly conducted. Mutation is occasionally (based on the R_m value) performed by randomly selecting a single child, then exchanging two of its disassembly tasks while ensuring precedence is preserved. The $R_x \cdot N$ least-fit parents are removed by sorting the entire parent population from worst-to-best based on fitness.

Since the GA saves the best parents from generation to generation and it is possible for duplicates of a solution to be formed using PPX, the solution set could contain multiple copies of the same answer resulting in the algorithm potentially becoming trapped in a local optima. This becomes more likely in a GA with solution constraints (such as precedence requirements) and small populations, both of which are seen in this study. To avoid this, DLBP GA can be modified to treat duplicate solutions as if they had the worst fitness performance (highest numerical value), relegating them to replacement in the next generation. With this new ordering, the best unique $(1 - R_x) \cdot N$ parents are kept along with all of the $R_x \cdot N$ offspring to make up the next generation then the process is repeated. To again avoid becoming trapped in local optima, DLBP GA—as is the case with many combinatorial optimization techniques—is run not until a desired level of performance was reached but rather for as many generations as deemed acceptable by the user. Since DLBP GA always keeps the best solutions thus far from generation to generation, there is no risk of solution drift or bias, and the possibility of mutation allows for a diverse range of possible solution space visits over time. See Figs. 14.2 and 14.3 for DLBP GA details.

14.4 DLBP-Specific Qualitative Modifications

DLBP GA is modified from a general GA in several ways. Instead of the worst portion of the population being selected for crossover, in DLBP GA all of the population is (randomly) considered for crossover. This better enables the selection of nearby solutions (i.e., solutions similar to the best solutions to-date) common in many scheduling problems. Also, mutation is performed only on the children, not the worst parents. This is done to address the small population used in DLBP GA and to counter PPX's tendency to duplicate parents. Finally, duplicate children are sorted to make their deletion from the population likely since there is a tendency for the creation of duplicate solutions (due to PPX) and due to the small population saved from generation to generation.

14.5 DLBP-Specific Quantitative Modifications

A small population has been used (20 versus the more typical 10,000 to 100,000) to minimize data storage requirements and simplify analysis [except for the personal computer (PC) case study where the generations are also varied as part of the study] while a large number of generations have been used (10,000 versus the more typical 10 to 1000) to compensate for this small population while not being so large as to take an excessive amount of processing time. This is also done to avoid solving all cases to optimality since it is desirable to determine the point at which the DLBP GA's performance begins to break

```
Procedure DLBP_GA {
    INITIALIZE_DATA            { Load data: part removal times, etc. }
    SET count := 1             { count is the generation counter }

    FOR N DO:                  { Randomly create an initial population }
        DO:
            RANDOMLY_CREATE_CHROMOSOME
        WHILE (PRECEDENCE_FAIL)
        CALC_FITNESS           { Establish the initial solution's fitness }
                               { Determine no. of parents for reproduction }
    SET num_parents := N · R_x   reproduction }
                               { Ensure an even number of parents }
    SET num_parents := 2 · (num_parents/2)
                               { Note: num_parents is typed as an integer }

    DO:                        { Run GA for MAX_GENERATIONS }
                               { Randomly order the parents for breeding }
        RANDOMIZE_PARENTS
                               { Perform crossover using PPX }
        FOR num_parents DO:
            CROSSOVER

        IF FLIP(R_m) DO:       { FLIP equals 1 R_m % of the time, else 0 }
                               { Randomly select and mutate a child }
            MUTATE_CHILD

                               { Sort: best parents to the last positions }
        REMOVE_LEAST_FIT
                               { Duplicate answers to the front and re-sort }
        REMOVE_REPEATS
                               { Add children to the population }
        CHILDREN_TO_POOL

        FOR N DO:              { Calculate each solution's fitness }
            CALC_FITNESS

                               { Next generation }
        count := count + 1
    WHILE (count< MAX_GENERATIONS)

    SAVE the last chromosome as the best_solution
}
```

Figure 14.2 Pseudocode of the DLBP GA procedure.

FIGURE 14.3
Flowchart of the DLBP GA procedure.

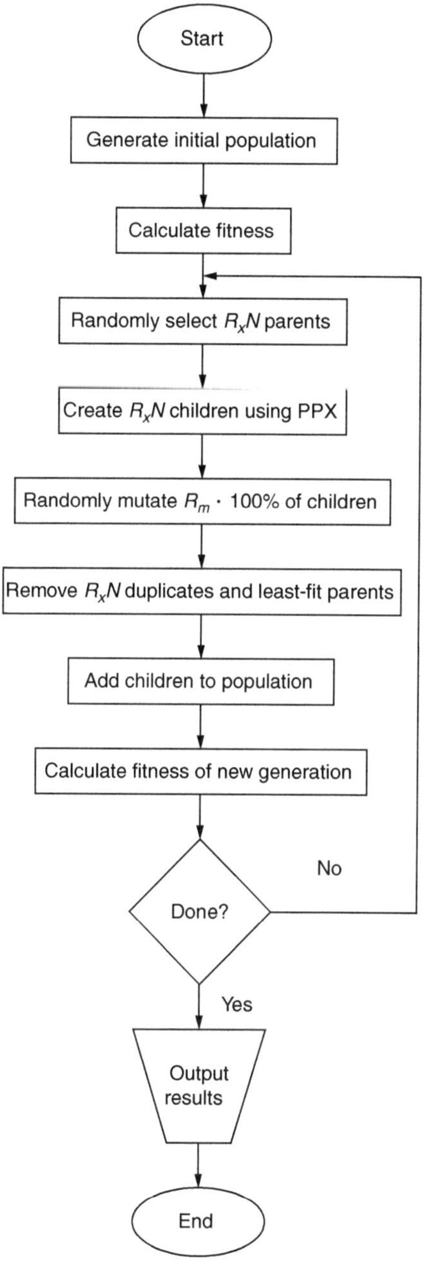

down and how that breakdown manifests itself. Lower than the recommended 90 percent (Koza, 1992), a 60 percent crossover is selected based on test and analysis. Developmental testing indicated that a 60 percent crossover provided better solutions and did so with one-third less processing time. Previous assembly-line balancing literature that indicated best results, have typically been found with crossover rates of 0.5 to 0.7, also substantiated the selection of this lower crossover rate. A mutation is performed about one percent of the time. Although some texts recommend 0.01 percent mutation while applications in journal papers have used as much as 100 percent mutation, 1.0 percent gave excellent algorithm performance for the DISASSEMBLY LINE BALANCING PROBLEM.

14.6 Numerical Results

The GA has been run on each of the four experimental instances from Chap. 10 with the results analyzed in this section.

14.6.1 Personal Computer Instance

DLBP GA is first used to solve the PC instance. It is known that there are four solutions to the PC instance that are optimal in F alone (McGovern and Gupta, 2004d). Since one purpose of the exercise in this section is to demonstrate DLBP GA performance with changes in generation size, the data was run varying the generations and without the use of hazard, demand, or direction data to avoid potential complications in the analysis. Due to GA's stochastic nature, it was run a minimum of five times with the results averaged to avoid reporting unusually favorable or unusually poor results. The DLBP GA converges to moderately good solutions very quickly. It is able to find at least one of the four solutions optimal in F after no more than 10 generations. 100 generations consistently provides three to four of the F-optimal solutions, while 1000 generations almost always result in the generation of all four optimal solutions.

DLBP GA is then applied to the traditional (complete) PC instance. The search of this instance (searching 10,000 generations of 20 chromosomes each) averaged 3.30 seconds over five runs with the optimal solution (Table 14.1) of four workstations and $F^* = 33$, $H^* = 7$, $D^* = 19{,}025$, and $R^* = 6$ being found in each of the runs.

14.6.2 The 10-Part Instance

DLBP GA finds feasible solutions to the 10-part instance in 3.16 seconds when averaged over five runs. Not only is this time slower than DLBP Exhaustive Search (which averaged 0.03 seconds) but it also resulted

Part ID	1	5	3	6	2	8	7	4
PRT	14	23	12	16	10	36	20	18
Workstation	1	1	2	2	2	3	4	4
Hazardous	0	0	0	0	0	0	1	0
Demand	360	540	620	750	500	720	295	480
Direction	1	2	1	4	0	4	1	1

TABLE 14.1 Typical GA Solution Using the Personal Computer Instance

Part ID	6	4	5	10	7	9	8	1	2	3
PRT	14	17	23	10	19	14	36	14	10	12
Workstation	1	1	2	2	3	3	4	5	5	5
Hazardous	0	0	0	0	1	0	0	0	0	0
Demand	750	0	0	0	295	360	0	0	500	0
Direction	5	2	5	3	2	5	1	2	0	0

TABLE 14.2 Typical GA Solution Using the 10-Part Instance

in only a near-optimal solution. The GA found the optimal solution (Table 13.2) of five workstations and $F^* = 211$, $H^* = 4$, $D^* = 9730$, and $R^* = 7$ during one of the five runs. The optimal number of workstations and the optimal balance was found on all of the runs while the hazardous part measure averaged $H = 4.80$ (with a best result being the optimal $H^* = 4$ and a worst result of $H = 5$), the high-demand part removal measure averaged $D = 9{,}054.00$ (with a best result of $D = 8885$—better than the optimal solution but at the expense of a worse performing hazard measure—and a worst result being the optimal measure of $D^* = 9730$), and the part removal direction measure averaged $R = 7.80$ (with a best result being the optimal $R^* = 7$ and a worst result of $R = 8$). A typical solution is shown in Table 14.2.

Although these results are not optimal, this is more a reflection of the challenges posed even by seemingly simple disassembly instances more than an indication of any limitations of this technique. Note that the inclusion of additional precedence constraints will increasingly move GA, and in fact all of the combinatorial optimization methodologies, toward optimal (due to a reduction in the feasible search space).

14.6.3 Cellular Telephone Instance

DLBP GA is then used to solve the cellular telephone instance. DLBP GA finds feasible solutions in 2.26 seconds when averaged over five runs. All of the GA solutions make use of 10 workstations and all had balance measures of $F = 81$. The hazardous part measure averaged $H = 78.40$ (ranging from a best result of $H = 78$ to a worst result of $H = 79$), the high-demand part removal measure averaged $D = 916.20$ (with a best result of $D = 914$ and a worst result of $D = 920$), and the part removal direction measure averaged $R = 10.60$ (with a best result of $R = 10$ and a worst result of $R = 12$). A typical solution is shown in Table 14.3.

GA was never able to generate the optimal number of workstations or the optimal balance (with the assumption that NWS = 9 and $F = 9$ is optimal; Gupta et al., 2004). Using the four manually generated solutions for a hot start, DLBP GA typically gives five different answers over the five runs, providing an indication that the hot-start solutions are adequate in number and diversity.

Part ID	2	1	4	5	10	3	9	6	8	7	11	12	14	15	18	13	17	16	19	21	25	22	20	23	24
PRT	2	3	10	10	2	3	15	15	15	15	2	2	2	2	3	2	2	2	18	1	2	5	5	15	2
Workstation	1	1	1	2	2	2	3	4	5	6	6	7	7	7	7	7	7	7	8	9	9	9	9	10	10
Hazardous	1	1	0	0	0	0	0	0	0	0	0	1	0	0	0	0	0	0	1	0	1	0	0	1	0
Demand	7	4	1	1	2	1	1	1	1	1	1	4	1	1	2	1	2	1	8	4	4	6	1	7	1
Direction	3	2	5	5	4	5	5	5	5	5	4	4	2	0	5	5	4	2	5	4	4	4	4	4	4

Table 14.3 Typical GA Solution Using the Cellular Telephone Instance

Notice that as the instances grow in size, the runtime actually decreases. This is attributed to the precedence constraints possessed by these three instances, the GA process, and the fixed population and number of generations. Since the population and number of generations are fixed and do not grow with instance size, any growth in runtime is due only to the additional processing time related to larger instances (which is minimal as is seen later in Fig. 14.11), or challenges in finding feasible chromosomes after mutation is performed and during the initial population generation. Since the cellular telephone instance used a hot-started initial population, this helped to minimize the total runtime of that instance. In addition, the number of precedence constraints has an effect on the number of feasible solutions as is seen in the longer runtime for the smaller PC instance. In effect, with only 16 solutions that are feasible in the search space (per Sec. 11.2), GA ends up making numerous unsuccessful attempts, first at randomly filling the initial population and then at randomly mutating the chromosomes in the pool during the run. The way DLBP GA is designed, when these attempts fail, the random generation or mutation process is repeated until a feasible solution is generated. The result of this is the observation that small instances with many precedence constraints (resulting in a smaller feasible search space) can actually run as long as or longer than some larger instances.

14.6.4 DLBP A Priori Instances

The developed GA is then used on the DLBP A Priori benchmark test cases to further measure its performance and demonstrate its limitations. All tests are performed using the previously described population size and mutation rates, and run for 10,000 generations using crossover rates of between 60 and 90 percent (with 60 percent reported here due to its more favorable time performance and efficacy). The population size and number of generations is intentionally selected to allow for performance decreases; all measures of efficacy are seen to slowly drop off with instance size.

An example of a GA metaheuristic solution can be seen below with $n = 12$ (Table 14.4). While there is more than one optimal solution for the A Priori instances (see Sec. 11.5.3), GA is the only technique reviewed here (other than, obviously, Exhaustive Search) that

Part ID	12	9	5	3	8	11	6	2	4	10	7	1
PRT	11	7	5	3	7	11	5	3	5	11	7	3
Workstation	1	1	1	1	2	2	2	2	3	3	3	3
Hazardous	1	0	0	0	0	0	0	0	0	0	0	0
Demand	0	1	0	0	0	0	0	0	0	0	0	0
Direction	0	0	0	0	0	0	0	0	1	1	1	1

TABLE 14-4 Typical GA Solution Using the A Priori Instance at $n = 12$

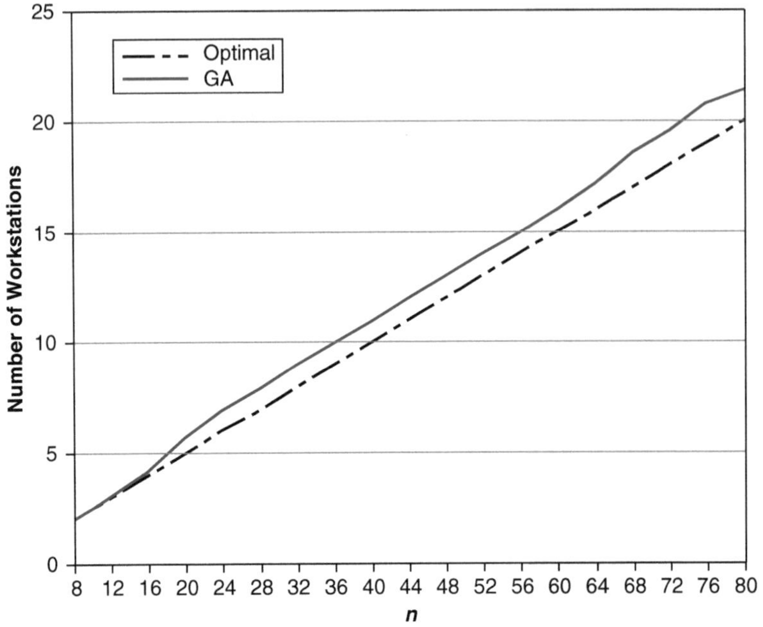

FIGURE 14.4 GA workstation calculation.

is able to regularly find one of the multiple optimum extreme points (three workstations and $F^* = 0$, $H^* = 1$, $D^* = 2$, and $R^* = 1$).

When considering all data sets of $n = \{8, 12, 16, \ldots, 80\}$, DLBP GA's efficacy index [from Eq. (12.1)] in number of workstations ranges from a high of 100 percent, down to 94 percent then stabilizes between 97 and 98 percent (Fig. 14.4). Overall, as given by Eq. (12.1), DLBP GA shows an efficacy index sample mean in number of workstations of 97 percent.

Figure 14.5 allows a detailed study of the decrease in performance with increased instance size, demonstrating optimal performance through $n = 12$ with a rapid decrease in performance through $n = 24$, followed by minor performance improvements until $n = 60$ with a similar steep decrease in performance through $n = 76$, followed by another period of minor performance improvements. Balance appears to grow almost in a *sawtooth-wave* fashion.

The normalized balance efficacy index starts as high as 100 percent then drops to 93 to 95 percent; it is never lower than 89 percent (Fig. 14.6) and has a sample mean of 94 percent. An instance size of $n = 16$ is seen to be the point at which the optimally balanced solution is not consistently found for the selected N and number of generations. Although DLBP GA's performance decreases with instance size, it can be seen in Fig. 14.6 that the solution found, while not

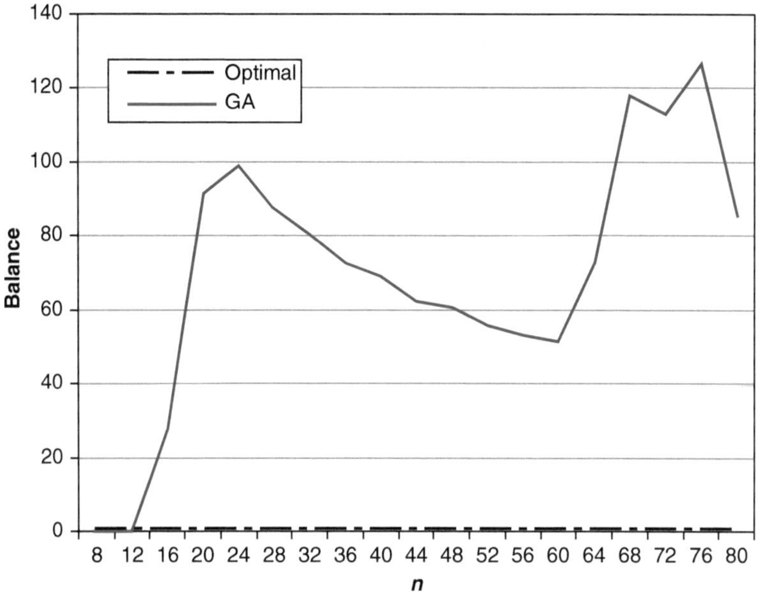

FIGURE 14.5 Detailed GA balance measure.

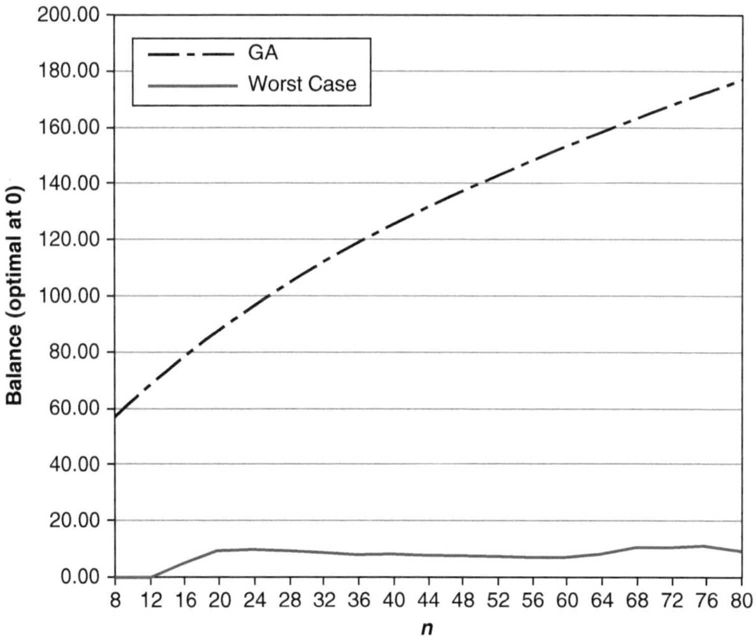

FIGURE 14.6 Normalized GA balance measure.

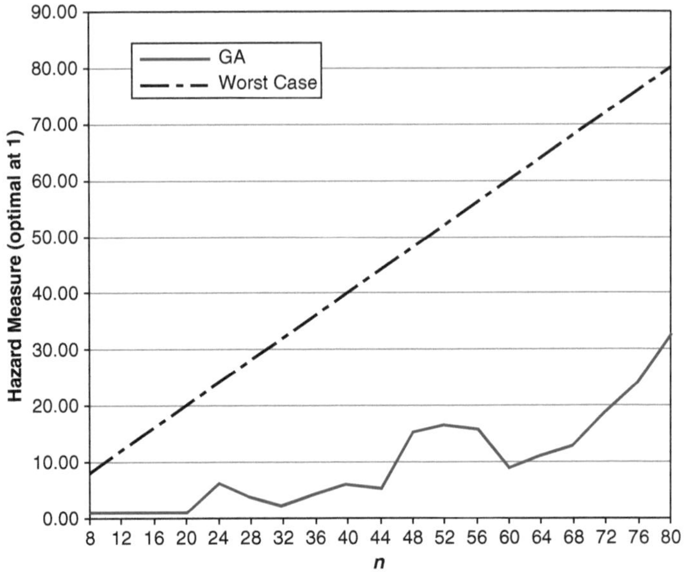

FIGURE 14.7 GA hazard measure.

optimal, is very near optimal and when normalized (as described in Sec. 12.4), roughly parallels the optimal balance curve.

The hazard measure tends to get worse with problem size (Fig. 14.7) with its efficacy index dropping relatively constantly from 100 to 60 percent and having a sample mean of $\overline{\text{EI}}_H = 84\%$.

The demand measure gets worse at a slightly more rapid rate than the hazardous part measure—as is expected due to the multicriteria priorities—with its efficacy index dropping from 100 to 45 percent (with a low of 42 percent) and having a sample mean of $\overline{\text{EI}}_D = 78\%$ (Fig. 14.8).

Finally, the part removal direction measure (Fig. 14.9) gets worse at a more rapid rate than the demand measure, again attributed to the multicriteria priorities, with its efficacy index dropping as low as 17 percent (at data set 15 where $n = 64$) and having a sample mean of $\overline{\text{EI}}_R = 49\%$.

Runtime increases very slowly with instance size. Based on the results of Fig. 14.10, a linear model is used to fit the curve with the linear regression equation calculated to be $T(n) = 0.0327n + 1.5448$ (Fig. 14.11) with a coefficient of determination of 0.9917 indicating 99.17 percent of the total variation explained by the calculated linear regression curve.

With a growth of $0.0327n + 1.5448$, we then list the average-case time complexity of DLBP GA on the DLBP A Priori data sets as $O(n)$ or *linear complexity* (Rosen, 1999). This is in agreement with the calculations from Sec. 14.2.

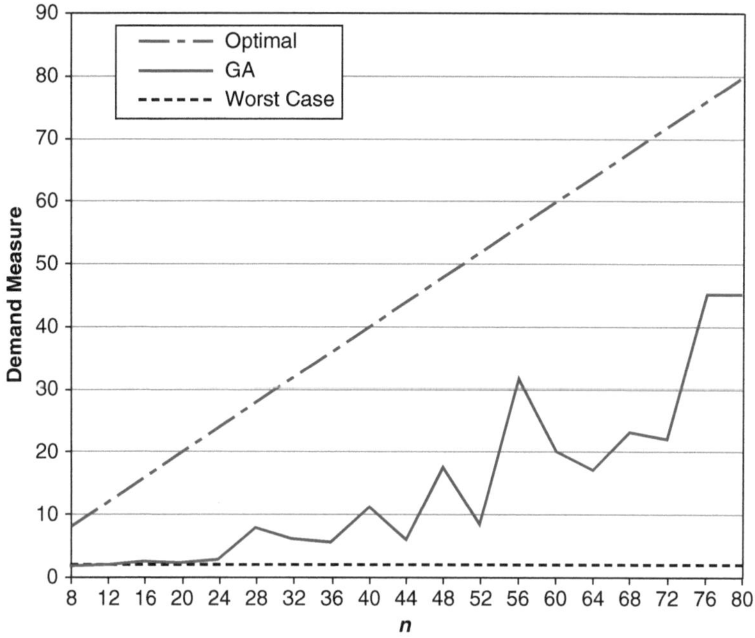

FIGURE 14.8 GA demand measure.

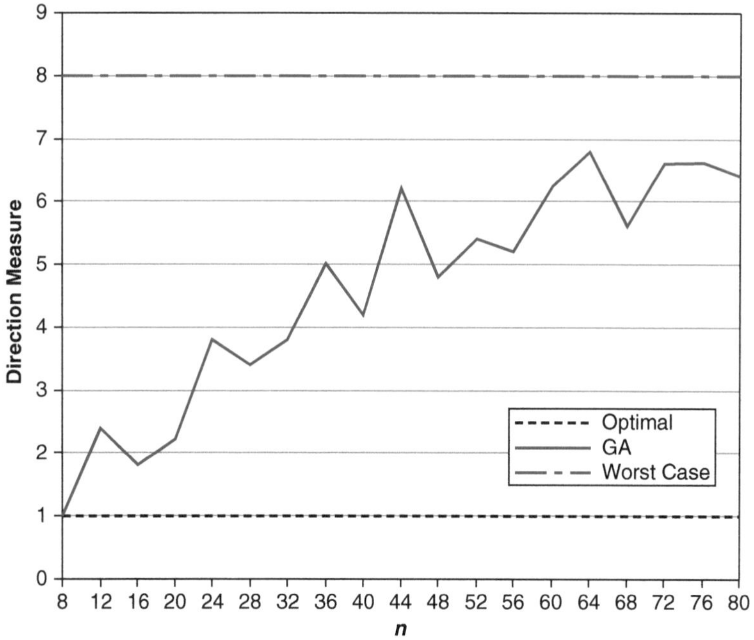

FIGURE 14.9 GA part removal direction measure.

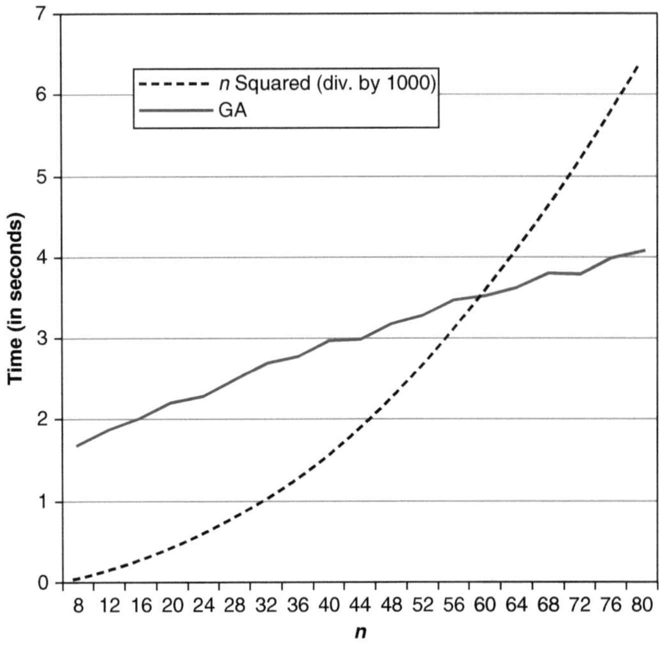

FIGURE 14.10 GA time complexity compared to second order.

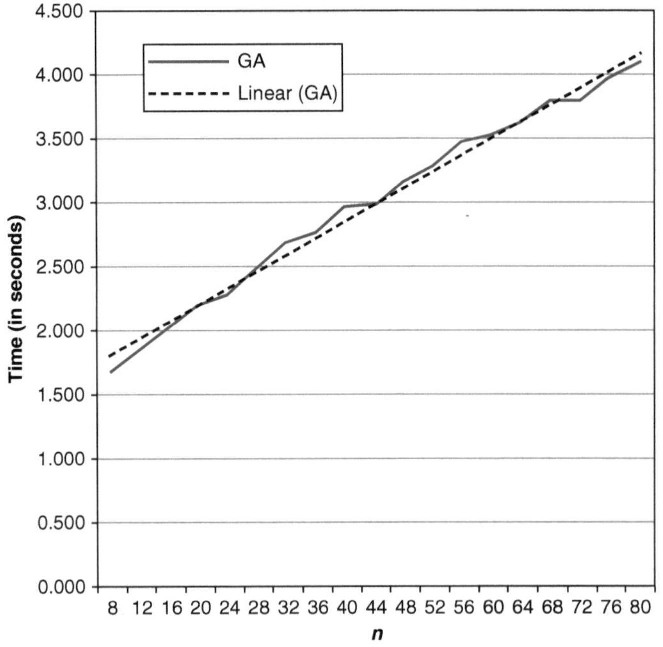

FIGURE 14.11 GA time complexity and first-order polynomial regression line.

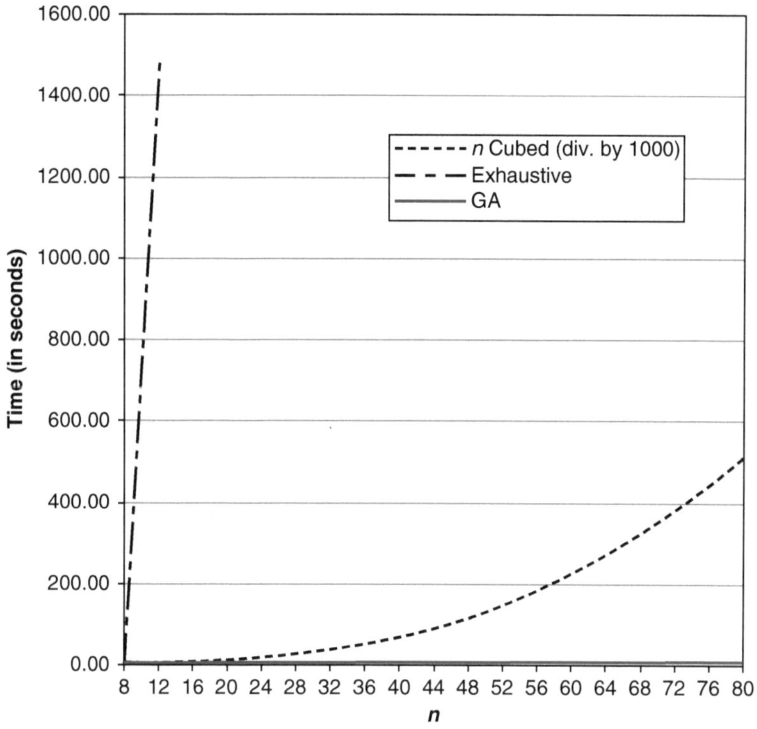

FIGURE 14.12 GA time complexity compared to exhaustive search.

Although larger instances can be expected to take longer to solve with DLBP GA, it is a very shallow, linear increase with significantly slower growth than the exhaustive search algorithm (Fig. 14.12).

DLBP GA maintained the normalized measure of balance within 6 percent of optimum on average. Since hazardous-part removal (Fig. 14.7), early removal of high-demand parts (Fig. 14.8), and minimizing part removal directions (Fig. 14.9) are all subordinate to balance and to the precedence constraints (and to each other respectively), the performance of each of these is seen to degrade at rates proportionate to their priority.

Although the preceding results are suboptimal, this is more a reflection of the especially designed DLBP A Priori data set than the GA search technique. Suboptimal solutions are not an atypical result when this data set is evaluated. The data has been designed to pose challenges—even at relatively small n values—to a variety of combinatoric solution-generating techniques.

Runtime performance can be improved by reducing the number of generations or reducing the population, while the opposite is true for improving other measures of performance; that is, increase the number of generations or increase the population to improve efficacy indices.

14.7 Conclusions

Although a near-optimum combinatorial optimization technique, the DLBP version of the GA metaheuristic quickly finds optimally balanced solutions, or solutions on average within single-digit percentages of optimal, in a variety of exponentially large search spaces, with the level of optimality increasing with the number of constraints. It is able to generate a feasible sequence with an optimal or near-optimal measure of balance while maintaining or improving the hazardous-materials, part-demand, and part-removal-direction measures and does so only at a linear growth cost in time complexity. DLBP GA can easily be structured for distributed computing for order-of-magnitude increases in processing speed.

CHAPTER 15

Ant Colony Optimization

15.1 Introduction

In this chapter the DISASSEMBLY LINE BALANCING PROBLEM (DLBP) is solved using an ant colony optimization (ACO) metaheuristic (also see Sec. 10.5). The version of ACO used here is an ant-cycle model, ant-system algorithm (Dorigo et al., 1999) enhanced for the DLBP. As is the case with the DLBP Genetic Algorithm technique (DLBP GA), the DLBP ACO method seeks to preserve precedence relationships within the product being disassembled and is further modified for multiple objectives, working to minimize the number of workstations and balance the part removal sequence while attempting to remove hazardous and high-demand product components as early as possible and remove parts with similar part removal directions together.

ACO is the second of the two stochastic methods studied in Part II and is also the second of the two methodologies classified here as metaheuristics. This chapter introduces the ant colony optimization metaheuristic and the DLBP variant. Section 15.2 provides background on the ACO model. Section 15.3 introduces qualitative enhancements made to the ACO for application to DLBP as well as the theoretical time complexity calculations. Section 15.4 describes the quantitative assignments made to the variables that guide the metaheuristic, while Sec. 15.5 reviews the ACO search results when run using the instances from Chap. 10. Finally, Sec. 15.6 summarizes the efforts of this chapter.

15.2 Model Description

ACO agents work cooperatively toward an optimal problem solution. Multiple agents are placed at multiple starting nodes. Each of the m ants is allowed to visit all remaining (unvisited) edges as indicated by a tabu-type list. Each ant's possible subsequent steps; that is, from a node p to node q giving edge $[p, q]$ (vertices and arcs in the DLBP), are evaluated for desirability and each is assigned a proportionate probability as shown (modified for the DLBP) in Eq. (15.1). Based on these probabilities, the next step in the tour is randomly selected for each ant. After completing an entire tour, all feasible ants are given the equivalent of additional pheromone (in proportion to tour desirability), which is added to each step it has taken on its tour. All paths are then decreased in their pheromone strength according to a measure of evaporation (equal to $1 - \rho$). This process is repeated for NC_{max} or until stagnation behavior is demonstrated. The ant system/ant-cycle model has been claimed to run in $O(NC \cdot n^3)$ (Dorigo et al., 1996). Since the number of cycles is considered to remain constant with instance size, the worst-case time complexity is then properly listed as $O(n^3)$. A time complexity analysis is made for the DLBP version of ACO in the next section.

15.3 DLBP-Specific Qualitative Modifications and the Metaheuristic

DLBP ACO is a modified ant-cycle algorithm (McGovern and Gupta, 2006a). It is designed around the DLBP by accounting for feasibility constraints and addressing multiple objectives. In DLBP ACO, each part is described by a vertex on the tour with the number of ants being set equal to the number of parts and having one ant uniquely on each part as the starting position. Each ant is allowed to visit all parts not already in the solution. (In the sequence example from Sec. 7.3—the eight-tuple $\langle 5, 2, 8, 1, 4, 7, 6, 3\rangle$—at time $t = 3$ component p would represent component 8 while component $q \in \{1, 4, 7, 6, 3\}$ and possible partial solution sequences $\langle 5, 2, 8, 1\rangle$, $\langle 5, 2, 8, 4\rangle$, $\langle 5, 2, 8, 7\rangle$, $\langle 5, 2, 8, 6\rangle$, and $\langle 5, 2, 8, 3\rangle$ would be evaluated for feasibility and balance.) Each ant's possible subsequent steps are evaluated for feasibility and current measure of balance, and then are assigned a proportionate probability as given by Eq. (15.1). Infeasible steps receive a probability of zero and so any ants having only infeasible subsequent task options are effectively ignored for the remainder of the cycle. The best solution found in each cycle is saved if it is better than the best found in all the cycles thus far (or if it is the first cycle; this is done in order to initialize the best solution measures for later comparison) and the process is repeated for a maximum designated number of cycles; it does not terminate for stagnation in order to allow for potential better solutions (due to the ACO's stochastic edge-selection capabilities). The best solution found is evaluated as described in Sec. 12.3.

Due to the nature of the DLBP (i.e., whether or not a task should be added to a workstation depends on that workstation's idle time and precedence constraints), the probability in Eq. (15.1) is not calculated once, at the beginning at each tour as is typical with ACO, but is calculated dynamically, generating new probabilities at each increment of t. The ACO probability calculation formula is modified for the DLBP with the probability Pr of ant r taking arc (p, q) at time t during cycle NC calculated as

$$\mathrm{Pr}^{r}_{p,q}(t) = \begin{cases} \dfrac{[\tau_{p,q}(\mathrm{NC})]^{\alpha} \cdot [\eta_{p,q}(t)]^{\beta}}{\sum_{r \in \mathrm{allowed}_r} [\tau_{p,r}(\mathrm{NC})]^{\alpha} \cdot [\eta_{p,r}(t)]^{\beta}} & \text{if } q \in \mathrm{allowed}_r \\ 0 & \text{otherwise} \end{cases} \tag{15.1}$$

where, for example, the notation "p, q" is used to represent either arc (p, q) or edge $[p, q]$. Also modified for the DLBP, the amount of trail on each arc is calculated after each cycle using

$$\tau_{p,q}(\mathrm{NC}+1) = \rho \cdot \tau_{p,q}(\mathrm{NC}) + \Delta\tau_{p,q} \tag{15.2}$$

where (unchanged from the general ACO formulation)

$$\Delta\tau_{p,q} = \sum_{r=1}^{m} \Delta\tau^{r}_{p,q} \tag{15.3}$$

and [also unchanged, other than the use of arcs versus edges and the recognition that arc $(p, q) \neq (q, p)$ in the DLBP; this is discussed further at the end of this section]

$$\Delta\tau^{r}_{p,q} = \begin{cases} \dfrac{Q}{L_r} & \text{arc } (p, q) \text{ used} \\ 0 & \text{otherwise} \end{cases} \tag{15.4}$$

As in other DLBP combinatorial optimization methodologies, DLBP ACO seeks to preserve precedence while not exceeding CT in any workstation. As long as the balance is at least maintained, DLBP ACO then seeks improvements in hazard measure, demand measure, and direction measure, but never at the expense of precedence constraints. For DLBP ACO, the visibility $\eta_{p,q}$ is also calculated dynamically at each increment of t during each cycle and is defined as

$$\eta_{p,q}(t) = \frac{1}{F_{t,r}} \tag{15.5}$$

where $F_{t,r}$ is the balance of ant r at time t (i.e., ant r's balance thus far in its incomplete solution sequence generation). The divisor for the change in trail is defined for DLBP ACO as

$$L_r = F_{n,r} \tag{15.6}$$

where a small final value for ant r's measure of balance at time n (i.e., at the end of the tour) provides a large measure of trail added to each arc. Though L_r and $\eta_{p,q}$ are related in this application, this is not unusual for ACO applications in general; for example, in the TRAVELING SALESPERSON PROBLEM (TSP), L_r is the tour length while $\eta_{p,q}$ is the reciprocal of a the distance between cities p and q (Dorigo et al., 1996). However, this method of selecting $\eta_{p,q}$ (effectively a short-term greedy choice) may not always translate into the best long-term (i.e., final tour) solution for a complex problem like the DLBP. Also note that, although this is a multicriteria problem, only a single criterion (the measure of balance F) is being used in the basic ACO calculations and trail selection. The other objectives are only considered after balance and then only at the completion of each cycle, not as part of the probabilistic vertex selection. That is, an ant's tour solution is produced based on F, while at the end of each cycle the best overall solution is then updated based on F, H, D, and R in that order per the routine in Chap. 12. This is done because the balance is considered (for the purpose of this demonstration) to be the overriding requirement. (One way to consider other criteria in the n greedy steps taken in the selection of a solution would be a weighting scheme in the probabilistic step selection.)

In addition to this new, dynamic nature of the probability calculation in Eq. (15.1), three procedural changes to the algorithm in Dorigo et al. (1996) are made (Fig. 15.1). First, the time counter is reset to zero in each cycle. This has no effect on the metaheuristic's performance and is done only for computer software readability. Secondly, "Place the m ants on the n nodes" is moved from step 1 to step 2 (and "nodes" is changed to "vertex" in recognition of the DLBP's formulation as a digraph). This is done so that the resetting of each ant to each vertex is repeated in each cycle, prohibiting the potential accumulation of ants on a single (or few) ending vertex (vertices) resulting in the subsequent restarting of all (or many) of the ants from that one (or few) vertex (vertices) in successive cycles; resetting the ants ensures solution diversity. This is also necessary due to the nature of the DLBP where the sequence is a critical and unique element of any solution. A part with numerous precedence constraints will typically be unable to be positioned in the front of the sequence (i.e., removed early) and will often be found toward the end. In the DLBP, not resetting each ant could potentially result in a situation where, for example, all ants choose the same final part due to precedence constraints; when each ant attempts to initiate a subsequent search (i.e., the next cycle) from

Procedure DLBP_ACO {

1. Initialize:

    SET NC := 0 { NC is the cycle counter }
    FOR every arc (p, q) **SET** an initial $\tau_{p,q}(\text{NC}) := c$ for trail intensity and $\tau_{p,q} := 0$

2. Format problem space:

    SET $t := 0$ { t is the time counter }
    Place the m ants on the n vertices
    SET $s := 1$ { s is the tabu list index }
    FOR $r := 1$ to m **DO**
        Place the starting part of the rth ant in $\mathbf{tabu}_r(s)$

3. Repeat until tabu lists are full { this step will be repeated $(n - 1)$ times }

    SET $s := s + 1$
    SET $t := t + 1$
    FOR $r := 1$ to m **DO**
        Choose the part q to move to, with probability $pr^r_{p,q}(t)$ as given by Eq. (15.1)
            { at time t the rth ant is on part $p = \mathbf{tabu}_r(s - 1)$ }
        Move the rth ant to the part q
        Insert part q in $\mathbf{tabu}_r(s)$

4. **FOR** $r := 1$ to m **DO**

    Move the rth ant from $\mathbf{tabu}_r(n)$ to $\mathbf{tabu}_r(1)$
    Compute F, H, D, and R for the sequence described by the rth ant
    SAVE the best solution found per procedure BETTER_SOLUTION

  FOR every arc (p, q)
    FOR $r := 1$ to m **DO**

$$\Delta\tau^r_{p,q} := \begin{cases} \dfrac{Q}{L_r} & \text{arc}(p,q) \text{ used} \\ 0 & \text{otherwise} \end{cases}$$

$$\Delta\tau_{p,q} := \Delta\tau_{p,q} + \Delta\tau^r_{p,q}$$

5. **FOR** every arc (p, q) compute $\tau_{p,q}(\text{NC} + 1)$ according to $\tau_{p,q}(\text{NC} + 1) := \rho \cdot \tau_{p,q}(\text{NC}) + \Delta\tau_{p,q}$

    SET NC := NC + 1
    FOR every arc (p, q) **SET** $\Delta\tau_{p,q} := 0$

6. **IF** $(\text{NC} < \text{NC}_{max})$

    THEN
        Empty all tabu lists
        GOTO STEP 2
    ELSE
        PRINT *best_solution*
        STOP

  }

FIGURE 15.1 DLBP ACO procedure.

that last vertex, all fail due to the infeasibility of attempting to remove a final part first, effectively terminating the search in just one cycle. Finally, step 6 normally could also terminate for stagnation behavior but (as further discussed in the next section) this is not desired for the DLBP and has been deleted.

It can be inferred from Dorigo et al. (1996) that p, q is equivalent to q, p. Although this is acceptable and desirable in a TSP-type problem, in the DLBP, sequence is an essential element of an n-tuple and of a DLBP solution (due to precedence constraints and size constraints at each workstation). For this reason, in DLBP ACO, edges p, q and q, p are directed (i.e., arcs) and therefore distinct and unique and as such, when trail is added to one, it is not added to the other.

If the number of cycles is considered to remain constant with instance size, one of the factors affecting the time complexity is the number of ants m. In the best case, ACO would make use of the minimum number of ants (i.e., one). In this situation the lone ant agent would initially (only in the DLBP or a similar BIN-PACKING-type process) be required to calculate the probabilities concerning the $n - 1$ vertices available to add to the tour. Once it did this, it would then review the $n - 1$ options and randomly select one based on the probabilistic weights. These events would take in proportion to $2(n - 1)$ time. The process is then repeated at the newly selected second vertex, but this time for $n - 2$ vertices, taking in proportion to $2(n - 2)$ time. This continues until the next to the last remaining vertex, which would take time in proportion to 2(2) (the vertex not selected becomes the last vertex by default). The resulting sequence of $(n - 2)$ loops can be formulated as a runtime $T(n)$ and reduced as follows

$$T(n) \propto 2(n-1) + 2(n-2) + \cdots + 2(2)$$

$$T(n) \propto 2[(n-1) + (n-2) + \cdots + 2]$$

$$T(n) \propto 2\{(n-1) + (n-2) + \cdots + [n-(n-2)]\}$$

$$T(n) \propto 2\{n(n-2) - [1 + 2 + \cdots + (n-2)]\}$$

$$T(n) \propto 2n(n-2) - \sum_{p=1}^{n-2} p$$

$$T(n) \propto 2n(n-2) - \frac{2(n-2)[(n-2)+1]}{2}$$

$$T(n) \propto 2n(n-2) - (n-2)[(n-2)+1]$$

$$T(n) \propto (2n^2 - 4n) - [(n^2 - 4n + 4) + (n-2)]$$

$$T(n) \propto n^2 - n - 2$$

This gives a best-case time complexity of $\Omega(n^2 - n - 2)$, which is properly written as $\Omega(n^2)$. As can be seen in this discussion, the added requirements posed by the DLBP's structure, while doubling the time required, do not increase the order of the search. Again it must be emphasized that this is a best case for the DLBP when using one ant; one ant is not used here nor is it recommended by Dorigo et al. (1996).

The theoretical worst case (and by its usage, expected to reflect the experimentally derived average case) would make use of the recommended number of ants $m = n$ as is done in this book. Keeping the number of cycles constant with instance size, the resulting runtime is formulated as

$$T(n) \propto \mathrm{m}(n^2 - n - 2)$$

and when $m = n$ this reduces to

$$T(n) \propto n^3 - n^2 - 2n$$

giving a resulting worst-case time complexity calculated to be $O(n^3)$, which is in agreement with Dorigo et al. (1996). No tight bound (big-theta) exists.

15.4 Quantitative Assignments

In DLBP ACO the maximum number of cycles is set at 300 (i.e., $\mathrm{NC}_{\mathrm{max}}$ = 300) since larger problems than those demonstrated in this book have been shown to reach their best solution by that count (Dorigo et al., 1996) and as a result of previous experimentation (Gupta and McGovern, 2004). The process is not run until no new improvements are seen but, as is the norm with many combinatorial optimization techniques, is run continuously on subsequent solutions until $\mathrm{NC}_{\mathrm{max}}$ is reached. This also enables the probabilistic component of ACO an opportunity to leave potential local minima. Repeating the DLBP ACO method in this way provides improved balance over time. Per the best ACO metaheuristic performance experimentally determined by Dorigo et al. (1996), the weight of pheromone in path selection α is set equal to 1.00, the weight of balance in path selection β is set equal to 5.00, the evaporation rate $1 - \rho$ is set to 0.50, and the amount of pheromone added if a path is selected Q is set to 100.00. The initial amount of pheromone on all of the paths c is set equal to 1.00.

15.5 Numerical Results

The DLBP ACO has been run on each of the four experimental instances from Chap. 10 with the results analyzed in this section.

15.5.1 Personal Computer Instance

The ACO metaheuristic is used to provide a solution to the DLBP based on the personal computer (PC) disassembly instance. Having a stochastic nature and variable run length (both traits shared with genetic algorithms), ACO is also used with this data set to demonstrate changes in its performance over numerous runs.

Due to its probabilistic component, DLBP ACO is able to find all four solutions optimal in F alone on different runs. For the same reason, it is seen to occasionally (though rarely) select a suboptimally balanced solution. It also does not find the F-optimal solutions with the same frequency. This can be attributed to Eq. (15.1) and the ACO's requirement to evaluate partial solutions before making a solution element selection decision. In addition, further research into the solutions generated during each cycle indicate that the process needs significantly less than an $NC_{max} = 300$ to stabilize at the ultimate solution. This is attributed to the numerous precedence constraints possessed by the product under study. Though details are not provided here, it can also be observed that arcs not selected are diluted very rapidly. For example, it would not be unusual to see the bulk of the arcs, though all had been initialized at $c = 1$, terminating with trail values of 1×10^{-91} after the 300 cycles.

The DLBP ACO is then applied to the PC instance in the more typical fashion. The search of the PC instance averaged 0.65 seconds per run over five runs with the optimal solution (Table 15.1) of four workstations and $F^* = 33$, $H^* = 7$, $D^* = 19{,}025$, and $R^* = 6$ being found in each of the five runs.

The DLBP ACO algorithm is able to generate a feasible disassembly sequence while attaining the first two objectives (minimize the number of workstations and balance the disassembly line, from Sec. 8.2). The last three objectives (remove hazardous components early in the disassembly sequence, remove high-demand components before low-demand components, and remove parts with similar part removal directions together) are also demonstrated, while rigid enforcement of the precedence constraints prevents earlier removal of these parts.

15.5.2 The 10-Part Instance

Over multiple runs DLBP ACO is able to successfully find several (again, due to the stochastic nature of ACO) feasible solutions to the 10-part instance; Table 15.2 depicts a typical solution sequence. This solution demonstrates the minimum number of workstations while placing the hazardous part (part number 7) and high-demand parts

Part ID	1	5	3	6	2	8	7	4
PRT	14	23	12	16	10	36	20	18
Workstation	1	1	2	2	2	3	4	4
Hazardous	0	0	0	0	0	0	1	0
Demand	360	540	620	750	500	720	295	480
Direction	1	2	1	4	0	4	1	1

Table 15.1 Typical ACO Solution Using the Personal Computer Instance

Part ID	10	5	4	6	7	9	8	1	3	2
PRT	10	23	17	14	19	14	36	14	12	10
Workstation	1	1	2	2	3	3	4	5	5	5
Hazardous	0	0	0	0	1	0	0	0	0	0
Demand	0	0	0	750	295	360	0	0	0	500
Direction	3	5	2	5	2	5	1	2	0	0

TABLE 15.2 Typical ACO Solution Using the 10-Part Instance

(parts 6, 7, and 9) relatively early in the removal sequence (the exception being part 2, primarily due to precedence constraints and part 2 having a smaller part removal time that is not conducive to the ACO's iterative greedy process) but allowing eight direction changes (due to precedence constraints, the optimum is $R^* = 7$ while the minimum is $R_{lower} = 5$ and worst case is $R_{upper} = 8$).

DLBP ACO finds feasible solutions to the 10-part instance in 1.14 seconds when averaged over five runs. Not only is this time slower than the DLBP Exhaustive Search (which averaged 0.03 seconds) but it also results in only near-optimal solutions (with the optimal solution sequence known to give five workstations and $F^* = 211$, $H^* = 4$, $D^* = 9730$, and $R^* = 7$). The optimal number of workstations and the optimal balance is found on all of the runs, while the hazardous part measure averaged $H = 4.40$ (with a best result of $H^* = 4$ and a worst result of $H = 5$), the high-demand part removal measure averaged $D = 10{,}736.00$ (with a best result of $D = 10{,}090$ and a worst result of $D = 11{,}635$), and the part removal direction measure averaged $R = 7.00$ (with a best result of $R = 6$ and a worst result of $R_{upper} = 8$).

15.5.3 Cellular Telephone Instance

DLBP ACO finds feasible solutions to the cellular telephone data set in 7.38 seconds on average. Due to its probabilistic component, DLBP ACO is seen to select a solution optimal to the number of workstations and in balance about 80 percent of the time (with the assumption that nine workstations and $F = 9$ is optimal; Gupta et al., 2004). This can again be attributed to Eq. (15.1) and ACO's requirement to evaluate partial solutions before making a solution element selection decision.

All of the DLBP ACO solutions in the runs conducted make use of nine workstations and have balance measures averaging $F = 9.80$ with a low of $F = 9$ and a high of $F = 13$. The hazardous part measure averaged $H = 86.40$ (ranging from a best result of $H = 82$ to a worst result of $H = 90$), the high-demand part removal measure averaged $D = 921.80$ (with a best result of $D = 868$ and a worst result of $D = 952$), and the part removal direction measure averaged $R = 11.60$ (with a best result of $R = 11$ and a worst result of $R = 12$). A typical solution is shown in Table 15.3.

Part ID	2	6	7	1	8	3	9	15	4	14	18	16	5	13	10	11	19	17	20	12	21	22	25	23	24
PRT	2	15	15	3	15	3	15	2	10	2	3	2	10	2	2	2	18	2	5	2	1	5	2	15	2
Workstation	1	1	2	2	3	3	4	4	5	5	5	5	6	6	6	6	7	8	8	8	8	8	8	9	9
Hazardous	1	0	0	1	0	0	0	0	0	0	0	0	0	0	0	0	1	0	0	1	0	0	1	1	0
Demand	7	1	1	4	1	1	1	1	1	1	2	1	1	1	2	1	8	2	1	4	4	6	4	7	1
Direction	3	5	5	2	5	5	5	0	5	2	5	2	5	5	4	4	5	4	4	4	4	4	4	4	4

TABLE 15.3 Typical ACO Solution Using the Cellular Telephone Instance

Part ID	6	5	9	8	7	10	1	2	11	4	3	12
PRT	5	5	7	7	7	11	3	3	11	5	3	11
Workstation	1	1	1	1	2	2	2	2	3	3	3	4
Hazardous	0	0	0	0	0	0	0	0	0	0	0	1
Demand	0	0	1	0	0	0	0	0	0	0	0	0
Direction	0	0	0	0	1	1	1	0	0	1	0	0

TABLE 15.4 Typical ACO Solution Using the A Priori Instance at $n = 12$

15.5.4 DLBP A Priori Instances

The DLBP ACO metaheuristic can be seen above with $n = 12$ (Table 15.4). In the five runs conducted for data collection, DLBP ACO is never able to structure the solution having the minimum number of workstations ($NWS^* = 3$); each run results in $NWS = 4$. As a result, the balance ranges from $F = 234$ to $F = 252$. The hazard measure remains constant at $H = 2$, the demand measure varies between $D = 1$ and $D = 4$, and the part removal direction measure varies between $R = 2$ and $R = 6$ (Sec. 11.5.2 contains information relative to the optimal values).

The DLBP A Priori data set forced the DLBP ACO algorithm to obtain a near-optimal number of workstations and other measures. This data also allows the metaheuristic to demonstrate hazardous material and high-demand part sequencing, as well as part removal direction selection.

On the full range of data ($n = \{8, 12, 16, \ldots, 80\}$), DLBP ACO regularly finds solutions with $NWS^* + 1$ workstations (Fig. 15.2). From

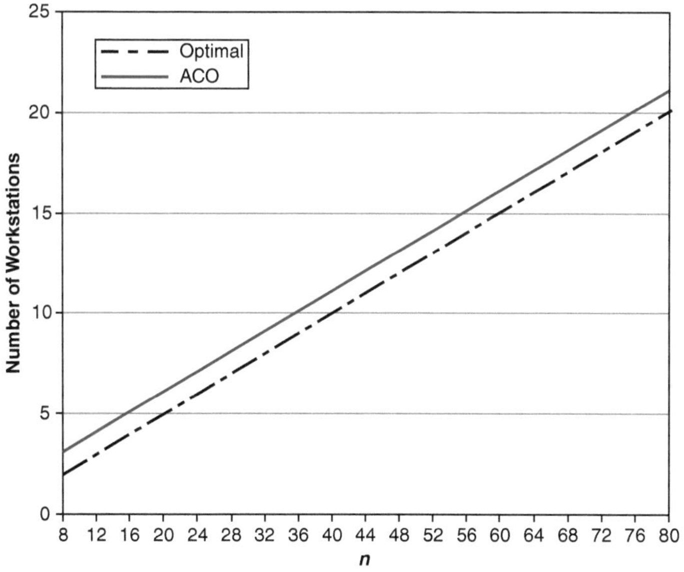

FIGURE 15.2 ACO workstation calculation.

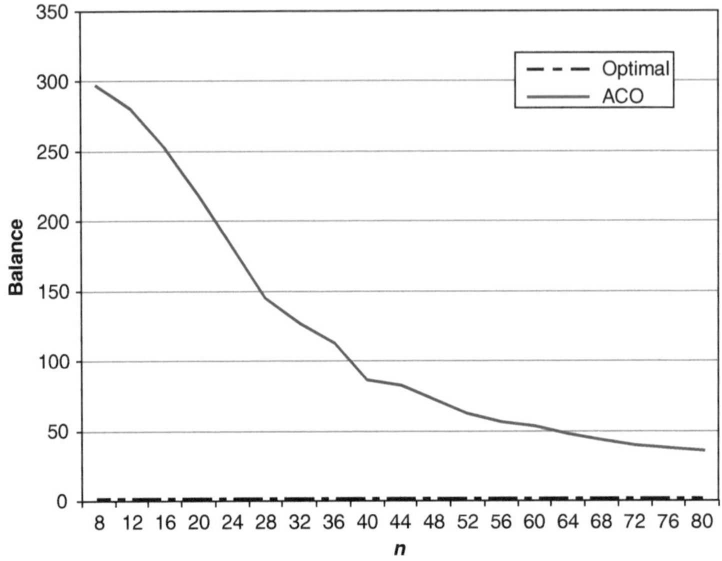

FIGURE 15.3 Detailed ACO balance measure.

Eq. (12.1), DLBP ACO's efficacy index in number of workstations moves from a low of 83 percent to a high of 98 percent. Overall, as given by Eq. (12.2), DLBP ACO shows an efficacy index sample mean in number of workstations of 95 percent.

Consistent improvement in the balance measure is seen with increases in data set size. Figure 15.3 allows a detailed study of the increase in performance with increased instance size, demonstrating poor performance with smaller instances steadily improving through to the last data set at $n = 80$.

The normalized balance efficacy index starts as low as 69 percent then increases through to 97 percent (Fig. 15.4) with a sample mean of 90 percent. While the DLBP ACO starts out poorly (in terms of balance), its performance smoothly improves with instance size. It can be seen in Fig. 15.4 that the normalized solution gets closer and closer to the optimal balance curve as the problem instance grows.

The hazardous part and the demanded part are both regularly suboptimally placed. Hazardous part placement stays relatively consistent with problem size (though effectively improving as compared to the worst case, as illustrated by Fig. 15.5). These results are as expected since hazard performance is designed to be deferential to balance and effected only when a better hazard measure can be attained without adversely affecting balance. The hazard measure's efficacy index starts out poorly at 23 percent, but by $n = 20$ it is able to maintain between 68 and 99 percent. Overall, the DLBP ACO's hazard measure has a sample mean of $\overline{EI}_H = 74\%$.

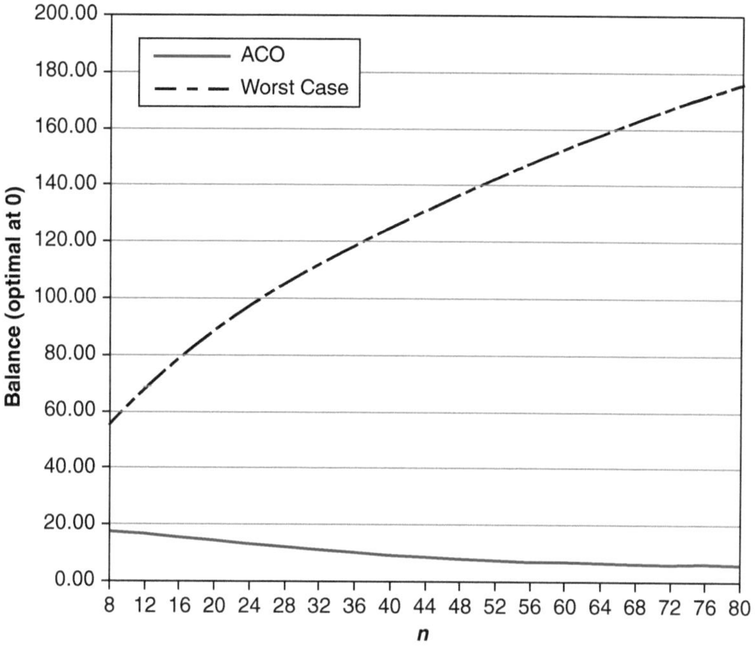

FIGURE 15.4 Normalized ACO balance measure.

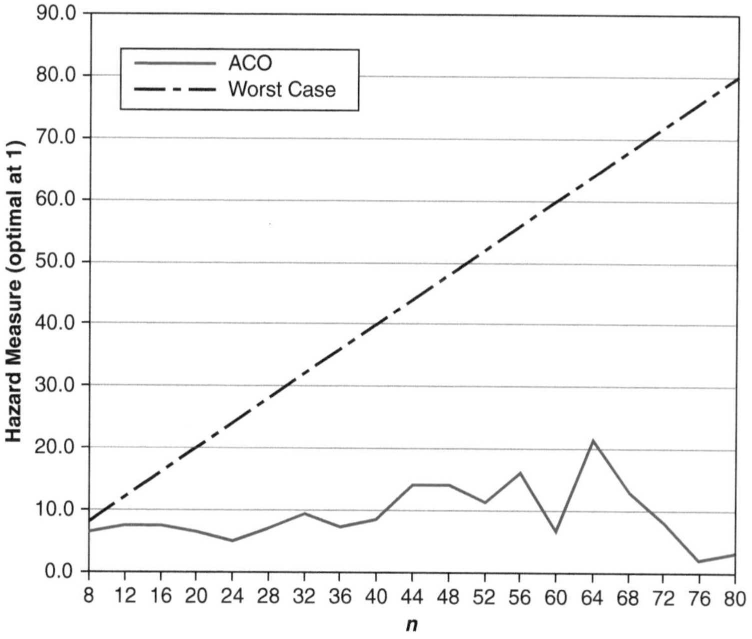

FIGURE 15.5 ACO hazard measure.

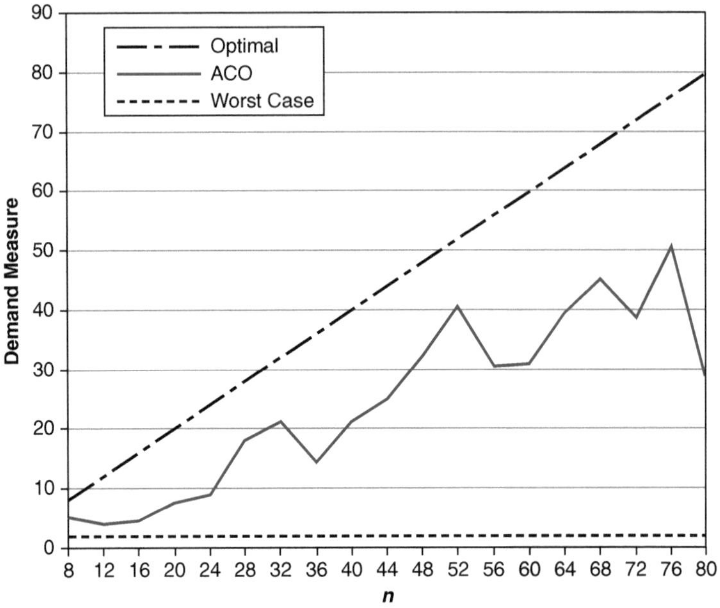

FIGURE 15.6 ACO demand measure.

The demand measure decreases in performance when compared to best case and maintains performance when compared to worst case (Fig. 15.6). These results too are as expected since demand performance is designed to be deferential to balance and hazardous part placement and effected only when a better demand measure can be attained without adversely affecting balance or hazardous part placement. The demand measure gets worse at a slightly more rapid rate than the hazardous part measure (as is expected due to the multicriteria priorities) with its efficacy index fluctuating between a low at $n = 76$ of 34 percent and $n = 16$ at 81 percent. The demand measure sample mean is calculated to be $\overline{\text{EI}}_D = 51\%$.

With part removal direction structured as to be deferential to balance, hazard, and demand, it is seen to decrease in performance when compared to the best case and when compared to the worst case (Fig. 15.7). Again, these results are as expected due the prioritization of the multiple objectives. The part removal direction measure's efficacy index starts at its highest value of 57 percent then drops to between 0 and 17 percent by $n = 24$ with a sample mean of $\overline{\text{EI}}_R = 13\%$.

Runtime increases moderately quickly with instance size. Based on Figs. 15.8 and 15.9 and the calculations in Sec. 15.3, a third-order polynomial regression model is used to fit the curve, with the regression equation calculated to be $T(n) = 0.0005n^3 - 0.0043n^2 + 0.3714n + 2.4412$. The degree is also in agreement with the time complexity notes in Sec. 15.2 based on previous studies done in this area.

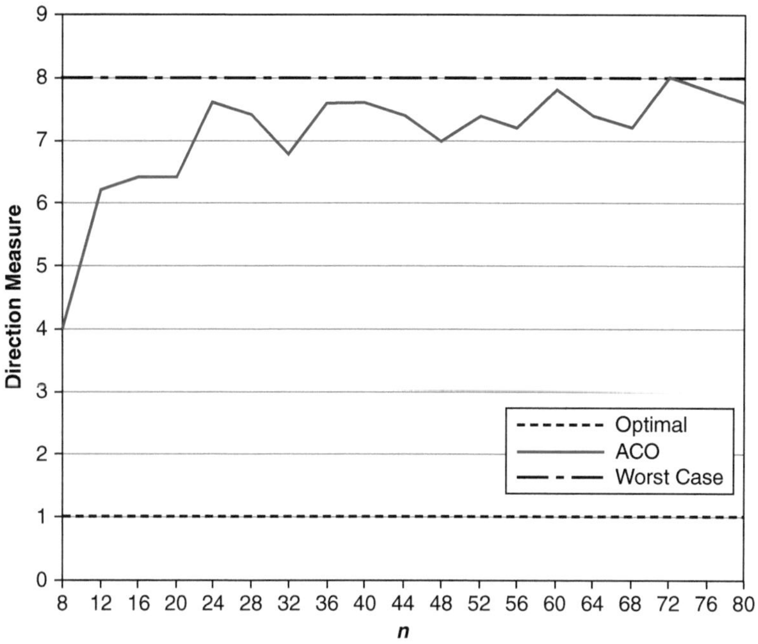

Figure 15.7 ACO part removal direction measure.

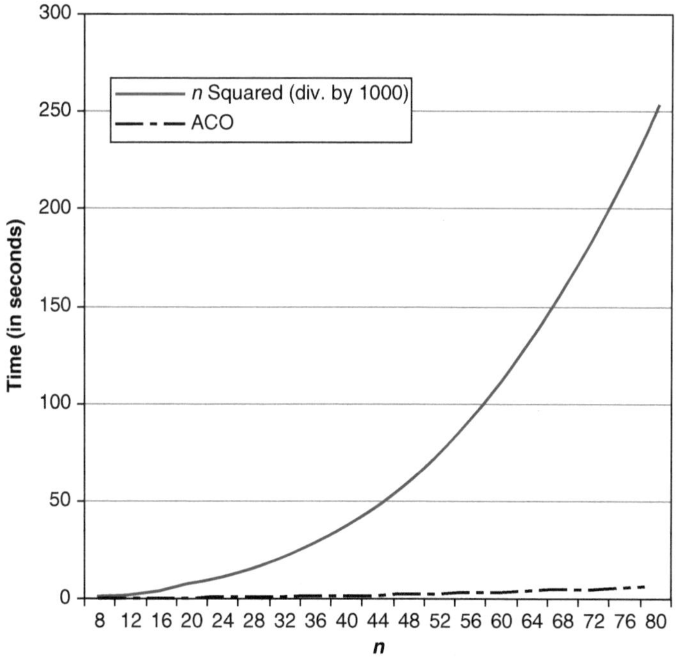

Figure 15.8 ACO time complexity compared to second order.

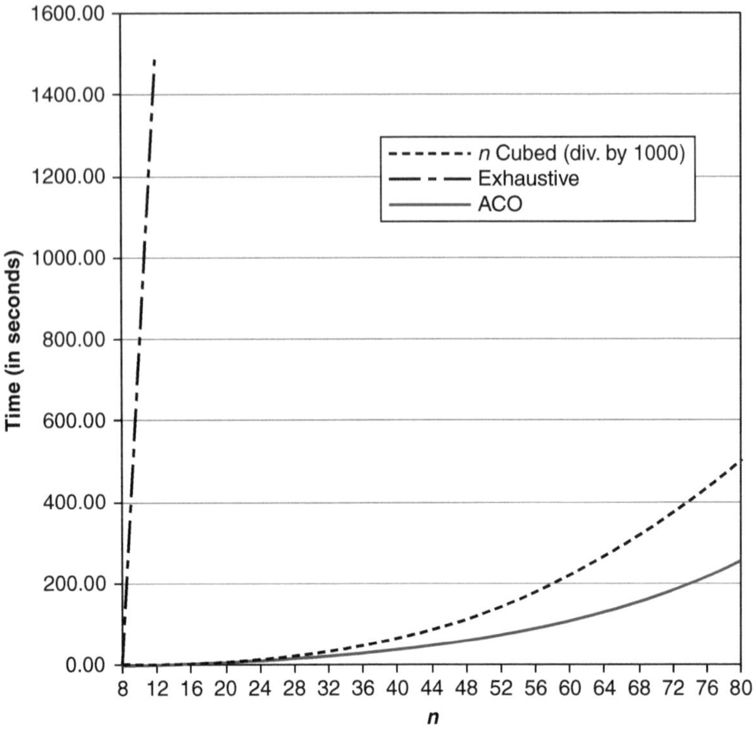

FIGURE 15.9 ACO time complexity compared to exhaustive search.

The coefficient of determination is calculated to be 1.0, indicating 100.00 percent of the total variation is explained by the calculated linear regression curve. As seen in Fig. 15.10, this regression gives the impression of being an exact match. With a growth of $0.0005n^3 - 0.0043n^2 + 0.3714n + 2.4412$, the average-case time complexity of DLBP ACO using the DLBP A Priori data is listed as $O(n^3)$ or *polynomial complexity* [defined to be $O(n^b)$ where $b > 1$; Rosen, 1999]. These experimental results are in agreement with the theoretical time complexity results given in Secs. 15.2 and 15.3.

Although it is slower than all of the other nonoptimal solution-generating methodologies demonstrated in this book, some of this can be attributed to the DLBP modification requiring the probability in Eq. (15.1) to be calculated dynamically, generating new probabilities at each increment of t instead of just once, at the beginning at each tour, as is typical with ACO.

Runtime performance can be improved in ACO by decreasing the number of cycles or decreasing the number of ant agents, while all other measures of performance can be expected to improve with an increase in the number of cycles or an increase in the number of ants. As with other search techniques, a larger n or the inclusion of

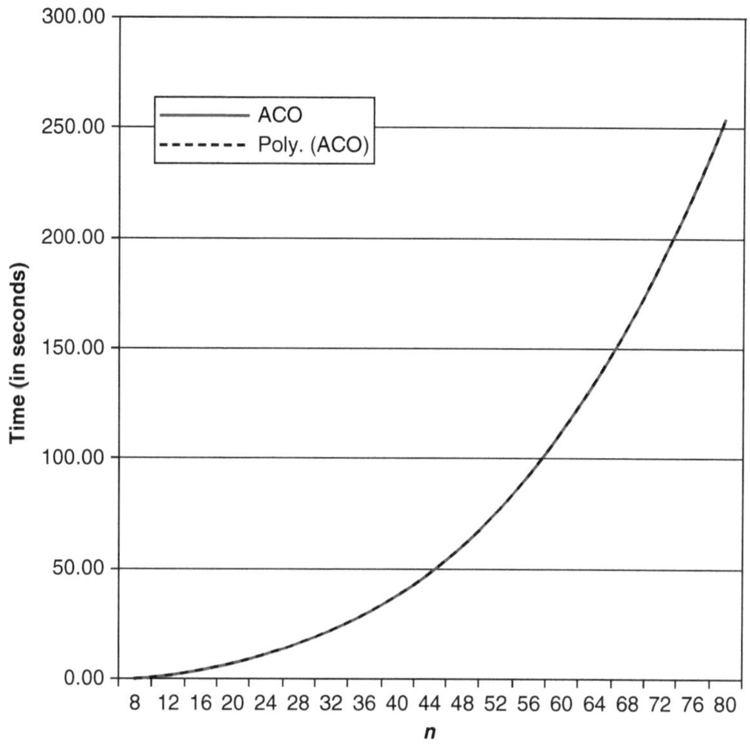

FIGURE 15.10 ACO time complexity and third-order polynomial regression line.

precedence constraints will move the DLBP ACO method toward the optimal solution. As shown in Figs. 15.9 and 15.10, the time complexity performance of DLBP ACO provides the trade-off benefit with the technique's near-optimal performance, demonstrating the moderate increase in time required with problem size that grows markedly slower than the exponential growth of exhaustive search.

15.6 Conclusions

A fast, near-optimal ant colony optimization approach to the multi-objective deterministic DLBP is developed and presented in this chapter. The DLBP ACO rapidly provides a feasible solution to the DLBP using an ant system/ant-cycle model algorithm based on the ant colony optimization metaheuristic and modified to meet multiple objectives. The DLBP ACO provides a near-optimal minimum number of workstations, with the level of optimality increasing with the number of constraints. It generates a feasible sequence with a near-optimal measure of balance while maintaining or improving the hazardous materials measure, the demand measure, and the part removal direction measure.

Several observations were able to be made through the in-depth study of the process required by the writing and test of the DLBP ACO software. The first of these, as discussed in Sec. 15.3, is the ease of which the ant colony optimization process is applied to the TRAVELING SALESPERSON PROBLEM and the challenges posed in applying it to different combinatorial optimization problems that are not formulated similarly to the TSP. This would seem to indicate that the process (as appears to be the case with *k*-opt; see Chap. 18) might have been developed specifically as a solution methodology for this problem (i.e., TSP) then proposed for use with other problems. Secondly, the two unique items in the ant colony optimization metaheuristic are the number of agents and the probabilistic weighting and selection of tours. Due to the greedy nature of the search, negating these two components (i.e., making use of one ant that always make the best deterministic choice for each step in the tour with no knowledge of previous tours) relegates the ant colony optimization process to a greedy search (Dorigo et al., 1996). That being said, ant colony optimization is a methodology that has gained a fairly wide following due to Dorigo's creative observation and application of successful searches performed in nature. The ACO work of Ding et al. (2010) is representative of the efforts of several DLBP research groups that have provided significant extensions to the work in this chapter.

Although a near-optimum technique, the DLBP ACO method quickly finds solutions within, on average, 10 percent of the optimal normalized balance in the exponentially large search spaces. While the DLBP ACO metaheuristic is difficult to fully apply to a multicriteria decision-making problem format and difficult to adapt to a BIN-PACKING-related problem, it is easily and effectively implemented in the solution of problems with nonlinear objectives as well as being suited to combinatoric problems.

CHAPTER 16
Greedy Algorithm

16.1 Introduction

In this chapter, the DISASSEMBLY LINE BALANCING PROBLEM (DLBP) is solved using a greedy algorithm. The greedy algorithm considered here is based on first-fit-decreasing (FFD) rules that are enhanced to preserve precedence relationships within the product being disassembled. FFD is further modified to a multiobjective greedy algorithm that seeks to minimize the number of workstations while attempting to remove hazardous and high-demand parts as early as possible and remove parts having similar part removal directions together. Examples are then used to illustrate the methodology and measure its performance. While the process is fast and is able to find the near-optimum number of workstations, it has no ability to equalize workload between the workstations. Section 10.6 provides some background into the greedy algorithm.

This chapter introduces the FFD greedy algorithm and the DLBP-specific variations. Section 16.2 provides background on the greedy model, reviews the qualitative enhancements made for application to the DLBP, and provides the calculations of the processes' theoretical time complexity. Section 16.3 describes the greedy search results when run using the instances from Ch. 10 and Sec. 16.4 summarizes the chapter.

16.2 Model Description

A greedy strategy always makes the choice that looks the best at the moment. That is, it makes a locally optimal choice in the hope that this choice will lead to a globally optimal solution. The DLBP Greedy algorithm (McGovern and Gupta, 2003a, 2003b, 2005a) is built around FFD rules. *First-fit-decreasing* rules require looking at each element in a list, from largest to smallest (part removal times in the DLBP) and putting that element into the first workstation in which it fits without violating precedence constraints. When all of the work elements have been assigned to a workstation, the process is complete. The greedy FFD algorithm in this chapter has been further modified with priority rules to meet multiple objectives. During the sorting process, the hazardous parts are prioritized, greedy-ranked large removal time to small. The remaining nonhazardous parts are greedy ranked next, large removal times to small. In addition, selecting the part with the larger demand ahead of those with lesser demands breaks any ties for parts with equal part removal times and selecting the part with an equivalent part removal direction breaks any ties for parts also having equal part removal directions. This is done to prevent damage to these more desirable parts. The DLBP Greedy algorithm is able to provide an optimal or near-optimal (minimum) number of workstations. As with the other processes, the more the constraints the more likely the optimal solution is found; that is, the level of performance will generally improve with the number of precedence constraints.

The specific details for this implementation are as follows. The DLBP Greedy algorithm first sorts the list of parts. The sorting is based on part removal times, whether or not the part contains hazardous materials, the subsequent demand for the removed part, and the part removal direction. Hazardous parts are put at the front of the list for selection into the solution sequence. The hazardous parts are ranked from largest to smallest part removal times. The same is then done for the nonhazardous parts. Any ties (i.e., two parts with equal hazard typing and equal part removal times) are not randomly broken, but rather ordered based on the demand for the part, with the higher-demand part being placed earlier on the list. Any of these parts also having equal demands is then selected based on their part removal direction being the same as the previous part on the list (i.e., two parts compared during the sorting that only differ in part removal directions are swapped if they are removed in different directions—the hope being that subsequent parts and later sorts can better place parts having equal part removal directions).

Once the parts are sorted in this multicriteria manner, the parts are placed in workstations in FFD greedy order while preserving precedence. Each part in the sorted list is examined from first to last. If the part had not previously been put into the solution sequence (as described by ISS_k tabu[1] list data structure), the part is put into the current workstation if idle time remains to accommodate it and as long as putting it into the sequence at that position will not violate any of its precedence constraints. ISS_k is defined as

$$\text{ISS}_k = \begin{cases} 1, \text{ assigned} \\ 0, \text{ otherwise} \end{cases} \quad \forall k \in P \tag{16.1}$$

---

[1]Although the name recalls tabu search as proposed by Glover (1989, 1990), the ISS_k tabu list is similar to that used by the ant system (Dorigo et al., 1996) including, for example, the absence of any *aspiration function*.

If no workstation can accommodate it at the given time in the search due to precedence constraints, the part is maintained on the sorted list (i.e., its ISS_k value remains zero) and the next part (not yet selected) on the sorted list is considered. If all parts have been examined for insertion into the current workstation on the greedy solution list, a new workstation is created and the process is repeated. Figure 16.1 shows the DLBP Greedy procedure. The two main processes (sort the data, then FFD greedy-assign the parts to workstations) are seen in step 2 and step 3 of Fig. 16.1.

The DLBP Greedy heuristic process is run once to generate its solution. Since the Greedy process makes use of *bubble sort*, which has a known time complexity of $T(n) \propto n(n-1)/2$ or $\Theta(n^2)$ (Rosen, 1999), the entire process will be no faster than that. As can be seen in studying Fig. 16.1, the best case for the FFD portion of the process (step 3) would be an instance and a sorted sequence that never violated the partial ordering imposed by the precedence constraints and required only one workstation. This would require only one run through the sorted data resulting in a best-case time complexity of $\Omega(n)$. When this is combined with the bubble sort algorithm's

Procedure DLBP_GREEDY {

1. **SET** $j := 0$

2. $\forall$ $PRT_k \mid 1 \le k \le n$, generate sorted part sequence PSS based on:
   $h_k := 1$ parts (i.e., hazardous), large PRT_k to small; then $h_k := 0$ parts, large PRT_k to small; if more than one part has both the same hazard rating and the same PRT_k, then sort by demand, large d_k to small; if more than one part has the same hazard rating, PRT_k, and demand rating, then sort by equal part removal direction

3. $\forall$ $p \in PSS \mid p := PSS_k$, $1 \le k \le n$, generate greedy part removal sequence PSG by:

IF	$ISS_p == 1$ (i.e., part p already included in PSG) $\vee$ PRECEDENCE_FAIL $\vee$ $I_j < PRT_p$ (i.e., part removal time required exceeds idle time available at ST_j)		
THEN	**IF**	$p == n$ (i.e., at last part in sequence)	
	THEN	Increment j (i.e., start new workstation) Start again at $p := 1$	
	ELSE	Try next p	
ELSE	Assign p to ST_j and set $ISS_p := 1$		

4. Using PSG:
   Calculate F, H, D, and R

}

FIGURE 16.1 DLBP Greedy procedure.

known number of comparisons, the two processes then require a best-case runtime of

$$T(n) \propto \frac{n(n-1)}{2} + n$$

$$T(n) \propto \frac{n(n-1)+2n}{2}$$

$$T(n) \propto \frac{n^2+n}{2}$$

$$T(n) \propto \frac{n(n+1)}{2}$$

giving a resulting best-case time complexity calculated to be $\Omega(n^2)$.

The worst-case sequencing for the FFD portion of the process (step 3) would give a reverse ordering of the parts (e.g., the largest hazardous part would be required by precedence constraints to be the last part removed while the smallest nonhazardous part would be the first part required to be removed with a similar lexicographic ordering for all parts in-between). Therefore, the first time the greedy routine would take time in proportion to n, the next time through $n - 1$, then $n - 2$, and so on until the last remaining sequence element is the first element in the sorted list which would take time in proportion to 1. The resulting sequence can be formulated as a runtime and reduced as follows

$$T(n) \propto n + (n-1) + (n-2) + \cdots + 1$$

$$T(n) \propto n + (n-1) + (n-2) + \cdots + [n - (n-1)]$$

$$T(n) \propto n(n) - [1 + 2 + \cdots + (n-1)]$$

$$T(n) \propto n(n) - \sum_{p=1}^{n-1} p$$

$$T(n) \propto n^2 - \frac{(n-1)[(n-1)+1]}{2}$$

$$T(n) \propto n^2 - \frac{(n^2-2n+1+n-1)}{2}$$

$$T(n) \propto n^2 - \frac{n^2-n}{2}$$

$$T(n) \propto n^2 - \frac{n(n-1)}{2}$$

When this is combined with the bubble sort algorithm's known number of comparisons, the two processes then require a worst-case runtime of

$$T(n) \propto \frac{n(n-1)}{2} + n^2 - \frac{n(n-1)}{2}$$

$$T(n) \propto n^2$$

giving a resulting worst-case time complexity calculated to be $O(n^2)$.

Note that since the best-case time complexity is $\Omega(n^2)$ and the worst-case time complexity is $O(n^2)$, an asymptotic tight bound (big-theta) exists as $\Theta(n^2)$.

While being very fast and generally efficient, the FFD-based greedy algorithm is not always able to optimally minimize the number of workstations. In addition, there is no capability to equalize the station times in the workstations; in fact, the FFD structure lends itself to filling the earlier workstations as much as possible, often to capacity, while later workstations may end up with progressively greater and greater idle times. This results in an extremely poor balance measure.

16.3 Numerical Results

The greedy algorithm is run on each of the four experimental instances from Ch. 10 with the results analyzed in this section.

16.3.1 Personal Computer Instance

The DLBP Greedy algorithm has been applied to the personal computer (PC) instance. Being a deterministic process, multiple runs have been made, but only made to provide time complexity averaging. The search of the PC instance always took less than 1/100th of a second (listed as 0.00 seconds each run per the original data output) with the optimal solution (Table 16.1) of four workstations and $F^* = 33$, $H^* = 7$, $D^* = 19{,}025$, and $R^* = 6$ being found.

Part ID	1	5	3	6	2	8	7	4
PRT	14	23	12	16	10	36	20	18
Workstation	1	1	2	2	2	3	4	4
Hazardous	0	0	0	0	0	0	1	0
Demand	360	540	620	750	500	720	295	480
Direction	1	2	1	4	0	4	1	1

TABLE 16.1 Greedy Solution Using the Personal Computer Instance

Part ID	5	4	6	7	8	9	1	10	3	2
PRT	23	17	14	19	36	14	14	10	12	10
Workstation	1	1	2	2	3	4	4	4	5	5
Hazardous	0	0	0	1	0	0	0	0	0	0
Demand	0	0	750	295	0	360	0	0	0	500
Direction	5	2	5	2	1	5	2	3	0	0

TABLE 16 2 Greedy Solution Using the 10-Part Instance

16.3.2 The 10-Part Instance

The DLBP Greedy algorithm is able to successfully find a feasible solution (Table 16.2) having the minimum number of workstations while placing the hazardous part (part number 7) and high-demand parts (parts 2, 6, 7, and 9) relatively early in the removal sequence (the exception being part 2, primarily due to precedence constraints and a smaller part removal time). The speed for the C++ implemented program on this problem is again less than 1/100th of a second.

DLBP Greedy finds a feasible solution but not the optimal solution (five workstations and $F^* = 211$, $H^* = 4$, $D^* = 9730$, and $R^* = 7$). Application of DLBP Greedy results in the optimal number of workstations and the optimal hazard measure $H^* = 4$, with a balance of $F = 393$, a demand measure of $D = 10{,}590$, and a part removal direction measure of $R_{\text{upper}} = 8$.

16.3.3 Cellular Telephone Instance

DLBP Greedy finds feasible solutions to the cellular telephone data, again taking less than 1/100th of a second. DLBP Greedy requires 10 workstations [recall that the DLBP Ant Colony Optimization technique (DLBP ACO) regularly found 9; the highest found by any of these methodologies is 11], a balance measure of $F = 199$ (again, DLBP ACO has the best found—and presumed optimum—balance of $F = 9$ while the worst seen is $F = 399$), a hazard measure of $H = 89$ [a single run of the DLBP Genetic Algorithm (DLBP GA) gave $H = 78$ while one of the DLBP ACO runs found $H = 90$], a demand measure of $D = 973$ (one DLBP ACO run came up with $D = 868$), and a direction measure of $R = 12$ (the best seen was a DLBP GA run of $R = 10$). The DLBP Greedy solution is shown in Table 16.3.

The results shown in the table demonstrate how the sorting and FFD portions of DLBP Greedy operate. Per its design, this technique removes hazardous parts early on, though many of these parts have to wait due to precedence constraints. Four parts with larger part removal times (15 seconds) are removed very early on (although parts 19 and 23 take as long or longer, precedence constraints prevent earlier removal) but at the expense of using entire workstations, demonstrating a condition that limits DLBP Greedy's

Part ID	1	2	4	3	6	7	8	9	16	5	10	15	18	14	13	17	20	11	12	21	25	19	22	23	24
PRT	3	2	10	3	15	15	15	15	2	10	2	2	3	2	2	2	5	2	2	1	2	18	5	15	2
Workstation	1	1	1	1	2	3	4	5	5	6	6	6	6	7	7	7	7	7	7	7	7	8	9	10	10
Hazardous	1	1	0	0	0	0	0	0	0	0	0	0	0	0	0	0	0	0	1	0	1	1	0	1	0
Demand	4	7	1	1	1	1	1	1	1	1	2	1	2	1	1	2	1	1	4	4	4	8	6	7	1
Direction	2	3	5	5	5	5	5	5	2	5	4	0	5	2	5	4	4	4	4	4	4	5	4	4	4

TABLE 16.3 Greedy Solution Using the Cellular Telephone Instance

Part ID	12	11	3	10	9	8	7	5	6	4	2	1
PRT	11	11	3	11	7	7	7	5	5	5	3	3
Workstation	1	1	1	2	2	2	3	3	3	3	3	4
Hazardous	1	0	0	0	0	0	0	0	0	0	0	0
Demand	0	0	0	0	1	0	0	0	0	0	0	0
Direction	0	0	0	1	0	0	1	0	0	1	0	1

TABLE 16.4 Greedy Solution Using the A Priori Instance at $n = 12$

effectiveness. The methodology can be seen to perform well in terms of part removal direction even though this is the lowest of the prioritized objectives. Precedence relationships play a large role in determining the allowable solution sequences to this problem instance.

16.3.4 DLBP A Priori Instances

The DLBP Greedy technique can be seen above with $n = 12$ (Table 16.4). This data set forces the DLBP Greedy algorithm to obtain the near-optimal solution of NWS = 4 workstations versus the optimal NWS = 3.

The speed is again less than 1/100th of a second. Although not one of the multiple optimum extreme points, the single hazardous part has been optimally placed in position $k = 1$ (the first part to be removed in the sequence) with the single demanded part sub-optimally placed (though still correctly subordinate to the hazardous part) in position $k = 5$. Note that the first task listed in the greedy solution is part number 12 (part removal time of 11 seconds), which is the only part labeled as hazardous.

Various performance measures of the DLBP Greedy technique are demonstrated in the analysis of the DLBP A Priori benchmark data sets of $n = \{8, 12, 16, \ldots, 80\}$. This difficult data set results in the DLBP Greedy algorithm consistently obtaining the near-optimal solution of $\text{NWS} = \text{NWS}^* + 1$ workstations (Fig. 16.2). In terms of the number of workstations required by DLBP Greedy, from Eq. (12.1) the heuristic's efficacy index in number of workstations ranges from a low of $\text{EI}_{\text{NWS}} = 83\%$ to a high of $\text{EI}_{\text{NWS}} = 98\%$. Overall, as given by Eq. (12.2), DLBP ACO shows an efficacy index sample mean in number of workstations of $\overline{\text{EI}}_{\text{NWS}} = 95\%$.

In general, balance efficacy (with the previously described DLBP A Priori data sets) is actually seen to improve with problem size (Fig. 16.3) with the balance measure decreasing in a fashion approximating a *step function* (or *staircase function*). Consistent improvement in the balance measure is seen, with poor performance when dealing with smaller instances steadily improving through to the last data set having $n = 80$.

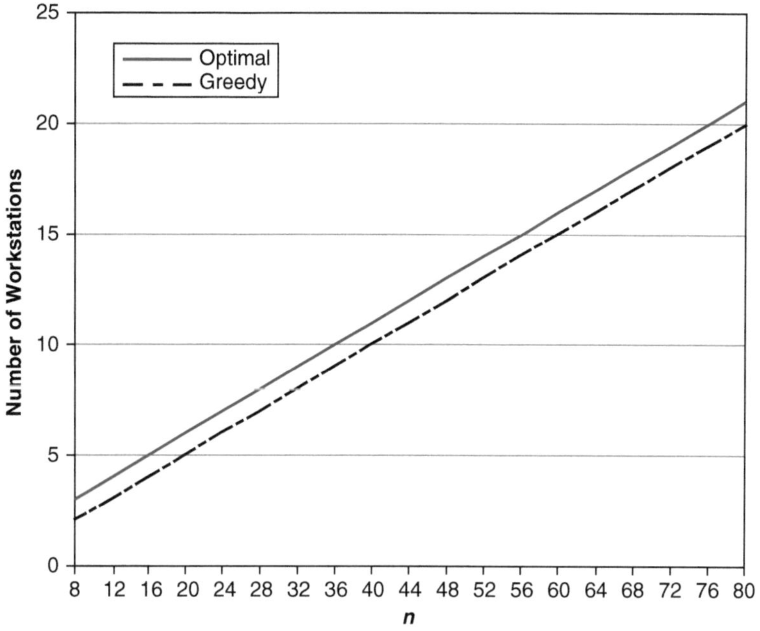

FIGURE 16.2 Greedy workstation calculation.

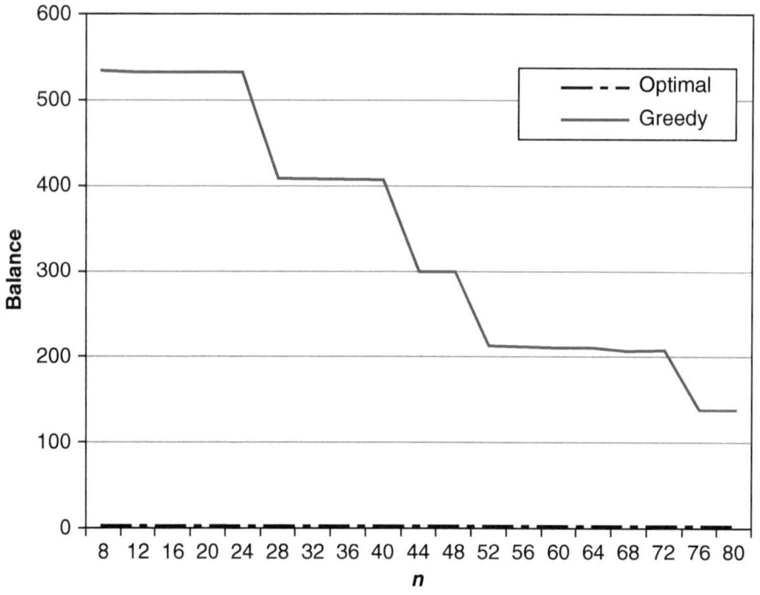

FIGURE 16.3 Detailed Greedy balance measure.

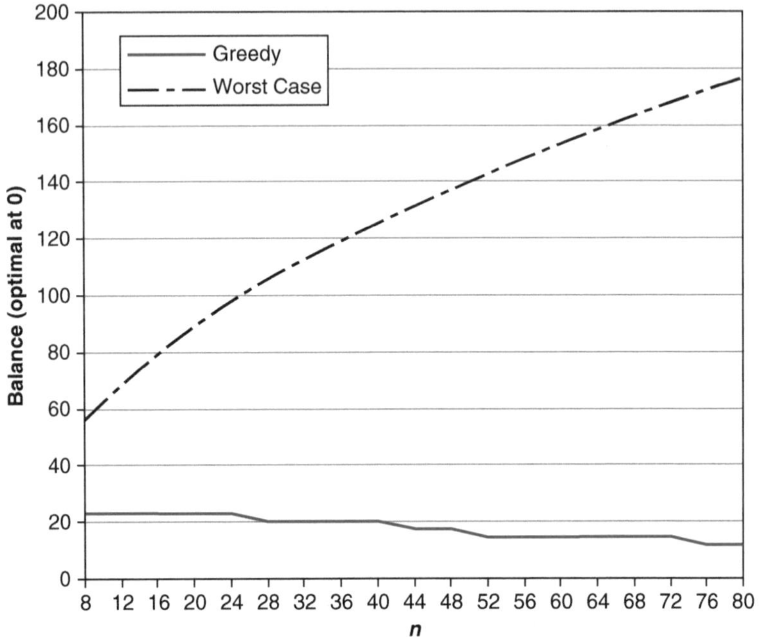

Figure 16.4 Normalized Greedy balance measure.

The normalized balance efficacy index starts as low as $EI_F = 59\%$ then increases through to $EI_F = 93\%$ (Fig. 16.4) and has a sample mean of $\overline{EI}_F = 83\%$. While DLBP Greedy starts out poorly in its normalized balance measure, its performance increasingly improves with instance size; the normalized balance solution moves closer and closer to optimal as the instance size grows.

The hazardous part is regularly optimally placed, as illustrated by Fig. 16.5. This is a direct consequence of the problem-specific sorting process that puts the hazardous parts at the front of the sorted list. The hazard measure's efficacy index starts out optimally at $EI_H = 100\%$ and remains there for all instances, giving a sample mean of $\overline{EI}_H = 100\%$.

The demand measure remains relatively consistent with instance size (Fig. 16.6). Its efficacy index stays between $EI_D = 70\%$ (at $n = 12$) and $EI_D = 63\%$ for all of the data, giving a demand measure sample mean of $\overline{EI}_D = 64\%$.

With part removal direction structured as to be deferential to balance, hazard, and demand, it was seen to drop to low performance levels and remain there for all data set sizes (Fig. 16.7)—a result of the prioritization of the multiple objectives. The part removal direction measure's efficacy index drops immediately from its highest value of $EI_R = 43\%$ to $EI_R = 14\%$ where it remains for the remainder of the instances, resulting in a sample mean of $\overline{EI}_R = 16\%$.

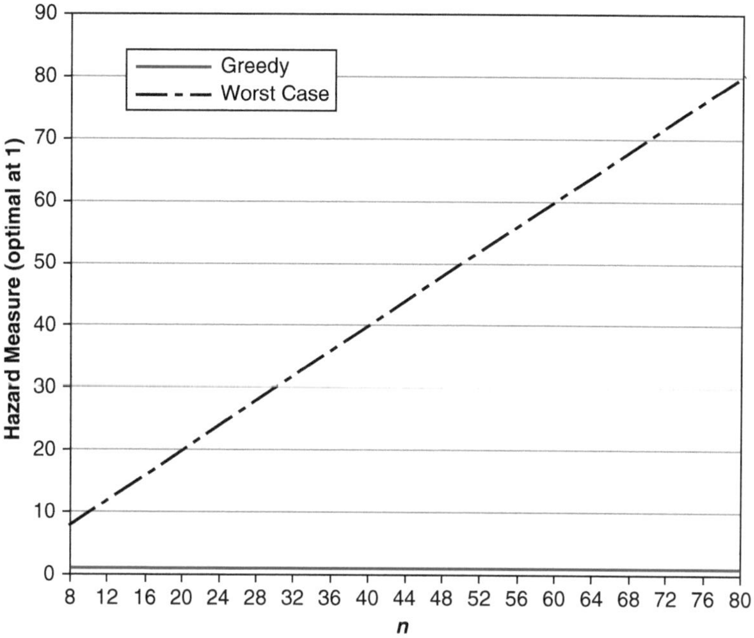

FIGURE 16.5 Greedy hazard measure.

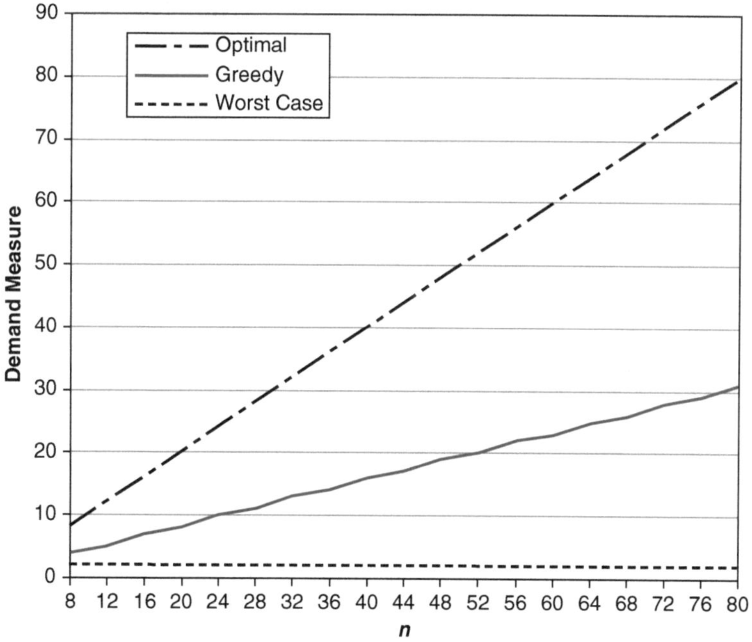

FIGURE 16.6 Greedy demand measure.

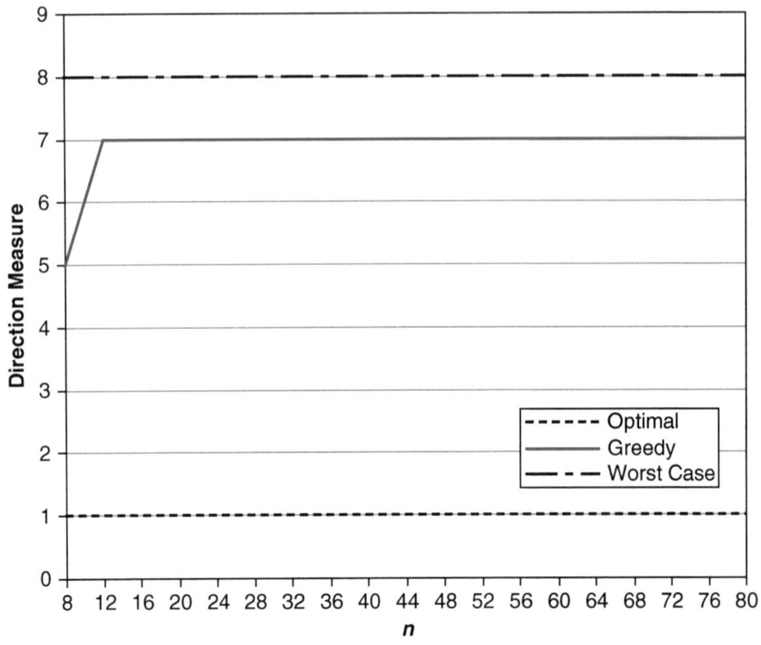

FIGURE 16.7 Greedy part removal direction measure.

In terms of the time complexity, except for a lone data point, each DLBP Greedy run for all $n = \{8, 12, 16, \ldots, 80\}$ data sets took less than 1/100th of a second. While this is recognized as being unusually fast, it does not allow for a detailed time complexity analysis due to the software's sampling rate. For this reason, the DLBP Greedy heuristic is then run to collect additional data, this time on data sets of $n = \{8, 12, 16, \ldots, 760\}$ (note that the only results used from the larger set of instances are the runtimes). With the new data range, runtime can be seen to increase slowly with instance size, with DLBP Greedy almost appearing to run in linear time. Based on the work performed in Sec. 16.2 and the worse coefficient of determination being generated when using a linear regression model, a second-order polynomial regression model is used to fit the curve, with the regression equation calculated to be $T(n) = 7 \times 10^{-8}n^2 + 2 \times 10^{-6}n + 0.00001$. The polynomial's extremely small coefficients are in keeping with the slow time complexity growth in instance size. The coefficient of determination is calculated to be 0.9571, indicating 95.71 percent of the total variation is explained by the calculated linear regression curve. As seen in Fig. 16.8, this regression provides an accurate fit.

With a growth of $7 \times 10^{-8}n^2 + 2 \times 10^{-6}n + 0.00001$, the average-case time complexity of DLBP Greedy using the DLBP A Priori data is listed as $O(n^2)$ or polynomial complexity. This is in agreement with the calculations performed in Sec. 16.2. With $n = 80$ running in less than 1/100th

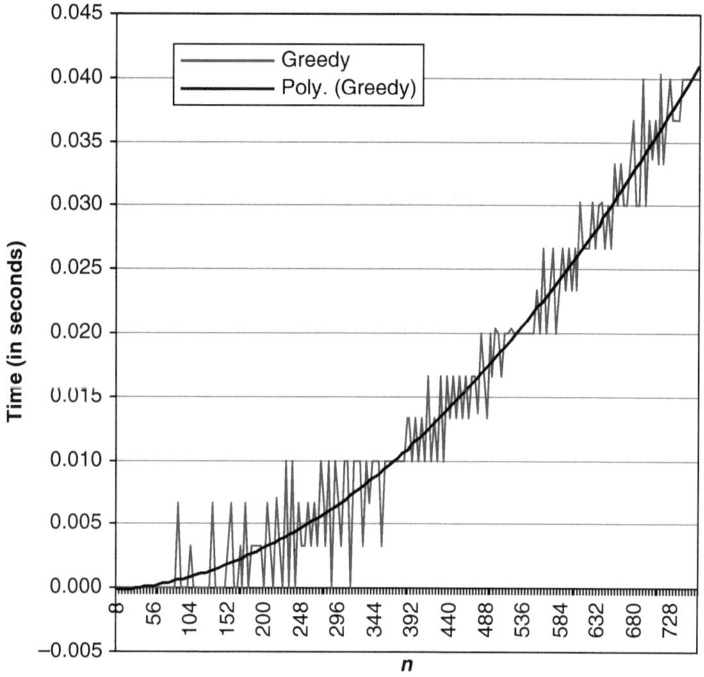

FIGURE 16.8 Greedy time complexity and second-order polynomial regression line.

of a second and $n = 760$ taking only 0.04 seconds, DLBP Greedy is by far the fastest methodology demonstrated in the Part II study on the data sets tested (note, however, that DLBP GA is superior in terms of the rate of growth). The main reason for this speed is due to the greedy process of only going through the data set once (actually twice due to the requirement to sort the data first—this is the slowest part of the operation and the reason for Greedy having polynomial complexity instead of linear complexity). Many of the other processes demonstrated in this book continue to go through the data, refining the solution using local search methodologies or over many generations/cycles.

Runtime performance can be improved by changing to a faster sorting algorithm [such as *quick sort*, which runs on average in $2n \log_2 n$, though it is still $O(n^2)$ worst case, or *merge sort*, which runs in $O(n \log_2 n)$], while other measures of performance cannot easily be improved without changing the first-fit-decreasing and greedy nature of the algorithm.

16.4 Conclusions

An exceptionally fast, near-optimal approach to the multiobjective deterministic DLBP was presented in this chapter. It rapidly provides a feasible solution to the DLBP using a greedy algorithm based on FFD rules and modified to meet multiple objectives. The heuristic provided excellent solutions very rapidly, an asset for large problems (such as automobile disassembly, where over 10,000 parts may be involved) or real-time problem solution. Although a near-optimum technique, the DLBP Greedy method quickly found solutions in exponentially large search spaces with solution performance improving with problem size and number of precedence constraints. This method appears moderately well suited to the multicriteria decision-making problem format as well as for the solution to problems with nonlinear objectives. It should also be noted that DLBP Greedy is a problem-specific heuristic and may not be expected to lend itself to similar success with different problems.

CHAPTER 17

Greedy/Adjacent Element Hill-Climbing Hybrid

17.1 Introduction

In this chapter the DISASSEMBLY LINE BALANCING PROBLEM (DLBP) is solved using a hybrid technique. This method sequentially finds, and then improves upon, an instance solution using a first-fit-decreasing (FFD)-based greedy algorithm followed by the application of a hill-climbing heuristic. The hybrid combination of a greedy algorithm and the hill-climbing heuristic is instrumental in rapidly obtaining near-optimal solutions to the DLBP's intractably large solution space. The greedy algorithm with the hill-climbing heuristic technique performs so well and is so rapid that it should lend itself to real-time solution of the DLBP on a dynamic disassembly line, generating solutions as components arrive for disassembly on mixed-model and *mixed-product* lines. The greedy algorithm is detailed in Chap. 16. While that process is exceptionally fast and is able to find the near-optimum number of workstations, it has no ability to balance the workstations. As a result, a hill-climbing heuristic has been applied to balance the part removal solution sequence, addressing the major weakness of applying an FFD algorithm to a line balancing problem. The adjacent element hill-climbing (AEHC) heuristic only compares tasks assigned in adjacent workstations. This is done both to conserve search time (by not investigating all tasks in all workstations) and to only investigate swapping tasks that will most likely result in a feasible sequence (since the farther apart the positional changes, the less likely that precedence will be preserved for both of the tasks exchanged and for all of the tasks between them). Examples are considered to illustrate the implementation of the hybrid methodology.

This chapter introduces the greedy/hill-climbing hybrid heuristic and the DLBP-specific variations. Section 17.2 provides background on the hill-climbing model, reviews the qualitative features specific to the DLBP, and demonstrates the theoretical complexity calculations. Section 17.3 describes the results when the greedy/hill-climbing hybrid is applied to the instances from Chap. 10 while Sec. 17.4 summarizes the chapter. An introduction to the adjacent element hill-climbing heuristic is provided in Sec. 10.7.

17.2 Model Description

A two-phase approach to the DLBP quickly provides a near-optimal and feasible balance sequence using a hill-climbing local search heuristic, AEHC (McGovern and Gupta, 2003b, 2005a). The adjacent element hill-climbing heuristic takes advantage of knowledge about the problem's format and constraints in order to provide a solution that is better balanced than DLBP Greedy alone but significantly faster than other methodologies (including DLBP Greedy/2-Opt, Chap. 18). Hill climbing makes use of an iterative greedy strategy, which is to move in the direction of increasing value. AEHC is designed to consider swapping each task in every workstation only with each task in the next adjacent workstation in search of improved balance. It does this while preserving precedence and not exceeding CT in any workstation. Only adjacent workstations are compared to enable a rapid search and since it is deemed unlikely that parts several workstations apart can be swapped and still preserve the precedence of all of the tasks in-between. As shown in the Fig. 17.1 example, a part has a limited number of other parts it can be considered with for exchange. In the 2-opt methodology presented in Chap. 18 and using the Fig. 17.1 data, part number 6, for example, would be considered for exchange by parts 1 through 5, and then would consider exchanges with parts 7 through 11; that is, any part could be exchanged with any other part. In AEHC, part 6 would be considered for exchange only by parts 4 and 5, and would consider exchanges only with parts 8, 9, and 10.

The neighborhood definition and search details of AEHC are as follows. After a minimum-NWS feasible solution is generated (by, e.g.,

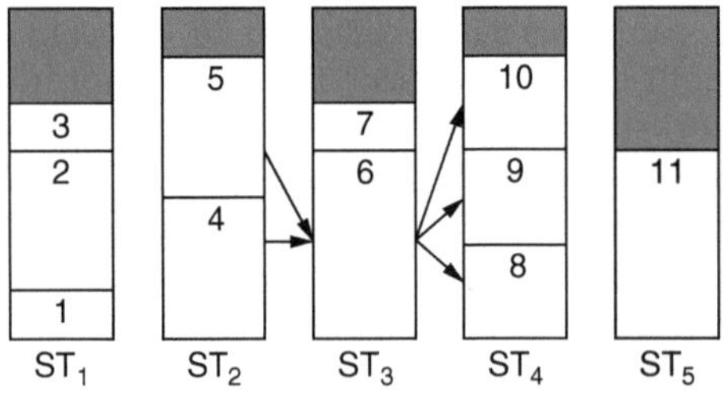

FIGURE 17.1 Adjacent element hill-climbing (AEHC) example.

the DLBP Greedy algorithm), the AEHC heuristic is applied to improve the balance. The AEHC does this by going through each task element (part) in each workstation and comparing it to each task element in the next adjacent workstation. If the two task elements can be exchanged while preserving precedence, without exceeding either workstation's available idle time, and with a resulting improvement in the overall balance, the exchange is made and the resulting solution sequence is saved as the new solution sequence. This process is repeated until task elements of the last workstation have been examined.

In a single run of AEHC, two additional situations (one best case, one worst case) present themselves due to the search's design.

The fastest of these would consist of either: two workstations with one of the workstations containing only one element, or alternatively n workstations with each containing only one element. The least number of comparisons is then defined as

$$\text{Exchanges} = n - 1 \tag{17.1}$$

The slowest of these would consist of two workstations with each of the workstations containing exactly $n/2$ elements. In this situation, each of the $n/2$ elements in workstation one compares itself to each of the $n/2$ elements in workstation two, resulting in the number of comparisons defined as

$$\text{Exchanges} = \frac{n}{2}\left(\frac{n}{2}\right) = \frac{n^2}{4} \tag{17.2}$$

While exhaustive search visits all $n!$ tours, from Eqs. (17.1) and (17.2) the total number of tours visited by an run of DLBP AEHC, including the original data set (such as ⟨1, 2, 3, 4⟩), can then be bound by

$$n \leq \text{tours} \leq \frac{n^2}{4} + 1 \tag{17.3}$$

The heuristic is given in *pseudocode* format in Fig. 17.2.

In the DLBP Greedy/AEHC hybrid heuristic, the DLBP Greedy process is run once to generate an initial solution. Hill climbing is typically continuously run on subsequent solutions for as long as is deemed appropriate or acceptable by the user or until it is no longer possible to improve, at which point it is assumed that the local optima has been reached (Hopgood, 1993). Repeating the AEHC method in this way provides improved balance with each run. During its development (McGovern and Gupta, 2003b) AEHC was tested both ways—run only once after the greedy solution was generated, as well as run repeatedly on each subsequent solution until the local optima was obtained (a single AEHC iteration has several benefits including the observation that AEHC can be seen to typically provide its largest single balance performance improvement in the very first iteration). Additionally, a single run of AEHC may be recommended for the real-time solution of a very large

Procedure DLBP_AEHC {

1. Initialize: { PSG is the greedy solution sequence }
    SET PST := PSG { PST is the current solution sequence }
    SET TMP := PSG { TMP is a temporary solution sequence }
    SET BST:= PSG { BST is the best solution sequence thus far }
    SET START := 1 { START is the position counter }

2. $\forall j \mid 1 \leq j \leq$ NWS – 1, attempt to generate better balanced AEHC part removal sequence PST by performing AEHC part swap (i.e., swap each TMP_p in workstation j with every TMP_q in workstation j + 1)
    { NPW_j is the number of parts in workstation j }
    $\forall\, p \in \text{TMP} \mid p = \text{TMP}_k,\ \text{START} \leq k \leq \text{START} + \text{NPW}_j - 1$

    $\forall\, q \in \text{TMP} \mid q = \text{TMP}_k,\ \text{START} + \text{NPW}_j \leq k \leq \text{START} + \text{NPW}_j - 1 + \text{NPW}_{j+1}$

    Calculate new ST_j and ST_{j+1}

    IF [$(\text{CT} - \text{new ST}_j \geq 0) \wedge$
    $(\text{CT} - \text{new ST}_{j+1} \geq 0)$]

    THEN Calculate new F, H, D, and R

    IF { [BETTER_SOLUTION(TMP, BST)]
    $\wedge$
    (PRECEDENCE_PASS) }

    THEN SET BST:= TMP

    SET START:= START+ NPW_j

3. Save results:
    SET PST := BST
    SET TMP := BST

4. Repeat STEP 2 until no further improvements in F, H, D, or R can be made

}

FIGURE 17.2 DLBP AEHC procedure.

DLBP on a dynamic mixed-model and mixed-product disassembly line, or on a line having a fast pace or a short cycle time. AEHC is run repeatedly here.

The best-case time complexity of the Greedy/AEHC hybrid can be no faster than the best-case time complexity of Greedy alone, calculated to be $\Omega(n^2)$. The best case for AEHC would take place when no improvement can be made over what Greedy has generated. In this situation, AEHC would still go through the entire list of greedy-sorted items but only once. Here, the two situations described by Eqs. (17.1) and (17.2) present themselves. Based on these, AEHC alone can be as slow as $T(n) \propto n^2/4$ and it can never be faster than $T(n) \propto n - 1$. When combined with the fastest DLBP Greedy results of

$$T(n) \propto \frac{n(n-1)}{2}$$

this gives us

$$T(n) \propto \frac{n(n-1)}{2} + n - 1$$

$$T(n) \propto \frac{n(n+1)}{2} - 1$$

which is formally $\Omega(n^2)$.

The worst-case time complexity of the Greedy/AEHC hybrid is at least as slow as the worst-case time complexity of Greedy alone, calculated to be $O(n^2)$. A single run of AEHC alone can be as slow as $T(n) \propto n^2/4$. When combined with the worst-case Greedy results of $T(n) \propto n^2$, this gives us $T(n) \propto n^2 + (n^2/4)$, which is formally $O(n^2)$. Due to the iterative nature of AEHC, the worst-case time complexity of the heuristic has not been calculated, while the work completed above indicate that it can be no better than $O(n^2)$. [Note that in Lawler et al. (1985) and using the example of the TRAVELING SALESPERSON PROBLEM, Johnson and Papadimitriou provide a theorem by Papadimitriou and Steiglitz showing that a local search cannot be guaranteed to find a tour whose length is bounded by a constant multiple of the optimal tour length even if an exponential number of iterations is allowed.]

With a best-case time complexity of $\Omega(n^2)$ and the worst-case time complexity of at least $O(n^2)$, a tight bound for the hybrid may exist at $\Theta(n^2)$ but this is unlikely [from Eqs. (17.1) and (17.2), the best-case time complexity of AEHC alone is $\Omega(n)$ and its worst-case time complexity is at least $O(n^2)$].

17.3 Numerical Results

The greedy/hill-climbing hybrid is run on each of the four experimental instances from Ch. 10 with the results analyzed in this section.

17.3.1 Personal Computer Instance

The DLBP Greedy/AEHC hybrid has been applied to the personal computer (PC) instance. As DLBP Greedy alone successfully found the optimal solution, the subsequent AEHC heuristic could do no better and terminated itself. The hybrid search of the PC instance also took less than 1/100th of a second with the optimal solution (Table 17.1) of four workstations and $F^* = 33$, $H^* = 7$, $D^* = 19{,}025$, and $R^* = 6$ being found.

17.3.2 The 10-Part Instance

On the 10-part instance, DLBP Greedy generates a feasible solution having the minimum number of workstations while placing the hazardous part (part number 7) and high-demand parts (parts 2, 6, 7, and 9)

Part ID	1	5	3	6	2	8	7	4
PRT	14	23	12	16	10	36	20	18
Workstation	1	1	2	2	2	3	4	4
Hazardous	0	0	0	0	0	0	1	0
Demand	360	540	620	750	500	720	295	480
Direction	1	2	1	4	0	4	1	1

TABLE 17.1 Greedy/AEHC Hybrid Solution Using the Personal Computer Instance

relatively early in the removal sequence (the exception being part 2, primarily due to precedence constraints and a smaller part removal time). DLBP AEHC is able to improve the overall balance and the demand measure while maintaining the original Greedy-optimized hazard measure and observing the precedence constraints (Table 17.2). The speed for the hybrid on this instance is again less than 1/100th of a second.

The DLBP Greedy/AEHC hybrid does not find the optimal solution (five workstations and $F^* = 211$, $H^* = 4$, $D^* = 9730$, and $R^* = 7$). DLBP Greedy is instrumental in calculating the optimal number of workstations and hazard measure, while the hill-climbing search improves the balance (from Greedy's $F = 393$ to $F = 369$) and demand (from Greedy's $D = 10{,}590$ to $D = 9{,}840$) measures and obtains a better than optimal direction measure ($R = 6$ versus $R^* = 7$, though at the expense of suboptimal balance and demand measures).

17.3.3 Cellular Telephone Instance

DLBP Greedy/AEHC finds feasible solutions to the cellular telephone data, taking less than 1/100th of a second on average. DLBP Greedy generates a solution of 10 workstations and AEHC is unable to improve on any of the other measures, ending with a balance measure of $F = 199$ [DLBP Ant Colony Optimization technique (DLBP ACO) has the best found and believed to be optimum balance of $F = 9$], a hazard measure of $H = 89$ [DLBP Genetic Algorithm (DLBP GA) gave the low of $H = 78$ while ACO had the high of $H = 90$], a demand measure of $D = 973$ (one ACO run came up with a low of $D = 868$), and a direction measure of $R = 12$ (the best seen was a GA run giving $R = 10$). The DLBP Greedy/AEHC solution is shown in Table 17.3.

Part ID	5	6	4	7	8	9	1	10	3	2
PRT	23	14	17	19	36	14	14	10	12	10
Workstation	1	1	2	2	3	4	4	4	5	5
Hazardous	0	0	0	1	0	0	0	0	0	0
Demand	0	750	0	295	0	360	0	0	0	500
Direction	5	5	2	2	1	5	2	3	0	0

TABLE 17.2 Greedy/AEHC Hybrid Solution Using the 10-Part Instance

Part ID	1	2	4	3	6	7	8	9	16	5	10	15	18	14	13	17	20	11	12	21	25	19	22	23	24
PRT	3	2	10	3	15	15	15	15	2	10	2	2	3	2	2	2	5	2	2	1	2	18	5	15	2
Workstation	1	1	1	1	2	3	4	5	5	6	6	6	6	7	7	7	7	7	7	7	7	8	9	10	10
Hazardous	1	1	0	0	0	0	0	0	0	0	0	0	0	0	0	0	0	0	1	0	1	1	0	1	0
Demand	4	7	1	1	1	1	1	1	1	1	2	1	2	1	1	2	1	1	4	4	4	8	6	7	1
Direction	2	3	5	5	5	5	5	5	2	5	4	0	5	2	5	4	4	4	4	4	4	5	4	4	4

TABLE 17.3 Greedy/AEHC Hybrid Solution Using the Cellular Telephone Instance

Per its design, DLBP Greedy removes hazardous parts early on, though many of these parts have to wait due to precedence constraints. Four parts with larger part removal times (15 seconds) are removed very early on (although parts 19 and 23 take as long or longer, precedence constraints prevent earlier removal) but at the expense of using an entire workstation, demonstrating a condition that limits DLBP Greedy/AEHC's effectiveness since the greedy portion works to take out large (part removal time) items early (and therefore, potentially adjacent to each other in the sequence) and AEHC is not allowed to look out far enough to add smaller time items into the sequence to better fill those workstations.

17.3.4 DLBP A Priori Problem Instances

The DLBP Greedy/AEHC hybrid technique can be seen below with $n = 12$ (Table 17.4). While the DLBP Greedy algorithm obtains the near-optimal solution of NWS = 4 workstations (a measure AEHC is not capable of improving on due to its design,) versus the optimal NWS = 3, this data allows AEHC to obtain a better-balanced solution nearby and demonstrate hazardous material and high-demand item sequencing. The time to complete the search is less than 1/100th of a second. The single hazardous part is optimally maintained by AEHC in position $k = 1$. The single demanded part still remains (suboptimally) in position $k = 5$. This is an artifact of the AEHC criteria; the concept of only considering parts in adjacent workstations in the interest of saving time means that swapping parts within a workstation is not considered. As a result, the obvious swap of parts 9 and 10—which would improve both the demand and the direction measures—is not considered by this hill-climbing search. Though the hazard measure was already optimal at $H^* = 1$ and the demand measure remained at $D = 5$, by swapping parts 4 and 8 and parts 1 and 7 from their sequence positions in the first-phase Greedy solution, AEHC improves the balance from $F = 532$ to $F = 380$ and the direction measure from $R = 7$ to $R = 5$. This provides an illustration of the intuitive search strategy of AEHC, with a small number of simple part swaps quickly giving large gains in performance.

Part ID	12	11	3	10	9	4	1	5	6	8	2	7
PRT	11	11	3	11	7	5	3	5	5	7	3	7
Workstation	1	1	1	2	2	2	3	3	3	3	3	4
Hazardous	1	0	0	0	0	0	0	0	0	0	0	0
Demand	0	0	0	0	1	0	0	0	0	0	0	0
Direction	0	0	0	1	0	1	1	0	0	0	0	1

TABLE 17.4 Greedy/AEHC Hybrid Solution Using the A Priori Instance at $n = 12$

Various performance measures of the DLBP Greedy/AEHC hybrid are then demonstrated with the analysis of the DLBP A Priori benchmark data sets of $n = \{8, 12, 16, \ldots, 80\}$. Since AEHC is not designed to minimize the number of workstations, the results determined by DLBP Greedy alone remain unchanged and are not repeated here (see Fig. 16.2 for graphical results and Sec. 16.3.4 for the quantitative analysis).

As with Greedy, balance efficacy is seen to improve with problem size (Fig. 17.3) though with the balance measure decreasing much more smoothly than the approximate step function seen with Greedy (note that these two graphs are not of the same scale, with the Greedy chart in Fig. 16.3 having a y-axis 50 percent taller than that of the AEHC chart in Fig. 17.3). Again, as with Greedy, poor performance is seen when dealing with smaller instances, thought the balance steadily improves through to the last data set of $n = 80$.

Providing improvements in balance is the main goal of the AEHC design and, as seen in this section, its improvements are significant with minor increases in time. The normalized balance efficacy index is improved from a Greedy low $EI_F = 59\%$ to an AEHC low of $EI_F = 70\%$, from a Greedy high of $EI_F = 93\%$ to an AEHC high of $EI_F = 95\%$, and from a Greedy sample mean of $\overline{EI}_F = 83\%$ to an AEHC sample mean of $\overline{EI}_F = 87\%$ (Fig. 17.4).

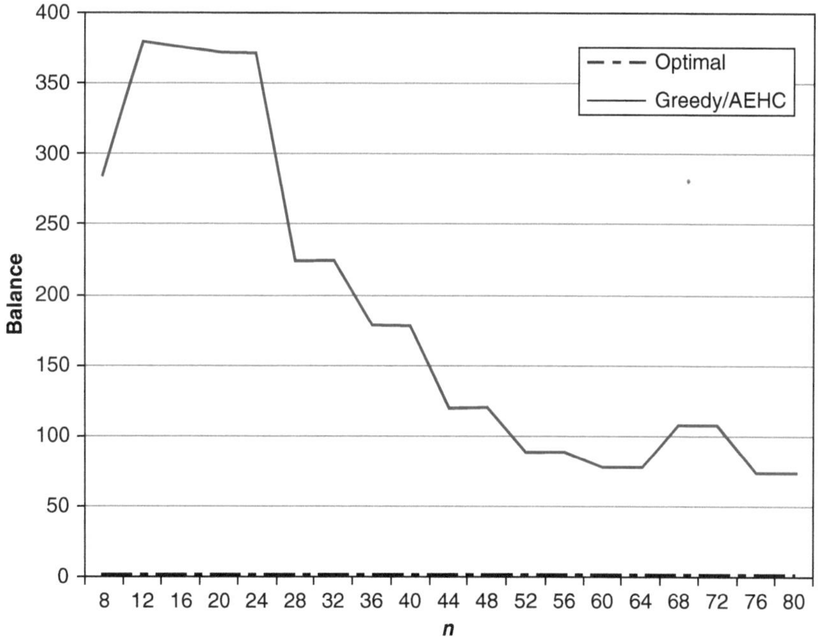

FIGURE 17.3 Detailed Greedy/AEHC hybrid balance measure.

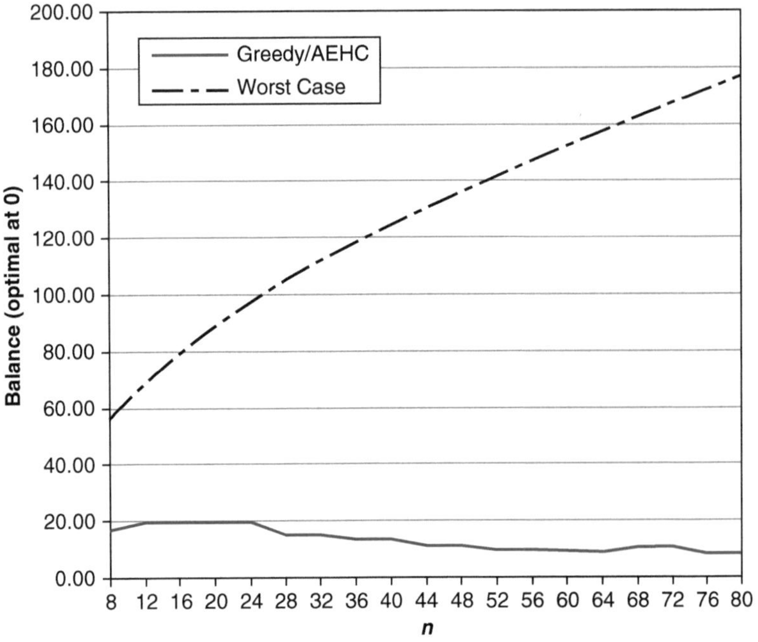

FIGURE 17.4 Normalized Greedy/AEHC hybrid balance measure.

The hazardous part is regularly optimally placed by Greedy (due to the sorting process that puts the hazardous parts at the front of the sorted list) and this position is maintained by AEHC, as illustrated by Fig. 17.5. The hazard measure's efficacy index starts out optimally at $EI_H = 100\%$ and remains there for all instances, giving a sample mean of $\overline{EI}_H = 100\%$.

The demand measure grows relatively consistent with instance size (Fig. 17.6). Its efficacy index starts at $EI_D = 100\%$ and then drops immediately to $EI_D = 70\%$ where it continues a very slow decline to $EI_D = 63\%$ (with a low of $EI_D = 53\%$ at $n = 36$). The resulting demand measure sample mean is $\overline{EI}_D = 65\%$.

AEHC is able to provide a significant improvement over Greedy alone when considering part removal direction (Fig. 17.7). While Greedy's part removal direction measure's efficacy index drops immediately from its highest value of $EI_R = 43\%$ to $EI_R = 14\%$ where it remains for the remainder of the instances giving a sample mean of $\overline{EI}_R = 16\%$, AEHC starts at $EI_R = 71\%$ and never drops lower than Greedy's high of $EI_R = 43\%$, resulting in a sample mean of $\overline{EI}_R = 48\%$.

Though AEHC is seen to fill one of its design goals of continuing the small time complexity of Greedy, this results in similar problems to those as seen in Sec. 16.3.4 where the times collected do not allow for analysis. For AEHC, even though the times collected included the time required by Greedy, the data points with values greater than

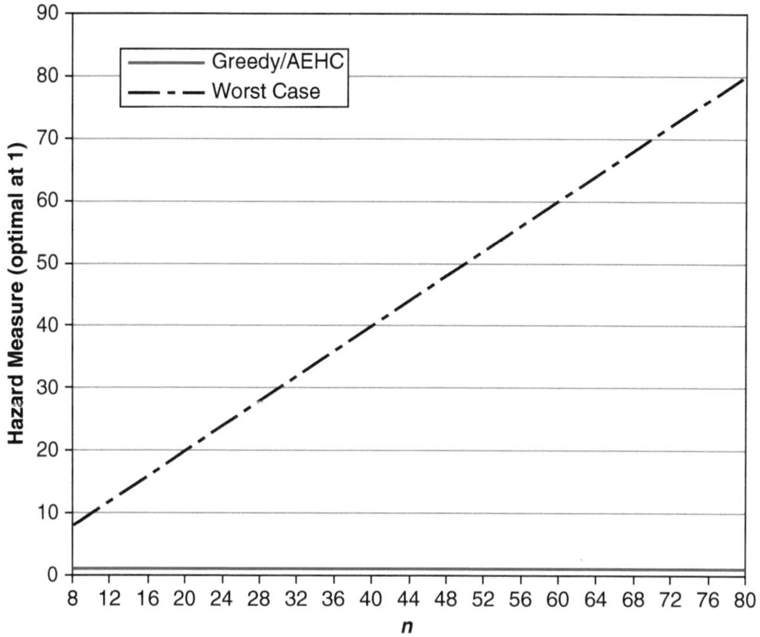

FIGURE 17.5 Greedy/AEHC hybrid hazard measure.

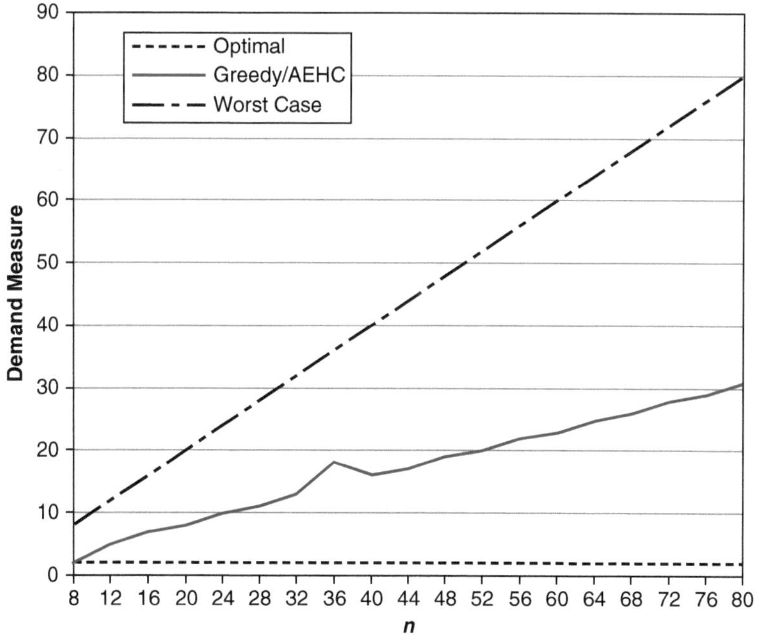

FIGURE 17.6 Greedy/AEHC hybrid demand measure.

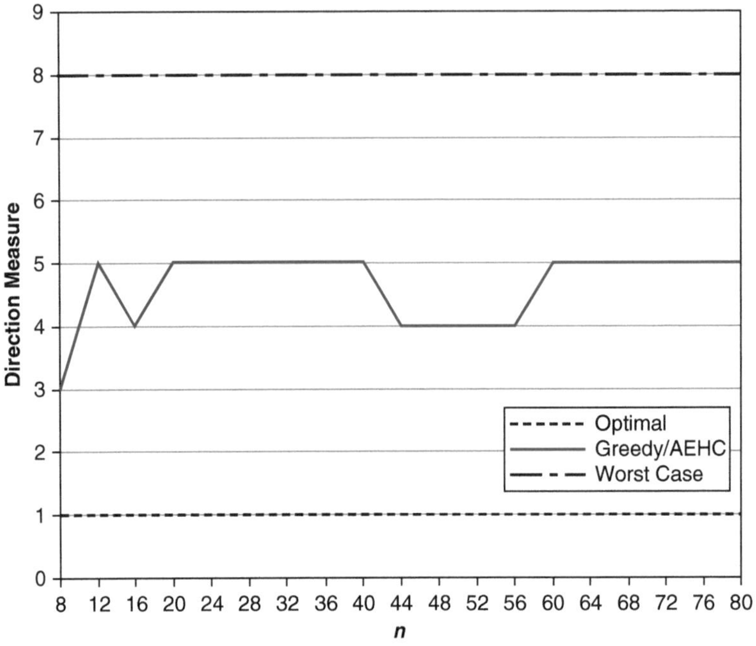

Figure 17.7 Greedy/AEHC hybrid part removal direction measure.

zero were few and those that were captured lacked the resolution required to allow for a true representation of the time complexity curve. For this reason, the Greedy/AEHC hybrid was run to collect additional data using the same data sets as in Sec. 16.3.4 of $n = \{8, 12, 16, \ldots, 760\}$. (Again, the only results used from the larger set of instances are the runtimes.)

Using this larger data set, runtime is only slightly more than that seen with Greedy alone. Since DLBP Greedy has been shown to have an average-case time complexity of $O(n^2)$, a second-degree polynomial regression model is used to fit the AEHC curve. The regression equation is calculated to be $T(n) = 3 \times 10^{-7}n^2 - 1 \times 10^{-5}n + 0.0094$ (only slightly slower than the $T(n) = 7 \times 10^{-8}n^2 + 2 \times 10^{-6}n + 0.00001$ regression model of Greedy alone). The polynomial's extremely small coefficients are indicative of the slow runtime growth in instance size. The coefficient of determination is calculated to be 0.9909, indicating a very high 99.09 percent of the total variation is explained by the calculated linear regression curve. As seen in Fig. 17.8, this regression provides an accurate fit.

With a growth of $3 \times 10^{-7}n^2 - 1 \times 10^{-5}n + 0.0094$, the average-case time complexity of DLBP Greedy/AEHC using DLBP A Priori data is listed as $O(n^2)$ or polynomial complexity. This is very much in agreement with the partial time complexity calculations performed

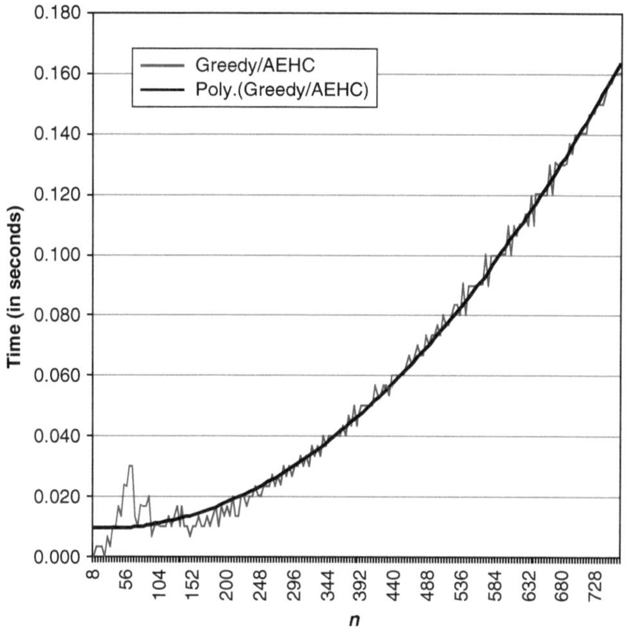

Figure 17.8 Greedy/AEHC hybrid time complexity and second-order polynomial regression line.

in Sec. 17.2; however, it should be noted that the worst-case time complexity is not known, though based on these experimental results and the earlier complexity calculations, it can be no less than $O(n^2)$.

With $n = 80$ running, on average, in 0.01 seconds and $n = 760$ taking, on average, only 0.16 seconds (just 0.12 seconds longer than Greedy alone at 0.04 seconds), AEHC appears to fulfill its design intent of improving upon a Greedy solution while not significantly impacting Greedy's speed advantage. The Greedy/AEHC hybrid is shown with a second-order curve in Fig. 17.9 to demonstrate just how close to linear this algorithm's growth is. Also, even though GA is linear in its time complexity, its growth is much more dependent on the number of generations and the size of the population chosen (hence the linear growth seen, since these variables were maintained constant in that study). Even with GA's linear time complexity growth in n, Greedy and the Greedy/AEHC hybrid are still significantly faster than DLBP GA, which takes 4.08 seconds on average to solve the DLBP A Priori instances at $n = 80$.

As with Greedy, runtime performance can be improved by changing to a faster sorting algorithm for the first phase, while other measures of performance cannot easily be improved without changing the original design of the heuristic.

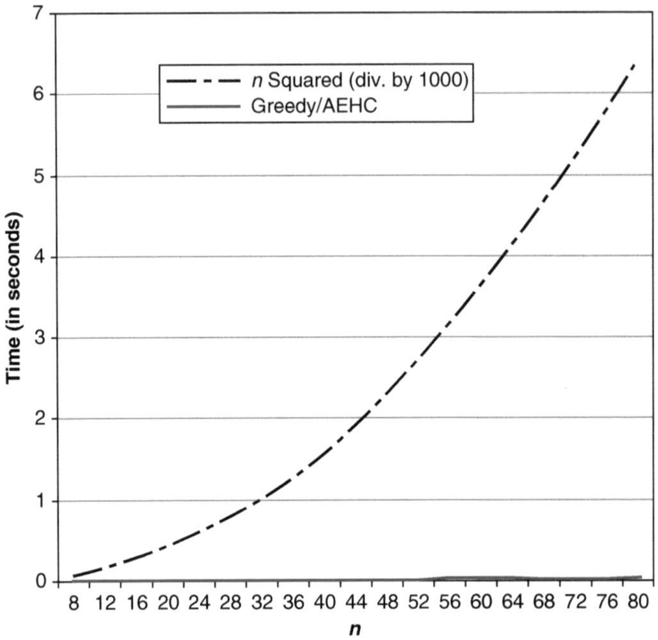

FIGURE 17.9 Greedy/AEHC hybrid time complexity compared to second order.

17.4 Conclusions

A very fast, two-phase hybrid approach to the DLBP is detailed in this chapter. This method combines the FFD greedy algorithm with a problem-specific hill-climbing heuristic to generate a near-optimal and feasible balance sequence almost as quickly as the Greedy method alone. AEHC only completes a pair-wise workstation comparison to allow for rapid search and since it is deemed unlikely that tasks several workstations out can be swapped and still preserve the precedence of all the tasks in-between. The Greedy/AEHC hybrid provides excellent solutions exceptionally rapidly, an asset not only for large problems but especially for real-time problem solution. A near-optimum technique, the DLBP Greedy/AEHC method is still able to quickly find solutions in exponentially large search spaces with solution performance improving with problem size and number of precedence constraints. As with Greedy alone, this specialized hybrid method appears moderately well suited to the multicriteria decision-making problem format. It is also very applicable to problems with nonlinear objectives, as well as for integer problems.

CHAPTER 18

Greedy/2-Opt Hybrid

18.1 Introduction

In this chapter, the DISASSEMBLY LINE BALANCING PROBLEM (DLBP) is solved using a second hybrid approach. This two-phase process is a combination of a first-fit-decreasing (FFD) greedy algorithm followed by application of a 2-optimal (2-opt) algorithm. The hybrid combination of the greedy algorithm and the 2-opt heuristic is instrumental in quickly obtaining near-optimal solutions to the DLBP. The greedy algorithm is based on first-fit-decreasing rules enhanced to preserve the precedence relationships within the product being disassembled. A 2-opt algorithm is then used to balance the part removal sequence and address other objectives per Sec. 12.3. The four experimental instances are then applied in order to demonstrate implementation of the two-phase hybrid and to generate performance measures.

The 2-opt algorithm, as is the case with all k-optimal neighborhood searches, is iterative in nature. The algorithm is reapplied to each new solution n-tuple generated in search of neighborhood improvements. 2-opt has had a great deal of success with TRAVELING SALESPERSON PROBLEM (TSP)-type problems (which a study of the algorithm's functionality would seem to indicate it was designed to solve), but—as a result—requires some degree of modification to allow a successful application to the DLBP. Also, the methodology considers many more part swaps than the adjacent element hill climbing (AEHC) heuristic, enabling a potential reduction in the number of workstations as well as better overall solutions, though at a time complexity cost.

This chapter introduces the greedy/2-opt hybrid heuristic and the DLBP-specific variations. Section 18.2 provides background on the 2-opt model and reviews the modifications required to enable application to a BIN-PACKING type of problem as well as enhancements specific to the DLBP. This section also provides the time complexity calculations for the hybrid. Section 18.3 describes the results when the DLBP Greedy/2-Opt hybrid is applied to the instances from Chap. 10, while Sec. 18.4 summarizes the chapter.

18.2 Model Description

A second phase for DLBP Greedy is implemented to compensate for that algorithm's inability to balance the workstation assignments. This second phase quickly provides a near-optimal and feasible balance sequence using a tour improvement procedure (McGovern and Gupta, 2003a, 2005a). The best-known tour improvement procedures are the edge exchange procedures. In an k-opt algorithm (see Sec. 10.8), all exchanges of k edges are tested until there is no feasible exchange that improves the current solution; this solution is said to be k-optimal, with $k = 2$ and $k = 3$ being the most commonly used due to the number of exchange operations increasing rapidly with increases in k. In addition, a greedy solution is generally considered the best starting point for k-opt (Lawler et al., 1985), though random solutions are commonly implemented as well (see the discussion next). The k-opt search is especially well suited to a TSP-type problem since it seeks to effectively "untwist" a given search route, resulting in a shorter overall tour; the intuitive design of 2-opt can be seen in Fig. 18.1. In this figure (with the four nodes arranged in a square, each node's two closest nodes having unity spacing, and all edges being Euclidean), we know from geometry that the Hamiltonian circuit $\langle 1, 3, 2, 4, 1\rangle$ will have a total distance of $1.41 + 1 + 1.41 + 1 = 4.82$, while after being "untwisted" by 2-opt, the Hamiltonian circuit $\langle 1, 2, 3, 4, 1\rangle$ has a much shorter total distance of $1 + 1 + 1 + 1 = 4.00$. In the TSP, the latter tour provides the lowest cost. This chapter applies a modified 2-opt local search algorithm to the DLBP since the problem format does not fully lend itself to mimicking the TSP.

In a 2-opt algorithm, two tours are neighbors if one can be obtained from other by deleting two edges, reversing one of the resulting two paths, and reconnecting. In the DLBP, the 2-opt neighbors of the tour are pair-wise part exchanges within the current solution sequence. For this version of 2-opt, nodes (more accurately, vertices since the DLBP is described by a directed graph) are exchanged instead of edges (arcs for the DLBP). Not only is this a more logical application of this technique for the DLBP, it is also more practical

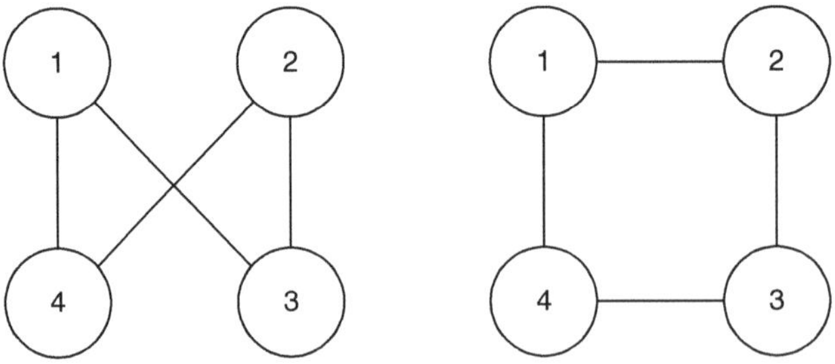

FIGURE 18.1 A 2-opt example.

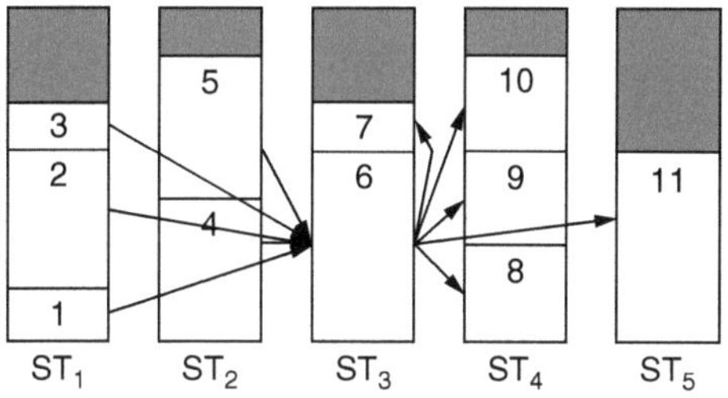

FIGURE 18.2 DLBP 2-Opt process.

since a true 2-opt will reverse large sections of the solution sequence, which is not conducive to precedence preservation. In addition, this method has been used in the past by other authors (Merkle and Middendorf, 2000).

The exchange mechanism used by the DLBP is demonstrated in Fig. 18.2 using part 6 as an example. As can be seen in the figure, part 6 can be exchanged with any other part; for example, parts 1 through 5 consider swapping themselves with part 6, then part 6 considers swapping itself with parts 7 through 11.

In DLBP 2-Opt with $n = 4$ and an original solution sequence four-tuple of $\langle 1, 2, 3, 4\rangle$, the six resulting pair-wise exchanges are $\langle 2, 1, 3, 4\rangle$, $\langle 3, 2, 1, 4\rangle$, $\langle 4, 2, 3, 1\rangle$, $\langle 1, 3, 2, 4\rangle$, $\langle 1, 4, 3, 2\rangle$, and $\langle 1, 2, 4, 3\rangle$. DLBP 2-Opt then makes comparisons defined as

$$\text{Exchanges} = \sum_{p=1}^{n-1} (n-p) \tag{18.1}$$

Exhaustive search visits all $n!$ tours, while the total number of tours visited by an iteration of DLBP 2-Opt, including the original data set (such as $\langle 1, 2, 3, 4\rangle$), can be calculated by the following formula

$$\text{Tours} = 1 + \sum_{p=1}^{n-1} (n-p) \tag{18.2}$$

Typically, a random starting tour is generated for the initial application of a 2-opt algorithm. In this chapter, the starting point is rapidly attained by the previously discussed DLBP Greedy algorithm, which provides a minimum-NWS feasible solution. This near-optimal starting point requires DLBP 2-Opt only to improve the remaining measures. DLBP 2-Opt tries switching the positions of any two parts in the current solution sequence in order to get the best tour in this neighborhood. It then repeats this process using this new tour (the best previous solution sequence) until no better solution (as described in Sec. 12.3) can be found. DLBP 2-Opt is designed to swap every task with each task in every subsequent workstation in search of improved balance. It does this while preserving precedence and not exceeding CT in any workstation. In addition, each DLBP 2-Opt neighbor under consideration recalculates the minimum number of workstations required for a given problem set, enabling the determination of more efficient part removal sequences that may exist in the 2-Opt neighborhood that did not exist in the DLBP Greedy solution. As long as the balance is at least maintained, DLBP 2-Opt then seeks improvements in hazard measure, demand measure, and part removal direction measure, but never at the expense of precedence constraints. The heuristic is given in pseudocode format in Fig. 18.3.

As is the norm with greedy algorithms, the DLBP Greedy process is run once to determine a solution. 2-Opt, however—like many combinatorial optimization techniques—is continuously run on subsequent solutions for as long as is deemed appropriate or acceptable by the user or until it is no longer possible to improve (which is the process used here), at which point it is assumed that the (local) optima has been reached. Repeating the DLBP 2-Opt method in this way provides improved performance over time.

The best-case time complexity of the Greedy/2-Opt hybrid can be no faster than the best-case time complexity of Greedy alone, calculated to be $\Omega(n^2)$. The best case for 2-Opt would take place when no improvement can be made over what Greedy has generated. In this

Procedure DLBP_2-OPT {

1. Initialize: { PSG is the greedy solution sequence }
   SET PST := PSG { PST is the current solution sequence }
   SET TMP := PSG { TMP is a temporary solution sequence }
   SET BST:= PSG { BST is the best solution sequence thus far }

2. $\forall\, p \in \text{PST} \mid p = \text{PST}_k,\ 1 \le k \le n$, attempt to generate better balanced 2-Opt part removal sequence PST by:

   $\forall\, q \in \text{TMP} \mid p + 1 \le q \le n$, perform 2-Opt part swap (i.e., swap TMP_p with every TMP_q)

   Calculate new F, H, D, and R

   IF { [BETTER_SOLUTION(TMP, BST)]
   $\wedge$
   (PRECEDENCE_PASS) }

   THEN SET BST:= TMP

3. Save results:
   SET PST := BST
   SET TMP := BST

4. Repeat STEP 2 until no more improvements in F, H, D, or R can be made

}

FIGURE 18.3 DLBP 2-Opt procedure.

situation, 2-Opt would still go through the entire list of greedy-sorted items, though only once. Equation (18.1) can be reformulated as

$$\text{Exchanges} = n(n-1) - \sum_{p=1}^{n-1} p$$

This can be written to reference the time complexity as

$$T(n) \propto n(n-1) - \sum_{p=1}^{n-1} p$$

$$T(n) \propto n(n-1) - \frac{(n-1)[(n-1)+1]}{2}$$

$$T(n) \propto n(n-1) - \frac{n(n-1)}{2}$$

$$T(n) \propto \frac{2n(n-1)}{2} - \frac{n(n-1)}{2}$$

$$T(n) \propto \frac{2n(n-1) - n(n-1)}{2}$$

$$T(n) \propto \frac{n(n-1)}{2}$$

When this is combined with Greedy's known number of comparisons [i.e., $T(n) \propto n(n+1)/2$] from Sec. 16.2, this gives

$$T(n) \propto \frac{n(n+1)}{2} + \frac{n(n-1)}{2}$$

$$T(n) \propto \frac{n^2 + n + n^2 - n}{2}$$

$$T(n) \propto \frac{2n^2}{2}$$

$$T(n) \propto n^2$$

resulting in a best-case time complexity calculated to be the asymptotic lower bound of $\Omega(n^2)$.

The worst-case time complexity of the Greedy/2-Opt hybrid is at least as slow as the worst-case time complexity of Greedy alone, calculated to be $O(n^2)$. A single iteration of 2-Opt alone is no faster than $T(n) \propto n^2$. When combined with the worst-case Greedy results of $T(n) \propto n^2$, this gives us $T(n) \propto n^2 + n^2$ or $T(n) \propto 2n^2$, which is formally $O(n^2)$. Similar to AEHC, due to the iterative nature of 2-Opt the worst-case time complexity of the heuristic has not been calculated, while the work completed above indicates that it can be no better than $O(n^2)$. In Lawler et al. (1985), Johnson and Papadimitriou discuss efforts using instances of second-best TSP tours having lengths an arbitrary multiple of the optimal length and never seeing any improvement in less than $3n/8$ edge exchanges while noting that there are an exponential number of potential edge exchanges of this sort and that there are an exponential number of these "second-best" tours.

With a best-case time complexity of $\Omega(n^2)$ and the worst-case time complexity of at least $O(n^2)$, a tight bound may exist at $\Theta(n^2)$, but this is unlikely.

18.3 Numerical Results

The greedy/2-opt hybrid is run on each of the four experimental instances from Chap. 10 with the results analyzed in this section.

18.3.1 Personal Computer Instance

The DLBP Greedy/2-Opt hybrid has been applied to the personal computer (PC) instance. As DLBP Greedy alone successfully found the optimal solution, the subsequent 2-Opt algorithm can do no better and terminates. The hybrid search of the PC instance took less than 1/100th of a second with the optimal solution (Table 18.1) of four workstations and $F^* = 33$, $H^* = 7$, $D^* = 19{,}025$, and $R^* = 6$ being found.

18.3.2 The 10-Part Instance

Using the 10-part instance, DLBP Greedy generates a feasible solution having the minimum number of workstations and then DLBP 2-Opt is able to improve the overall balance and the demand measure and maintain the optimal hazard measure, resulting in the optimal solution (five workstations and $F^* = 211$, $H^* = 4$, $D^* = 9730$, and $R^* = 7$) being found (Table 18.2). The speed for the hybrid on this instance is again less than 1/100th of a second.

Part ID	1	5	3	6	2	8	7	4
PRT	14	23	12	16	10	36	20	18
Workstation	1	1	2	2	2	3	4	4
Hazardous	0	0	0	0	0	0	1	0
Demand	360	540	620	750	500	720	295	480
Direction	1	2	1	4	0	4	1	1

TABLE 18.1 Greedy/2-Opt Hybrid Solution Using the Personal Computer Instance

Part ID	10	5	6	7	9	4	8	1	2	3
PRT	10	23	14	19	14	17	36	14	10	12
Workstation	1	1	2	2	3	3	4	5	5	5
Hazardous	0	0	0	1	0	0	0	0	0	0
Demand	0	0	750	295	360	0	0	0	500	0
Direction	3	5	5	2	5	2	1	2	0	0

TABLE 18.2 Greedy/2-Opt Hybrid Solution Using the 10-Part Instance

18.3.3 Cellular Telephone Instance

DLBP Greedy/2-Opt finds feasible solutions to the cellular telephone data, taking 0.02 seconds on average. While 2-Opt is not able to improve upon the first-phase Greedy solution of 10 workstations, it is able to improve on all of the other measures. The balance measure improves to $F = 93$ from the Greedy starting point of $F = 199$, the hazard measure improves from Greedy's $H = 89$ to $H = 87$, the demand measure improves from $D = 973$ to $D = 943$, and the part removal direction measure improves from $R = 12$ to $R = 10$. Also, all of these results are better than those obtained by Greedy/AEHC, though at the expense of time. The DLBP Greedy/2-Opt solution is shown in Table 18.3.

18.3.4 DLBP A Priori Problem Instances

The DLBP Greedy/2-Opt hybrid technique can be seen below as applied to the DLBP A Priori data with $n = 12$ (Table 18.4). DLBP 2-Opt is not able to improve upon Greedy's near-optimal solution of NWS = 4 workstations (optimal is NWS* = 3), though it is able to improve upon the balance (from $F = 532$ to $F = 174$) and the part removal direction measures (from $R = 7$ to $R = 4$) while maintaining the hazard (at $H^* = 1$) and demand (at $D = 5$) measures. The time to complete the search was less than 1/100th of a second.

DLBP 2-Opt maintains the optimum part removal sequence position for hazardous part 12 while improving the balance by increasing the number of parts in the last workstation from one to four. This improvement is seen in the Greedy idle times of $I_j = \langle 1, 1, 1, 23 \rangle$ improving to the more balanced $I_j = \langle 7, 8, 5, 6 \rangle$. Note that it would appear that parts 10 and 9 could be swapped to increase demand performance (at the acceptable cost of decreasing part removal direction performance); however, due to DLBP 2-Opt's ability to minimize the number of workstations, it would put parts 12, 4, 1, and 9 in workstation 1, decreasing the overall balance measure (because four workstations would still be required and now one would be filled to capacity; i.e., workstation 1 would have an idle time of $I_1 = 0$).

Various performance measures of the DLBP Greedy/2-Opt hybrid can be demonstrated with analysis using the DLBP A Priori benchmark

Part ID	2	1	4	3	6	7	8	9	5	10	15	13	18	14	17	20	16	19	21	25	11	12	22	23	24
PRT	2	3	10	3	15	15	15	15	10	2	2	2	3	2	2	5	2	18	1	2	2	2	5	15	2
Workstation	1	1	1	1	2	3	4	5	6	6	6	6	7	7	7	7	7	8	9	9	9	9	9	10	10
Hazardous	1	1	0	0	0	0	0	0	0	0	0	0	0	0	0	0	0	1	0	1	0	1	0	1	0
Demand	7	4	1	1	1	1	1	1	1	2	1	1	2	1	2	1	1	8	4	4	1	4	6	7	1
Direction	3	2	5	5	5	5	5	5	5	4	0	5	5	2	4	4	2	5	4	4	4	4	4	4	4

TABLE 18.3 Greedy/2-Opt Hybrid Solution Using the Cellular Telephone Instance

Part ID	12	4	1	10	9	11	6	5	7	8	2	3
PRT	11	5	3	11	7	11	5	5	7	7	3	3
Workstation	1	1	1	2	2	3	3	3	4	4	4	4
Hazardous	1	0	0	0	0	0	0	0	0	0	0	0
Demand	0	0	0	0	1	0	0	0	0	0	0	0
Direction	0	1	1	1	0	0	0	0	1	0	0	0

TABLE 18.4 Greedy/2-Opt Hybrid Solution Using the A Priori Instance at $n = 12$

data sets of $n = \{8, 12, 16, \ldots, 80\}$. Only once is 2-Opt able to improve upon Greedy's solution of NWS = NWS* + 1 workstations (Fig. 18.4). In terms of the number of workstations required by the DLBP Greedy/2-Opt hybrid, from Eq. (12.1) the hybrid's efficacy index in number of workstations begins at an optimal $EI_{NWS} = 100\%$, drops to a low of $EI_{NWS} = 89\%$, then continuously climbs through to $EI_{NWS} = 98\%$ with an efficacy index sample mean in number of workstations of $\overline{EI}_{NWS} = 96\%$.

As with Greedy and the Greedy/AEHC hybrid, balance efficacy is seen to improve with problem size (Fig. 18.5) though with the balance measure decreasing much more smoothly than the approximate step function seen with Greedy and even more so than Greedy/AEHC (again, note that the Greedy, Greedy/AEHC, and the Greedy/2-Opt

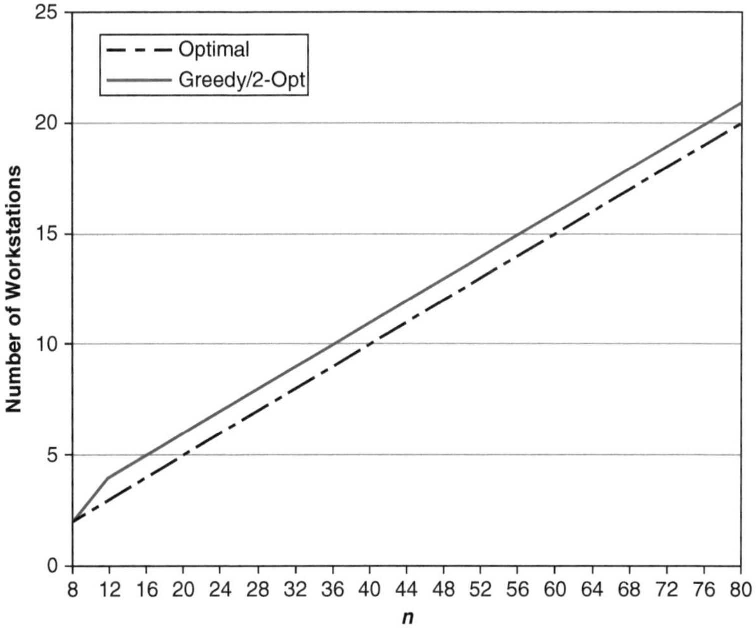

FIGURE 18.4 Greedy/2-Opt hybrid workstation calculation.

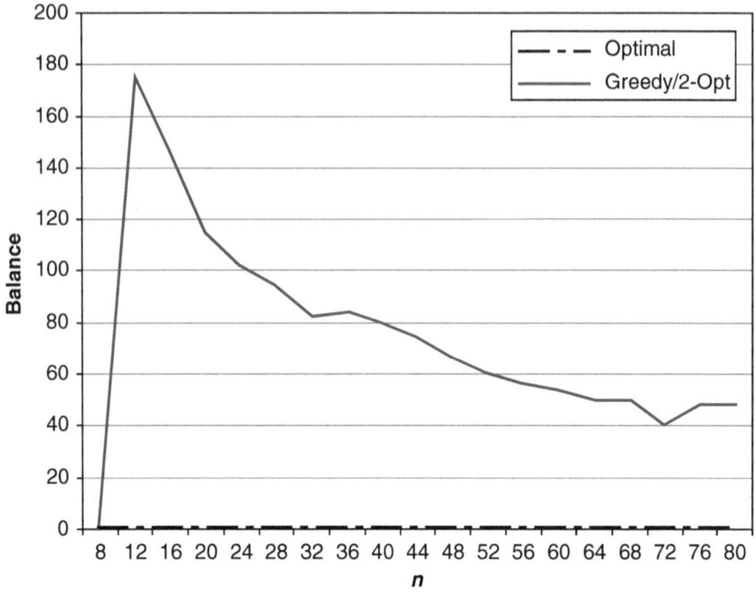

FIGURE 18.5 Detailed Greedy/2-Opt hybrid balance measure.

graphs are not of the same scale, with the Greedy chart in Fig. 16.3 having a *y*-axis that is three times taller than that of the 2-Opt chart in Fig. 18.5). As with Greedy and Greedy/AEHC, poor performance is seen when dealing with smaller instances (the exception being the optimal solution found with data set number 1), though the balance is seen to steadily improve through to the last data set of $n = 80$.

The normalized balance efficacy index improves from a Greedy low $EI_F = 59\%$ to a 2-Opt low of $EI_F = 81\%$, from a Greedy high of $EI_F = 93\%$ to a 2-Opt high of $EI_F = 100\%$, and from a Greedy sample mean of $\overline{EI}_F = 83\%$ to a 2-Opt sample mean of $\overline{EI}_F = 93\%$ (Fig. 18.6).

The hazardous part was regularly optimally placed by Greedy (due to the sorting process that puts the hazardous parts at the front of the sorted list) and this is maintained by 2-Opt (Fig. 18.7). The hazard measure's efficacy index starts out optimally at $EI_H = 100\%$ and remains there for all instances, giving a sample mean of $\overline{EI}_H = 100\%$.

The demand measure remains relatively consistent with instance size (Fig. 18.8). Its efficacy index fluctuates between $EI_D = 100\%$ and $EI_D = 70\%$ initially and then continues a very slow decline to $EI_D = 64\%$ (with a low of $EI_D = 61\%$ at $n = 20$). The resulting demand measure sample mean is $\overline{EI}_D = 74\%$.

Only slightly better than AEHC, 2-Opt is able to provide a similar improvement over Greedy alone when considering part removal direction (Fig. 18.9). Where Greedy's part removal direction measure's efficacy index drops immediately from its highest value of $EI_R = 43\%$ to $EI_R = 14\%$ where it remains for the remainder of the

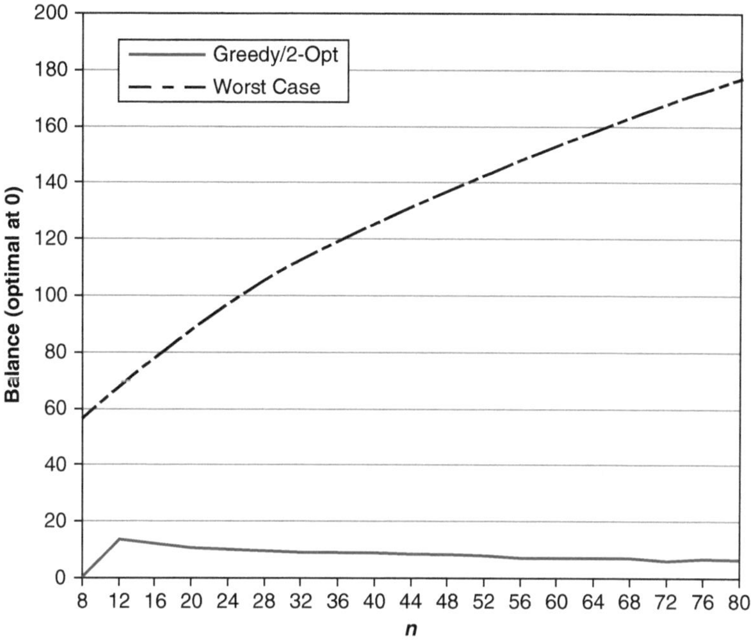

FIGURE 18.6 Normalized Greedy/2-Opt hybrid balance measure.

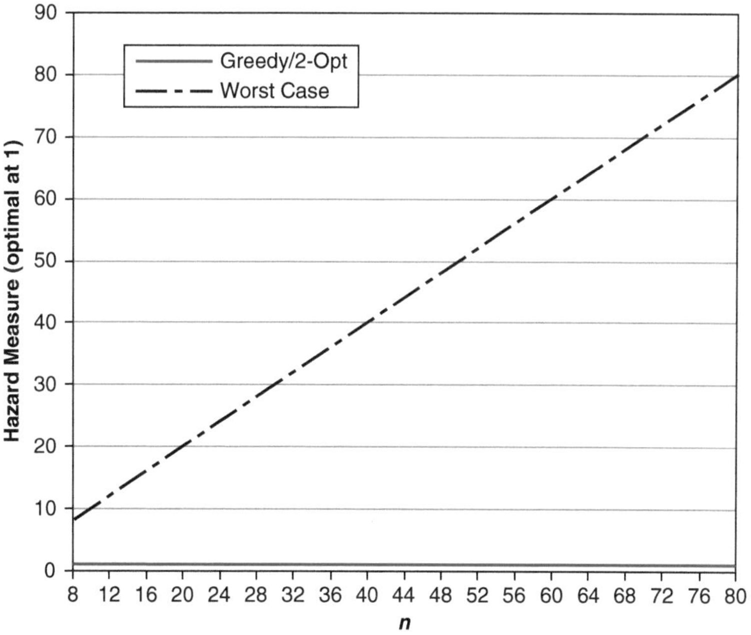

FIGURE 18.7 Greedy/2-Opt hybrid hazard measure.

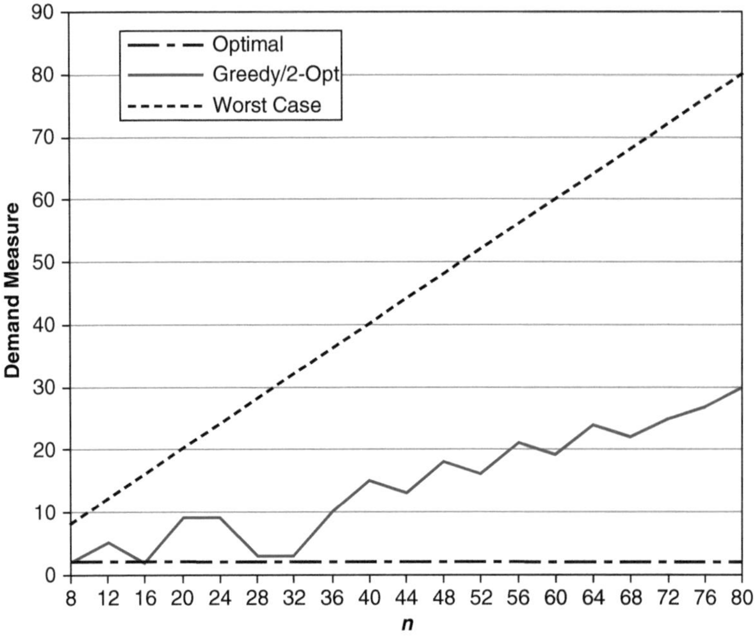

FIGURE 18.8 Greedy/2-Opt hybrid demand measure.

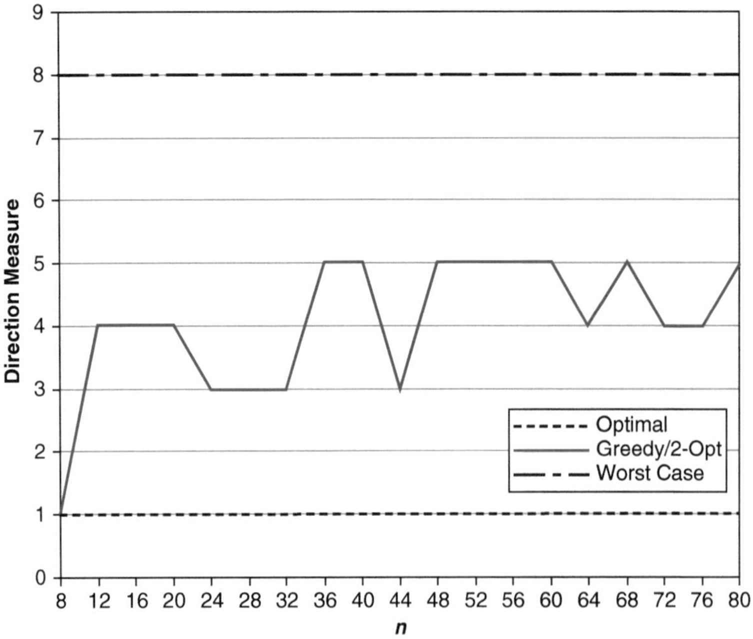

FIGURE 18.9 Greedy/2-Opt hybrid part removal direction measure.

instances (giving a sample mean of $\overline{\text{EI}}_R = 16\%$), 2-Opt starts at $\text{EI}_R = 100\%$ and never drops lower than Greedy's high of $\text{EI}_R = 43\%$, resulting in a sample mean of $\overline{\text{EI}}_R = 56\%$.

While DLBP Greedy/2-Opt is not as fast as Greedy/AEHC, it is still able to generate solutions very rapidly. 2-Opt did run slow enough to allow the use of the smaller A Priori instances of $n = \{8, 12, 16, \ldots, 80\}$ instead of $n = \{8, 12, 16, \ldots, 760\}$. Based on Figs. 18.10 and 18.11 as well as the fact that DLBP Greedy was shown to have an average-case time complexity of $O(n^2)$, a second-order polynomial regression model is used to fit the Greedy/2-Opt curve. Figures 18.10 and 18.11 are also valuable in demonstrating that even the slowest of the three greedy-based methodologies is tremendously fast when compared to faster growing second- or third-order curves.

The regression equation is calculated to be $T(n) = 5 \times 10^{-5}n^2 - 0.001n + 0.0033$ (as compared to the $T(n) = 7 \times 10^{-8}n^2 + 2 \times 10^{-6}n + 0.00001$ regression model of Greedy alone). The very small coefficients of n^2 are indicative of the slow runtime growth in instance size. The coefficient of determination is calculated to be 0.9195, with a percentage of 91.95 percent indicating that a moderately good fit is made using the calculated linear regression curve. However, this is attributed to the extreme speed of the hybrid resulting in runtime

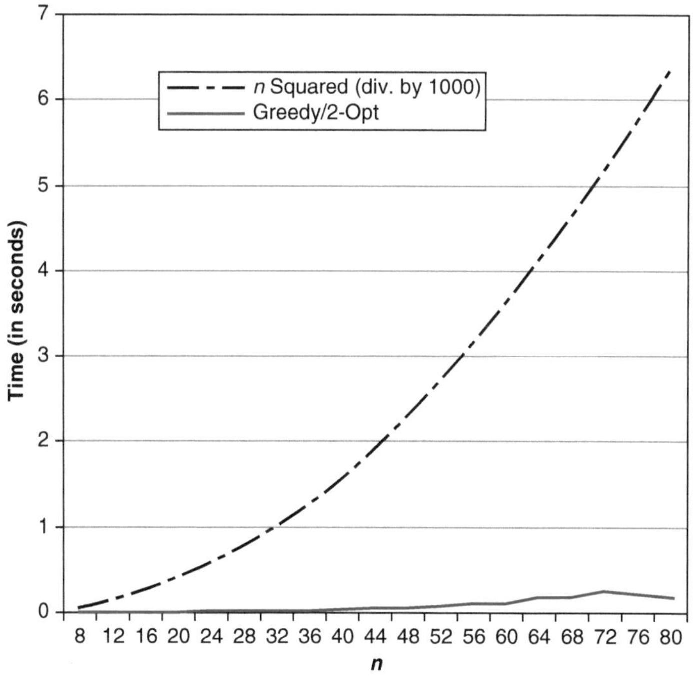

Figure 18.10 Greedy/2-Opt hybrid time complexity compared to second order.

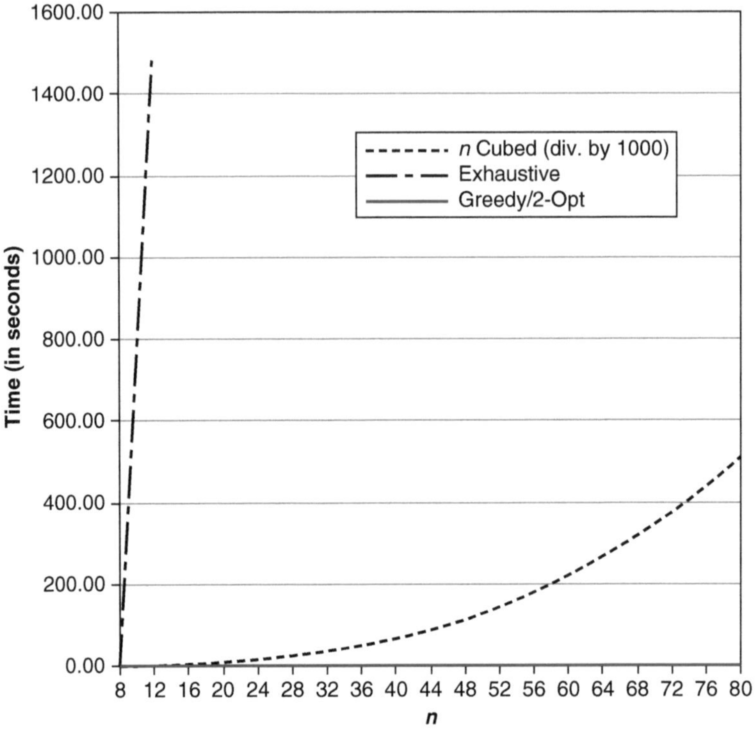

FIGURE 18.11 Greedy/2-Opt hybrid time complexity compared to exhaustive search.

data collected very near to the software's time resolution. A larger data set (as is required by Greedy and Greedy/AEHC) would be expected to result in a similar regression curve but with a higher coefficient of determination. As seen in Fig. 18.12, this regression provides an accurate fit. With a growth of $5 \times 10^{-5}n^2 - 0.001n + 0.0033$, the average-case time complexity of DLBP Greedy/2-Opt with the DLBP A Priori data is listed as $O(n^2)$ or polynomial complexity. This is in agreement with the partial time complexity calculations performed in Sec. 18.2; however, it should be noted that the worst-case time complexity has not been calculated, though based on these experimental results and the earlier complexity calculations, it can be no less than $O(n^2)$.

As with Greedy and Greedy/AEHC, runtime performance can be improved by changing to a faster sorting algorithm for the first phase. Other measures of performance can be improved by increasing the k value, for example, switching to a 3-opt algorithm.

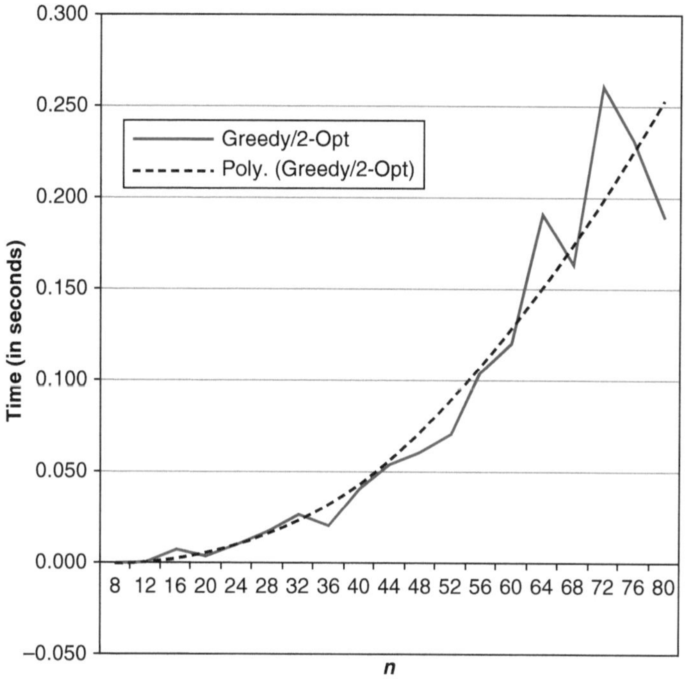

FIGURE 18.12 Greedy/2-Opt hybrid time complexity and second-order polynomial regression line.

18.4 Conclusions

An additional two-phase hybrid approach to the DLBP is presented in this chapter. This method combines the FFD greedy algorithm with a 2-opt algorithm modified for the BIN-PACKING problem by using arcs instead of edges and vertices instead of nodes, and by only allowing swaps of the vertices where normally edges would be exchanged. DLBP enhancements include the implementation of a multicriteria search objective. While not as fast as DLBP Greedy or the Greedy/AEHC hybrid, the DLBP Greedy/2-Opt hybrid is able to find solutions having higher efficacies in all measures and at an average-case time complexity cost that is significantly slower than the other, non-greedy-based methodologies considered in Part II of this book.

CHAPTER 19

H-K Heuristic

19.1 Introduction

The H-K heuristic rapidly provides a feasible solution using a modified exhaustive search technique that performs a data sampling of all solutions (Sec. 10.9). This general-purpose heuristic easily lends itself to the DISASSEMBLY LINE BALANCING PROBLEM (DLBP), seeking to provide a feasible disassembly sequence, minimize the number of workstations, minimize idle time, and ensure similar work content at each workstation, as well as addressing the placement of hazardous parts, high-demand parts, and parts with similar part removal directions per Sec. 12.3. The experimental instances are used to demonstrate the heuristic's implementation and measure its performance.

This chapter introduces the DLBP-specific variation of the H-K general-purpose heuristic. H-K is an uninformed search; it methodically moves through the search space regardless of prior, current, or forecast solution results. Section 19.2 allows a detailed look into the H-K process itself and reviews enhancements specific to the DLBP. Section 19.3 details the performance of the heuristic with changes in the search pattern, while Sec. 19.4 demonstrates the results when H-K is applied to the instances from Ch. 10. Section 19.5 summarizes the chapter.

19.2 DLBP Application

H-K (McGovern and Gupta, 2004b, 2008a) is modified for the DLBP. Applied to the permutation of the number of parts n and considering a product undergoing complete disassembly, the H-K process is modified to test the proposed part addition to the sequence for precedence constraints.

All of the parts are maintained in a tabu-type list ISS_k defined by Eq. (16.1). Each increment of the DLBP H-K-generated solution is considered for feasibility. If all possible parts for a given solution position fail these checks, the remainder of the positions are not further inspected, the procedure falls back to the previously successful solution addition, increments it by 1, and continues. If, however, it is feasible in its entirety, DLBP H-K then looks at each element in the solution and places that element using the next-fit rule (from the BIN-PACKING problem application; once a bin has no space for a given item attempted to be packed into it, that bin is never used again even though a later, smaller item may appear in the list and could fit in the bin, see Hu and Shing, 2002). DLBP H-K puts the element under consideration into the current workstation if it fits. If it does not fit, a new workstation is assigned and previous workstations are never again considered. Although next-fit is not considered to perform as well as first-fit, best-fit, first-fit-decreasing, or best-fit-decreasing when used in the general BIN-PACKING problem, it is the only one of these rules that will work with an H-K-generated DLBP solution sequence due to the existence of precedence constraints (see Ch. 16 or McGovern and Gupta, 2005a for a DLBP implementation of first-fit-decreasing). When all of the work elements have been assigned to a workstation, the process is complete and the balance, hazard, demand, and direction measures are calculated. The best of all of the inspected solution sequences is then saved as the problem solution. These processes are repeated until all items have been assigned to workstations.

Although the actual software implementation for the study in this book consisted of a very compact recursive algorithm, in the interest of clarity, the general DLBP H-K procedure is presented here as a series of nested loops (Fig. 19.1), where FS is the feasible sequence indication (FS = 1 if the solution is feasible, and ISS_k is a binary value: 1 if the kth part is already in the solution sequence).

19.3 DLBP A Priori Numerical Analysis for Varying Skip Size

The DLBP A Priori instances can be used to determine how skip size affects solution performance and time complexity (then, based on these results, how the heuristic's performance changes with problem size can be demonstrated, as can be seen in the next section). Because H-K is exceptionally deterministic and performs no preprocessing of the data, the order in which the data is presented can be expected to affect the solutions considered. For this reason, the analyses done in the following sections are first run with the data as given by Eq. (11.16) and then run again, but this time with the data presented in reverse. (Note that the algorithm in Fig. 19.1 actually generates the data as a permutation, so references to the data being given in reverse actually refer to the algorithm generating the permutations in reverse.) The better of the two results is used and the search times (forward and reverse) are then added together. Both forward and reverse results are used to ensure that unusually good (or bad) results are not artificially

```
Procedure DLBP_H-K {
        SET ISS_k := 0 ∀ k∈P              { equal to 1 if part k is in the solution sequence }
        SET FS := 1                       { feasible sequence value equal to 1 if feasible }
    ┌
    │   PS_1 := 1 to n, skip by ψ_1       { identification of kth part in a solution sequence }
    │         SET ISS_PS_1 := 1
    │     ┌
    │     │   PS_2 := 1 to n, skip by ψ_2
    │     │         WHILE     (ISS_PS_2 = = 1 ∨
    │     │                    PRECEDENCE_FAIL ∧
    │     │                    not at n)
    │     │               Increment PS_2 by 1
    │     │
    │     │         IF     ISS_PS_2 = = 1
    │     │         THEN SET FS := 0
    │     │         ELSE  SET ISS_PS_2 := 1
    │     │                :
    │     │       ┌        :
    │     │       │ IF     FS = = 1
    │     │       │              PS_n := 1 to n skip by ψ_n
    │     │       │                    WHILE     (ISS_PS_n = =1 ∨
    │     │       │                               PRECEDENCE_FAIL ∧
    │     │       │                               not at n)
    │     │       │                          Increment PS_n by 1
    │     │       │
    │     │       │                    IF     ISS_PS_n = = 0
    │     │       └                    THEN evaluate solution PS
    │     │                :
    │     │                :
    │     │   IF     FS = = 1
    │     │   THEN SET ISS_PS_2 := 0
    │     └   ELSE  SET FS := 1
    │
    │   SET ISS_PS_1 := 0
    └   SET FS := 1
}
```

FIGURE 19.1 The DLBP H-K procedure.

obtained due to some unknown structural feature of the DLBP A Priori data sets and since presenting data in both forward and reverse formats could be expected to be the most common application. This is not necessary where an instance contains a relatively large number of precedence constraints; individual parts in an instance do not bear a great deal of resemblance to each other; or where in an actual application one may expect a user of H-K to only run the data exactly as it was presented to them. Although not covered here in order to allow an introduction to the basic concepts of H-K, note that it could be expected that better results may be achieved with multiple runs of H-K consisting of the data being presented to H-K in differing orders.

Skip size is varied to determine if and how solution performance decreases with increases in ψ. All cases are the DLBP A Priori data set with $n = 12$, consideration of all skip sizes (including exhaustive search, i.e., $\psi = \{1, \ldots, n\}$), and using all smaller skip sizes per Eq. (10.3). This instance size is chosen because it is the largest multiple of $|PRT|$ that can be repeatedly evaluated by exhaustive search in a reasonable amount of time ($n = 12$ took just over 20 minutes, while $n = 16$ is calculated to take over 2 years; see Sec. 13.3). This is essential since exhaustive search provides the time complexity and solution performance baseline. A full range of skip sizes is used to provide maximum and minimum performance measures, and all intermediate skip sizes are used to allow the highest level of detail for the study. Although most of the following results are not optimal even though there are multiple optimum extreme points, this is again a reflection of the especially designed DLBP A Priori data set than any search technique.

The first analysis done is a time complexity study. As shown in Fig. 19.2, the solution time appears to grow slightly less than factorially to the inverse of ψ, so even a very small increase in the skip size will yield a significant speed up of any search. (The exhaustive search curve has no relationship to the H-K curve and is depicted only to provide a size and shape reference.)

A more detailed view of the H-K time-to-skip size relationship can be seen in Fig. 19.3. With an exponential curve (specifically n^n with

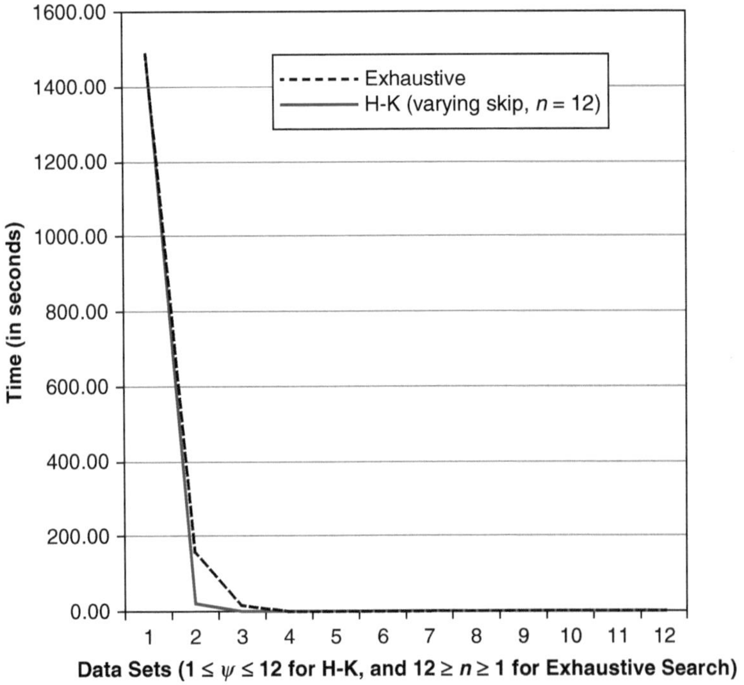

FIGURE 19.2 H-K time complexity.

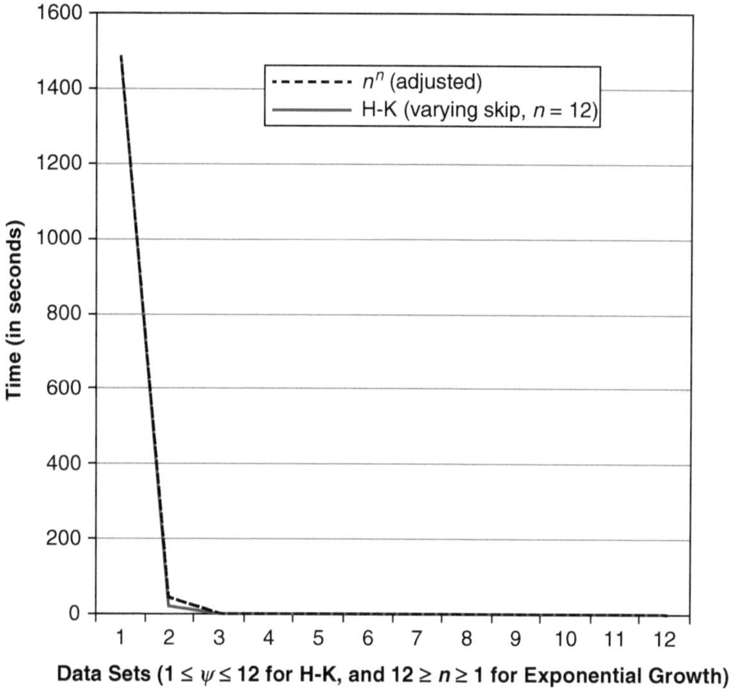

FIGURE 19.3 Detailed H-K time complexity.

$12 \geq n \geq 1$) adjusted for scale and overlaid, it can be seen that the rate of growth is actually exponential in the inverse of ψ. The average-case time complexity of H-K is then listed as $O(b^b)$ in skip size, where $b = 1/\psi$. Due to the nature of H-K, the number of commands executed in the software do not vary based on precedence constraints, data sequence, greedy or probabilistic decision making, improved solutions nearby, and the likes, so the worst case is also $O(b^b)$, as is the best case [i.e., big-omega of $\Omega(b^b)$], and, therefore, a tight bound exists, which is $\Theta(b^b)$.

Next, the workstation calculation performance is measured. The number of workstations calculated is seen to remain optimal through $\psi = 4$ (i.e., while ψ stays within 1/3 of n). As seen in Fig. 19.4, performance then decreases to NWS* + 1 and remains there; even at the maximum skip size ($\psi = 12$), NWS is never worse than NWS* + 1 (note that this is also indicative of the next-fit rule methodology). From Eq. (12.1) H-K's efficacy index in number of workstations while varying ψ starts at an optimal $EI_{NWS} = 100\%$ then drops to and remains at $EI_{NWS} = 89\%$ with an efficacy index sample mean in number of workstations of $\overline{EI}_{NWS} = 93\%$. Ignoring the $\psi = 1$ results (since this is effectively exhaustive search) and $\psi = 12$ results (since this is a trivial solution and one that is repeated for all ψ sizes), the efficacy index

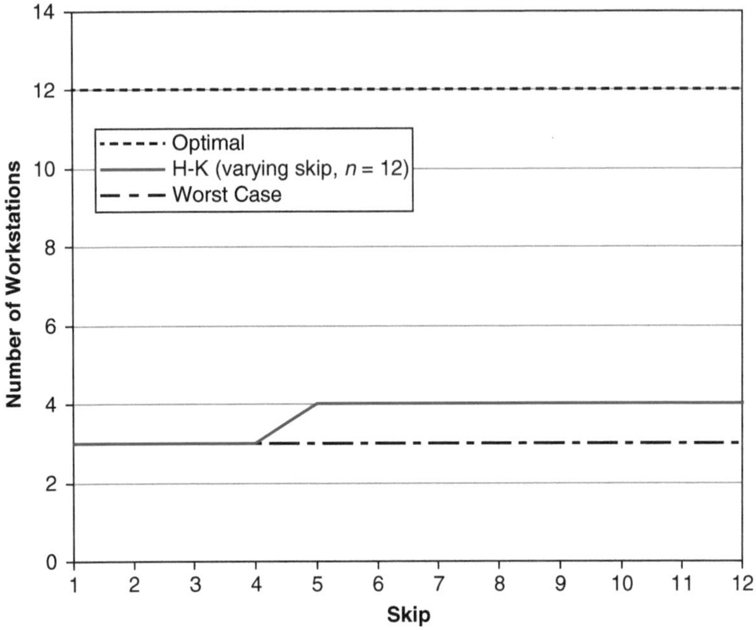

FIGURE 19.4 H-K workstation calculation with changes in ψ.

sample mean in number of workstations is more appropriately described by $\overline{EI}_{NWS} = 92\%$.

The balance performance is then measured. The measure of balance also remains optimal through $\psi = 4$ (i.e., while ψ stays within 1/3 of n). As seen in Fig. 19.5, performance again decreases as skip size increases but at a relatively slow rate. From Eq. (12.1) H-K's efficacy index is balanced while varying ψ starts at an optimal $EI_F = 100\%$ until $\psi = 5$ then drops to a low of $EI_F = 94\%$ giving an efficacy index sample mean in balance of $\overline{EI}_F = 97\%$ and remains the same when ignoring the $\psi = 1$ and $\psi = 12$ results.

The normalized balance gives a more scale-appropriate depiction of the balance performance. The normalized performance is seen in Fig. 19.6. The efficacy index is balanced while varying ψ starts at an optimal $EI_F = 100\%$ and drops to $EI_F = 76\%$, giving an efficacy index sample mean in normalized balance of $\overline{EI}_F = 85\%$; this remains the same when ignoring the $\psi = 1$ and $\psi = 12$ results.

The third objective is also seen to decrease in performance with increases in skip size, as illustrated by Fig. 19.7. The hazard measure remains optimal through $\psi = 3$ but eventually drops to worst case at $\psi = 9$. The hazard efficacy index while varying ψ starts at an optimal $EI_H = 100\%$ and worsens to $EI_H = 0\%$ giving an efficacy index sample mean in hazard of $\overline{EI}_H = 50\%$, which remains unchanged when ignoring the $\psi = 1$ and $\psi = 12$ results.

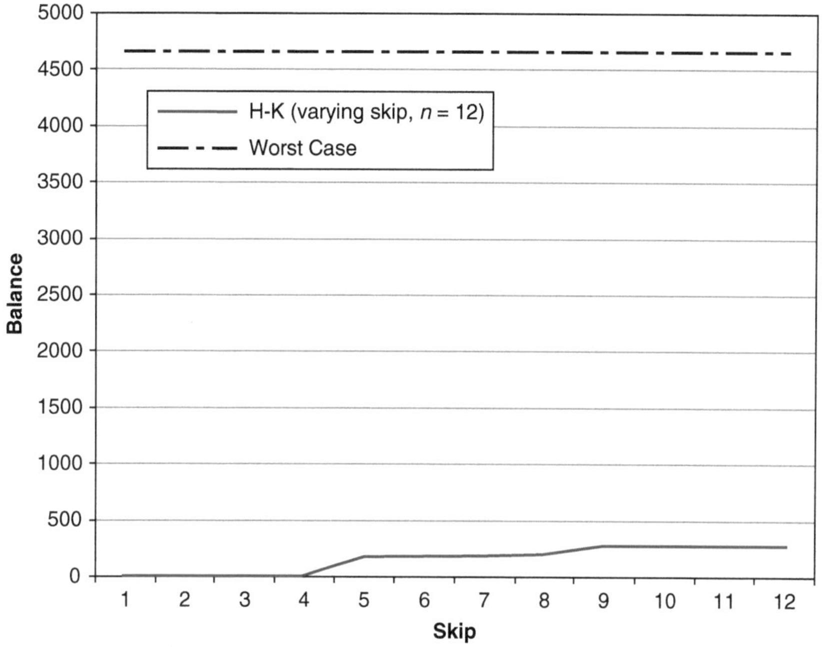

FIGURE 19.5 H-K F performance with changes in ψ.

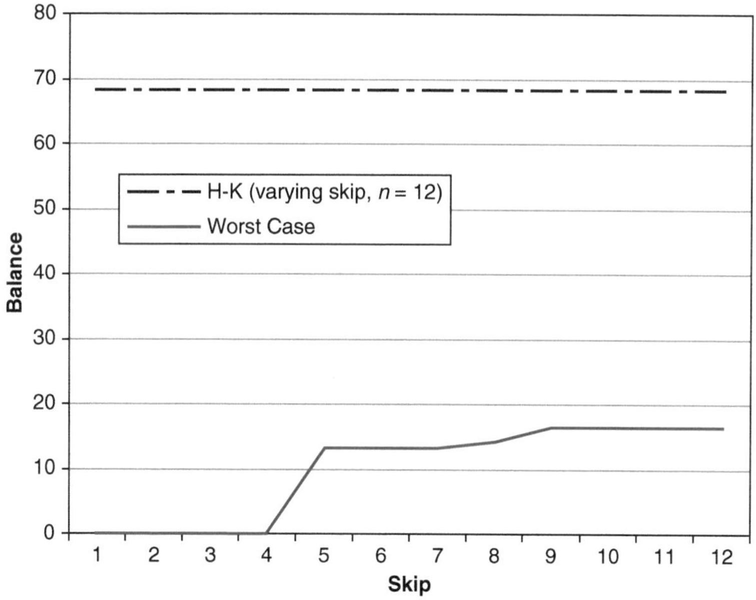

FIGURE 19.6 Normalized H-K F performance with changes in ψ.

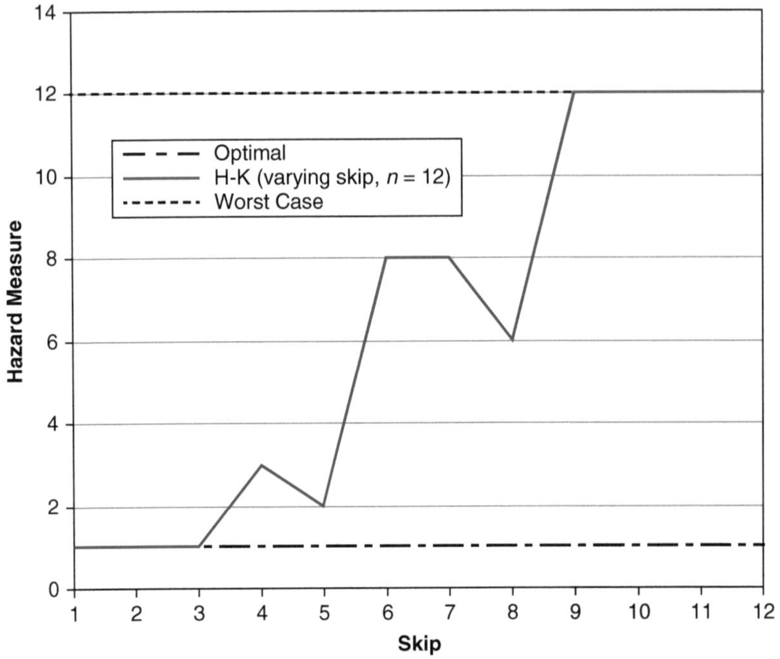

FIGURE 19.7 H-K H performance with changes in ψ.

The fourth and fifth objectives decrease in performance with increases in skip size as well, each at a slightly steeper angle (attributed to the prioritization of the objectives), as illustrated in Figs. 19.8 and 19.9. Optimal and better than optimal (i.e., removing the high-demand part first in the case of the hazardous part having not been removed first) D results are seen as far out as $\psi = 8$, while the optimal number of direction changes is only seen out to $\psi = 2$ (with the understanding that $\psi = 1$ is effectively exhaustive search). The demand efficacy index while varying ψ starts at an optimal $EI_D = 100\%$ then fluctuates between $EI_D = 20\%$ and $EI_D = 110\%$ (better than optimal, but at the expense of balance) giving an efficacy index sample mean in demand of $\overline{EI}_D = 58\%$. Ignoring the $\psi = 1$ and $\psi = 12$ results, the efficacy index sample mean is calculated to be $\overline{EI}_D = 56\%$.

The part removal direction measure reaches worst case by $\psi = 6$ but then improves slightly at $\psi = 9$. The part removal direction efficacy index while varying ψ starts at $\psi = 2$ at an optimal $EI_R = 100\%$, worsens to $EI_R = 0\%$, then climbs to $EI_R = 14\%$ giving an efficacy index sample mean in part removal direction of $\overline{EI}_R = 36\%$. Ignoring the $\psi = 1$ and $\psi = 12$ results, the efficacy index sample mean is calculated to be $\overline{EI}_R = 31\%$.

These studies show how the heuristic's performance decreases with skip size and provide at least a starting point for the selection of a skip

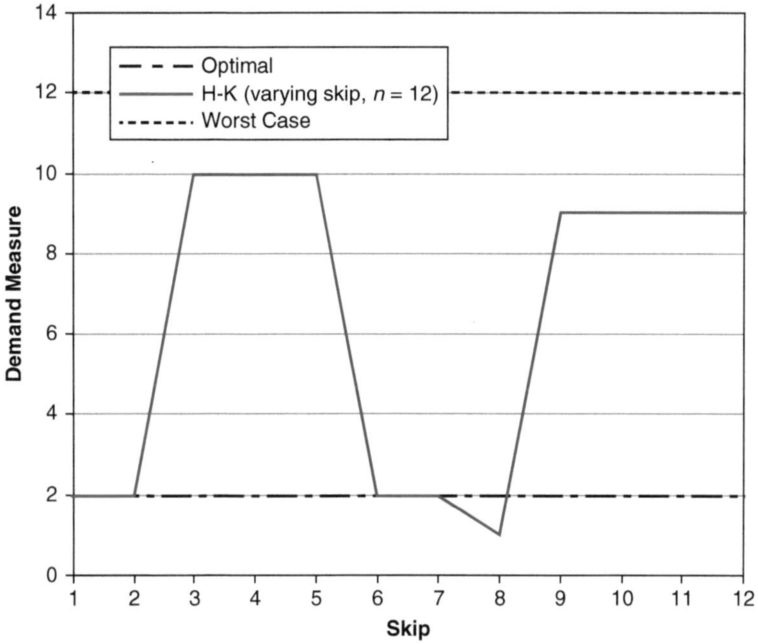

FIGURE 19.8 H-K *D* performance with changes in ψ.

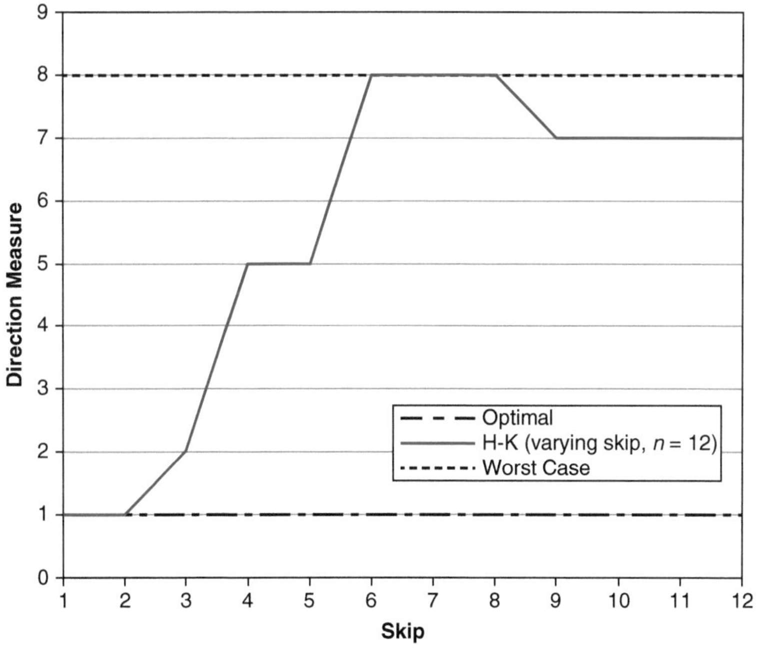

FIGURE 19.9 H-K *R* performance with changes in ψ.

size based upon the required level of performance as considered against time complexity. In large problems, time will quickly become the overriding consideration. These and other experiments indicate that a $\Delta\psi$ ranging from 10 to 12 (resulting in a ψ equal to 10 to 12 less than n) appears to give time complexity performance at par with many other heuristic techniques, regardless of hardware implementation.

19.4 Numerical Results

The H-K general-purpose heuristic has been run on each of the four experimental instances from Ch. 10 with the results analyzed in this section. Based upon the results in the previous section, all the instances are run on the forward data with $\Delta\psi$ varying from 1 to 10 (e.g., at $n = 25$, skip size would vary as $15 \leq \psi \leq 24$) with the best solution from these 10 searches kept. For small data sets the software has been set up so that it will not attempt any skip size smaller than $\psi = 3$ to avoid exhaustive or near-exhaustive searches (which would result in unrealistically large search times on small data sets). In addition, the DLBP A Priori data sets are run with the data in both forward and reverse.

19.4.1 Personal Computer Instance

DLBP H-K has been applied to the personal computer (PC) instance. The solution found is optimal in all measures except demand (Table 19.1). The H-K solution consists of the optimal four workstations and $F^* = 33$, $H^* = 7$, $D = 19{,}265$ ($D^* = 19{,}025$), and $R^* = 6$ being found and taking less than 1/100th of a second.

19.4.2 The 10-Part Instance

On the 10-part instance, DLBP H-K finds a near-optimal solution in less than 1/100th of a second. The solution shown in Table 19.2 is optimal in the number of workstations, balance, and hazard measures. It is suboptimal in demand but actually better than optimal in the part removal direction measure. With the optimal solution known to be five workstations and $F^* = 211$, $H^* = 4$, $D^* = 9730$, and $R^* = 7$,

Part ID	1	5	2	6	3	8	7	4
PRT	14	23	10	16	12	36	20	18
Workstation	1	1	2	2	2	3	4	4
Hazardous	0	0	0	0	0	0	1	0
Demand	360	540	500	750	620	720	295	480
Direction	1	2	0	4	1	4	1	1

TABLE 19.1 H-K Solution Using the Personal Computer Instance and $3 \leq \psi \leq 7$ (Forward Only)

Part ID	5	10	6	7	1	4	8	9	2	3
PRT	23	10	14	19	14	17	36	14	10	12
Workstation	1	1	2	2	3	3	4	5	5	5
Hazardous	0	0	0	1	0	0	0	0	0	0
Demand	0	0	750	295	0	0	0	360	500	0
Direction	5	3	5	2	2	2	1	5	0	0

Table 19.2 H-K Solution Using the 10-Part Instance and $3 \leq \psi \leq 9$ (Forward Only)

DLBP H-K's solution consists of $NWS^* = 5$, $F^* = 211$, $H^* = 4$, $D = 10{,}810$, and $R = 6$ (better than optimal, though at the expense of the demand measure).

19.4.3 Cellular Telephone Instance

The H-K solution to the cellular telephone instance is shown in Table 19.3. This solution results in $NWS = 11$, $F = 399$, $H = 82$, $D = 940$, and $R = 10$ and took 0.01 seconds on average.

19.4.4 DLBP A Priori Instances (Numerical Results for Varying *n*)

The DLBP H-K technique can be seen below (Table 19.4) on the DLBP A Priori data presented forward and reverse at $n = 12$ and $3 \leq \psi \leq 11$ (note that without the minimum allowed value of $\psi = 3$, skip values would include $2 \leq \psi \leq 11$). DLBP H-K is able to find a solution optimal in the number of workstations, balance, and hazard. The solution found came from the reverse set (both forward and reverse were used) and consisted of $NWS^* = 3$, $F^* = 0$, $H^* = 1$, $D = 10$ (optimal value is $D^* = 2$), and $R = 2$ (optimal value is $R^* = 1$). The time to complete the forward and reverse searches averaged 1.56 seconds.

Instance size is varied to demonstrate how solution performance changed with increases in n. All cases are the DLBP A Priori problem with sizes of $n = \{8, 12, 16, \ldots, 80\}$, $1 \leq \Delta\psi \leq 10$, and the resulting skip sizes of $n - 10 \leq \psi \leq n - 1$. On the full range of data (i.e., $n = \{8, 12, 16, \ldots, 80\}$), DLBP H-K finds solutions with NWS^* workstations up to $n = 12$, and then solutions with $NWS^* + 1$ workstations through data set 11 ($n = 48$), after which it stabilizes at $NWS^* + 2$ (Fig. 19.10). From Eq. (12.1) H-K's efficacy index in number of workstations starts at an optimal $EI_{NWS} = 100\%$, drops to a low of $EI_{NWS} = 92\%$, then continuously climbs through to $EI_{NWS} = 97\%$ with an efficacy index sample mean in number of workstations of $\overline{EI}_{NWS} = 96\%$.

Increases in the balance measure are seen with increases in data set size. A detailed view of balance performance with problem size can be seen in Fig. 19.11. The H-K curve approximates a step function similar to Greedy. However, H-K appears to differ from Greedy in

Part ID	1	2	3	4	5	6	7	8	9	10	11	12	13	14	15	16	17	18	19	20	21	22	23	24	25
PRT	3	2	3	10	10	15	15	15	15	2	2	2	2	2	2	2	2	3	18	5	1	5	15	2	2
Workstation	1	1	1	1	2	3	4	5	6	6	7	7	7	7	7	7	7	7	8	9	9	9	10	10	11
Hazardous	1	1	0	0	0	0	0	0	0	0	0	1	0	0	0	0	0	0	1	0	0	0	1	0	1
Demand	4	7	1	1	1	1	1	1	1	2	1	4	1	1	1	1	2	2	8	1	4	6	7	1	4
Direction	2	3	5	5	5	5	5	5	5	4	4	4	5	2	0	2	4	5	5	4	4	4	4	4	4

TABLE 19.3 H-K Solution Using the Cellular Telephone Instance and $15 \le \psi \le 24$ (Forward Only)

Part ID	12	2	5	8	11	1	4	7	10	9	6	3
PRT	11	3	5	7	11	3	5	7	11	7	5	3
Workstation	1	1	1	1	2	2	2	2	3	3	3	3
Hazardous	1	0	0	0	0	0	0	0	0	0	0	0
Demand	0	0	0	0	0	0	0	0	0	1	0	0
Direction	0	0	0	0	0	1	1	1	1	0	0	0

TABLE 19.4 H-K Solution Using the A Priori Instance at $n = 12$ and $3 \leq \psi \leq 11$ (Forward and Reverse)

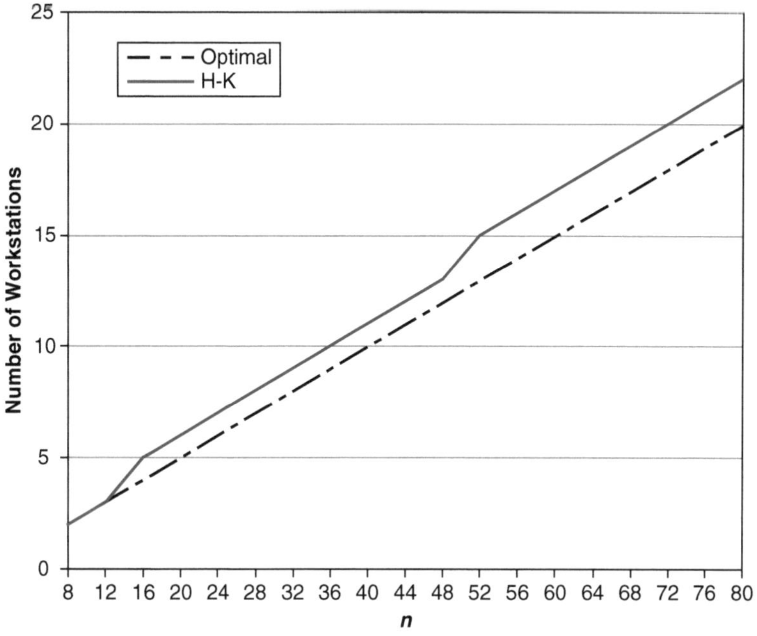

FIGURE 19.10 H-K workstation calculation.

the sense that good performance is actually seen more often when dealing with smaller instances (i.e., the step function of H-K is the reverse of that of Greedy).

While increases in the balance measure are seen with increases in data set size, as a percentage of the overall range from best case to worst case, the normalized balance measure tends to decrease (i.e., improve) with increases in the data set size (Fig. 19.12). The normalized balance efficacy index drops from a high of $EI_F = 100\%$ to a low of $EI_F = 85\%$ at $n = 16$ then slowly climbs to $EI_F = 92\%$ giving a sample mean of $\overline{EI_F} = 92\%$.

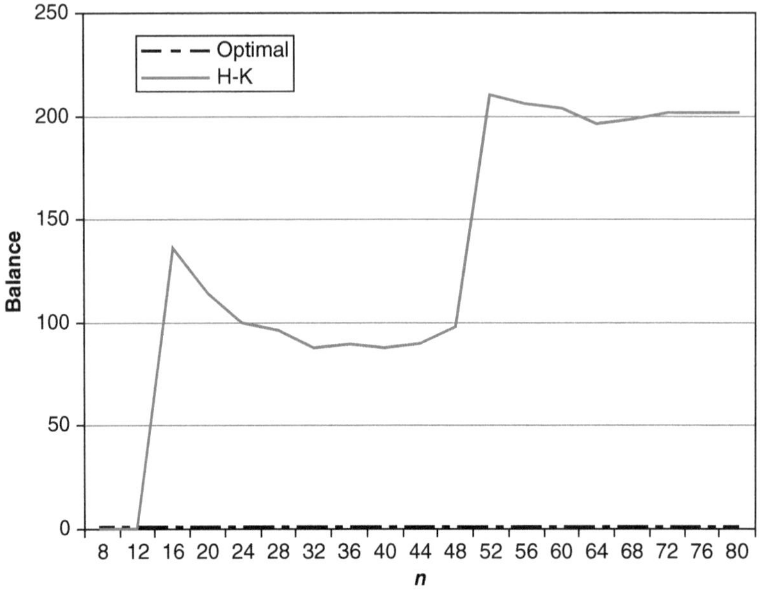

FIGURE 19.11 Detailed H-K balance measure.

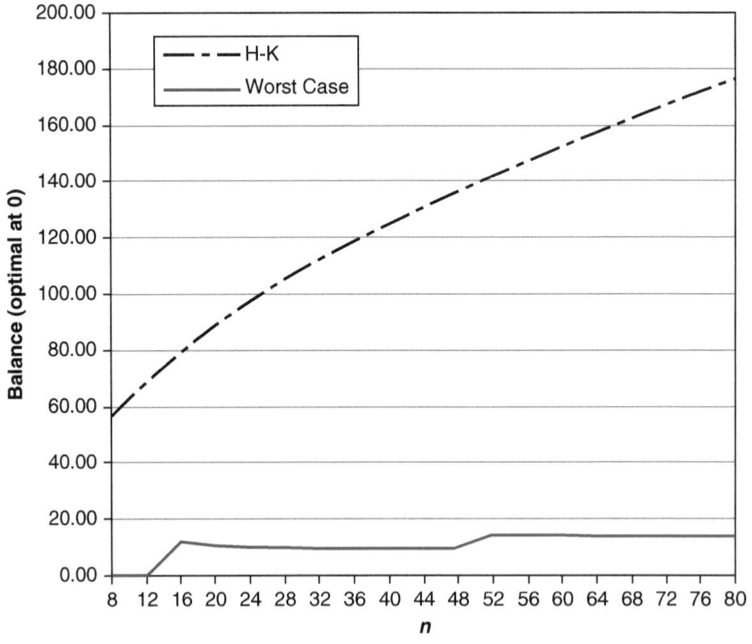

FIGURE 19.12 Normalized H-K balance measure.

The hazardous part is regularly suboptimally placed. Hazard part placement stays relatively consistent with problem size (though effectively improving as compared to the worst case, as illustrated by Fig. 19.13). These results are as expected since hazard performance is designed to be deferential to balance and affected only when a better hazard measure can be attained without adversely affecting balance. The hazard measure's efficacy index fluctuates, similar to a sawtooth wave function (or more accurately, a *triangle wave*), between $EI_H = 57\%$ and $EI_H = 100\%$, giving a sample mean of $\overline{EI}_H = 90\%$.

The high-demand part is also suboptimally placed, though at a higher rate than the hazardous part (Fig. 19.14). Its efficacy index fluctuates between $EI_D = 7\%$ and $EI_D = 103\%$ (due to better than optimal placement in position $k = 1$ at the expense of hazard placement). The resulting demand measure sample mean is $\overline{EI}_D = 49\%$.

With part removal direction structured as to be deferential to balance, hazard, and demand, it is seen to decrease in performance when compared to the best case and when compared to the worst case (Fig. 19.15). Again, these results are as expected due the prioritization of the multiple objectives. Though the part removal direction efficacy gets as high as $EI_R = 86\%$, by $n = 24$, it drops to $EI_R = 0\%$ and never rises higher again than $EI_R = 43\%$, resulting in a sample mean of $\overline{EI}_R = 20\%$.

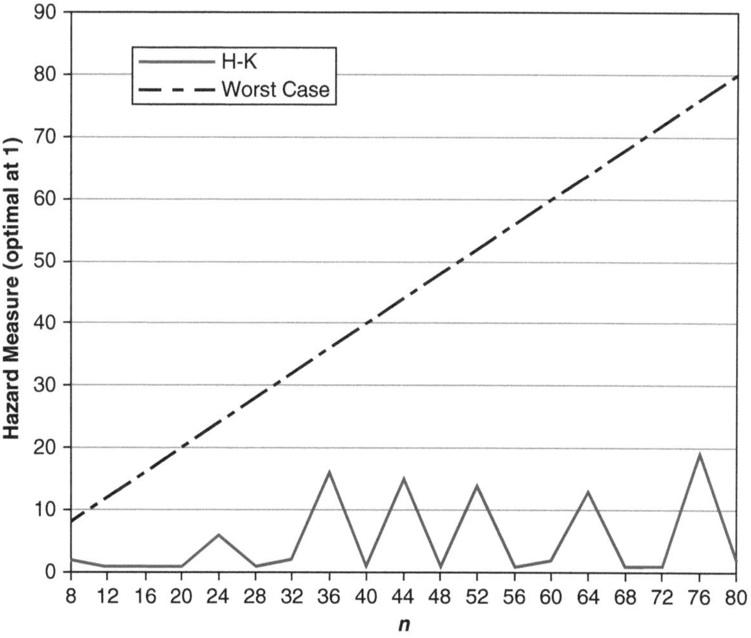

FIGURE 19.13 H-K hazard measure.

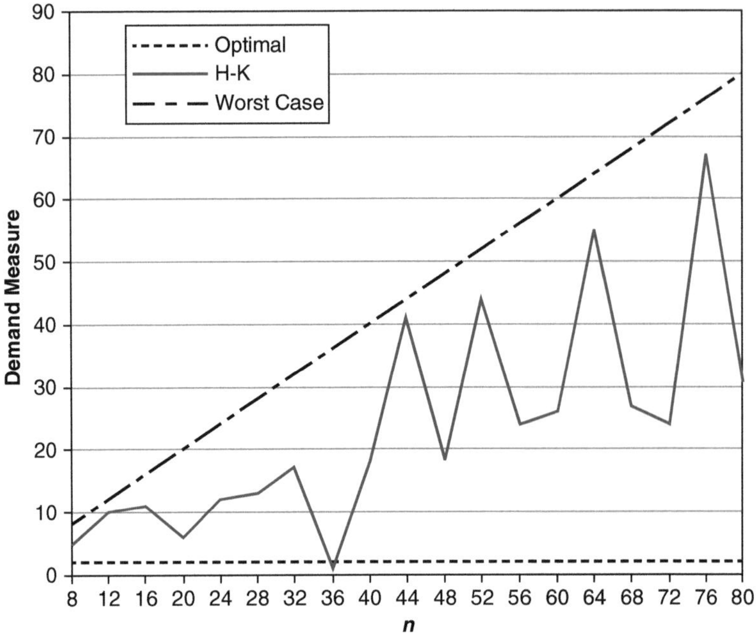

FIGURE 19.14 H-K demand measure.

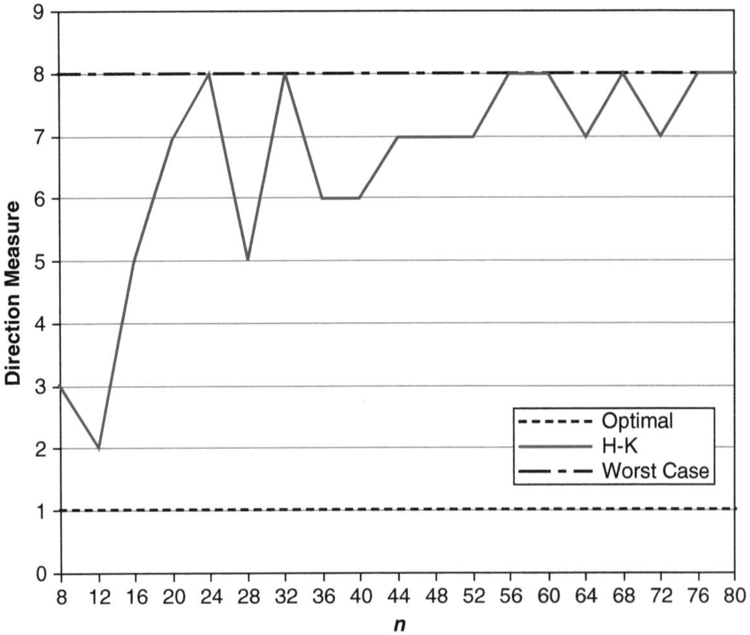

FIGURE 19.15 H-K part removal direction measure.

A smaller ψ or, as with other search techniques, the inclusion of precedence constraints will increasingly move the DLBP H-K method toward the optimal solution. As shown in Fig. 19.16, the time complexity performance of DLBP H-K provides the trade-off benefit with the technique's near-optimal performance, demonstrating the moderate increase in time required with problem size that grows markedly slower than the exponential growth of exhaustive search. Exhaustive and third-order curves are shown for comparison.

Based on Figs. 19.16 and 19.17, a second-order polynomial regression model is used to fit the H-K curve with $n = \{8, 12, 16, \ldots, 80\}$, $1 \leq \Delta\psi \leq 10$, and the resulting skip sizes of $n - 10 \leq \psi \leq n - 1$.

The regression equation is calculated to be $T(n) = 0.0033n^2 - 0.0002n + 0.2893$. The small n^2 coefficient is indicative of the slow runtime growth in instance size. The coefficient of determination is calculated to be 0.9974, indicating that 99.74 percent of the total variation is explained by the calculated linear regression curve (the anomaly seen in the H-K curve in Fig. 19.17 is due to the software rule that dictated that all ψ could be as small as $n - 10$, but no less than $\psi = 3$, to prevent exhaustive or near-exhaustive searches at small n). As seen in Fig. 19.18 this

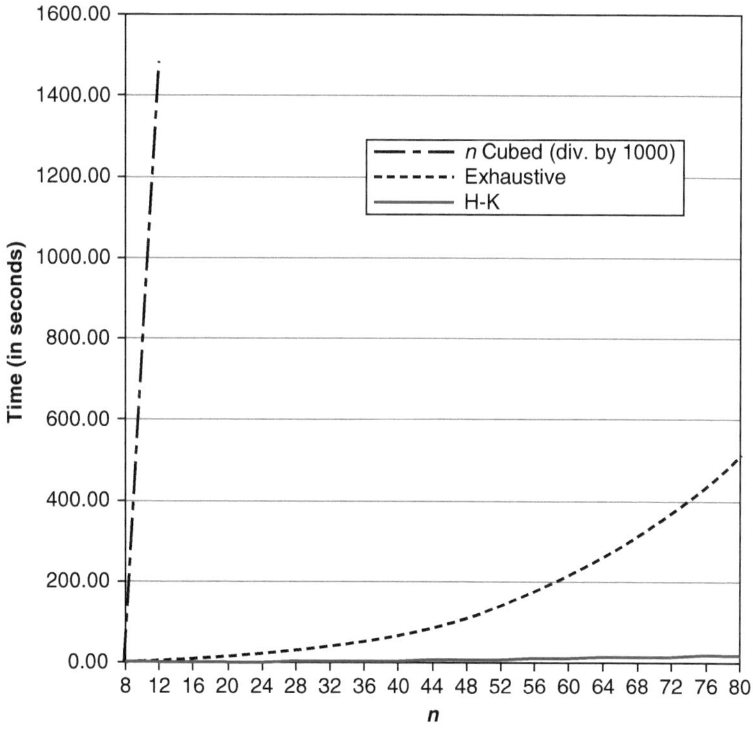

FIGURE 19.16 H-K time complexity compared to exhaustive search.

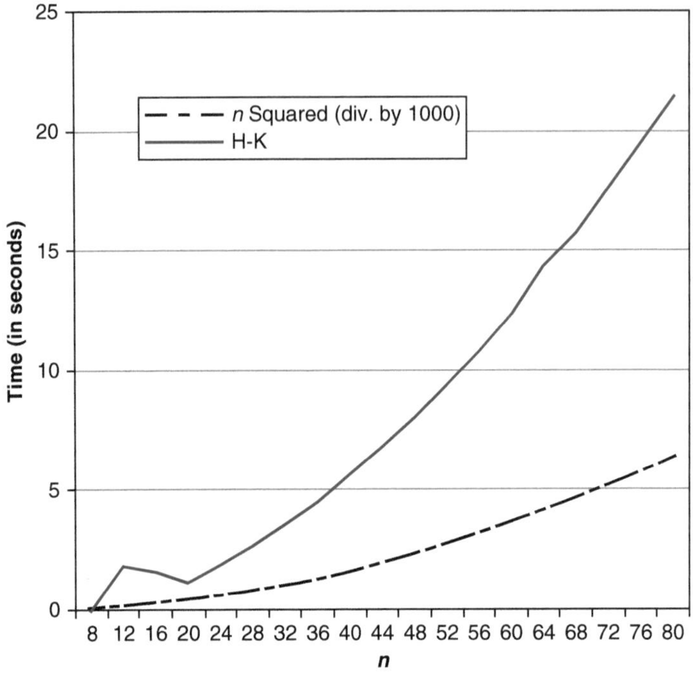

Figure 19.17 H-K time complexity compared to second order.

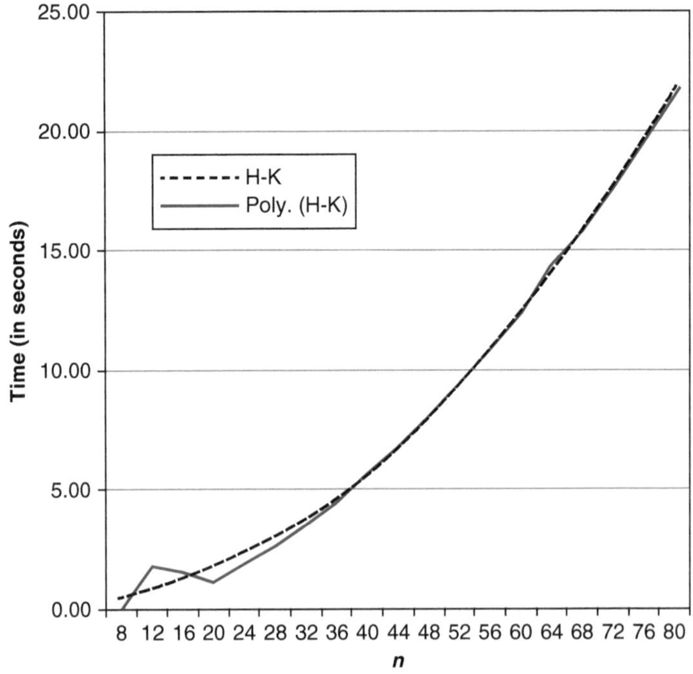

Figure 19.18 Detailed H-K time complexity and second-order polynomial regression line.

regression provides a very accurate fit. With a growth of $0.0033n^2 - 0.0002n + 0.2893$, the average-case time complexity of DLBP H-K curve (with forward and reverse data, $1 \leq \Delta\psi \leq 10$, and resulting skip sizes of $n - 10 \leq \psi \leq n - 1$) is listed as $O(n^2)$ or polynomial complexity. Per Sec. 19.3 the deterministic, single iteration nature of H-K also indicates that the process would be no faster than this so it is expected that the time complexity lower bound is $\Omega(n^2)$ and, therefore, the H-K appears to have an asymptotically tight bound of $\Theta(n^2)$ as configured here.

Average-case time complexity using the DLBP A Priori data then appears to be $O(b^b)$ in skip size (where $b = 1/\psi$), while H-K appears to grow at approximately $O(n^2)$ in instance size. Runtime performance can be improved by increasing skip size and by running the data in one direction only while the opposite (i.e., decreasing skip size and running different versions of the data, including running the data in forward and reverse order) would be expected to increase solution efficacy. Note that other techniques as discussed in Secs. 10.9.2 and 10.9.3 may decrease runtime and/or increase solution efficacy as well.

19.5 Conclusions

An uninformed deterministic search approach to combinatorial optimization problems and an application to the multiple-objective DLBP were presented in this chapter. The single-phase H-K general-purpose heuristic rapidly provides a feasible solution to the DLBP using a search technique that samples the entire solution space. The DLBP H-K heuristic provides a near-optimal minimum number of workstations, with the level of optimality increasing with the number of constraints. It generates a feasible sequence with an optimal or near-optimal measure of balance while maintaining or improving the hazardous materials measure, the demand measure, and the part removal direction measure. The H-K general-purpose heuristic appears well suited to the multicriteria decision-making problem format as well as for the solution of problems with nonlinear objectives. Of all the methodologies investigated in Part II, H-K provides the most natural application to distributed computing where the many networked computers would each work on a portion of the farmed-out search space and with each exponential increase in processing speed (attributed to some combination of increased CPU clock speed and/or increases in the number of computers in the grid), the skip size could be decremented resulting in improvements in solutions with no increase in perceived runtime. In addition, the H-K heuristic is especially applicable to generating diverse first-phase solutions for hybrids or hot-start solutions for local search methodologies.

CHAPTER 20

Quantitative and Qualitative Comparative Analysis

20.1 Introduction

In this chapter all of the combinatorial optimization methods are compared using the full range of the known-optimal solution data set (McGovern and Gupta, 2007b). They are compared to each other and to the known best-case and worst-case performance on the basis of the following: time complexity, total number of workstations, and measures of balance, hazard, demand, and part removal direction (McGovern and Gupta, 2007c). Each is then compared using the efficacy indices. The methods having probabilistic components [ant colony optimization (ACO) and genetic algorithm (GA)] have been run five times with the reported results being the mean of these runs while all other techniques—due to their deterministic nature—are run three times simply to provide mean runtime data.

This chapter provides a detailed side-by-side comparison of the various performance measures displayed by the six heuristic and metaheuristic methodologies (exhaustive search is not included in this comparison other than its use as a benchmark) when applied to the DISASSEMBLY LINE BALANCING PROBLEM (DLBP). Section 20.2 provides the quantitative and qualitative comparative analysis, demonstrating the results when the combinatorial optimization methodologies are applied to the instances from Ch. 10, while Sec. 20.3 summarizes the chapter.

20.2 Experimental Results

The four heuristics and two metaheuristics have been run on the DLBP A Priori experimental instances (with sizes of $n = \{8, 12, 16, \ldots, 80\}$) from Ch. 10 with the results comparatively analyzed in this chapter. Note that, due to the averaging of the results generated by the processes with stochastic characteristics, many of the reported results are not purely discrete.

For DLBP GA, all tests are performed using a population size of 20, mutation rates of 1.0 percent, 10,000 generations, and crossover rates of 60 percent.

For DLBP ACO the maximum number of cycles used is NC_{max} = 300, the weight of pheromone in path selection is $\alpha = 1.00$, the weight of balance in path selection is $\beta = 5.00$, the evaporation rate is $1 - \rho = 0.50$, the amount of pheromone added if a path is selected is $Q = 100.00$, and the initial amount of pheromone on all of the paths is $c = 1.00$.

For DLBP H-K, all cases are calculated using a varying delta skip of $1 \leq \Delta\psi \leq 10$ with the resulting skip sizes of $n - 10 \leq \psi \leq n - 1$. The software was configured so that it would not attempt any skip size smaller than $\psi = 3$ (to avoid exhaustive or near-exhaustive searches with small instances). In addition, the DLBP A Priori data sets were run with the data in forward and reverse order.

The first study performed is a measure of each of the technique's calculated number of workstations as compared to the optimal as given by Eq. (11.2). As shown in Fig. 20.1, all of the methods perform very well in workstation calculation, staying within two workstations of optimum for all data set sizes tested. Efficacy index sample means range from a low of $\overline{EI}_{NWS} = 95\%$ (DLBP ACO and DLBP Greedy) to a high of $\overline{EI}_{NWS} = 97\%$ (DLBP GA).

In terms of the calculated measure of balance, again, all of the methods are seen to perform very well and significantly better than the worst case (best case is found uniformly at $F = 0$, as illustrated in Fig. 20.2). However, examining the balance in greater detail provides some insight into the different techniques.

As seen in Fig. 20.2, DLBP ACO performs very well but—as seen later in the time complexity study—at the expense of runtime and extensive software modifications. Specifically, due to the nature of the DLBP (i.e., with precedence constraints, a part's placement in a solution sequence cannot be evaluated until what has come before it is known, and hence, how much time remains in the workstation under consideration) Eq. (15.1) is made to be dynamic; that is, Eq. (15.1) for the DLBP is calculated at each increment in time rather than just once at the start of each cycle. Although slightly more time consuming, this does not increase the overall time complexity of $O(n^3)$ and it results in excellent solution results. Also, DLBP ACO requires additional time since twice as many edges are evaluated than would be in a

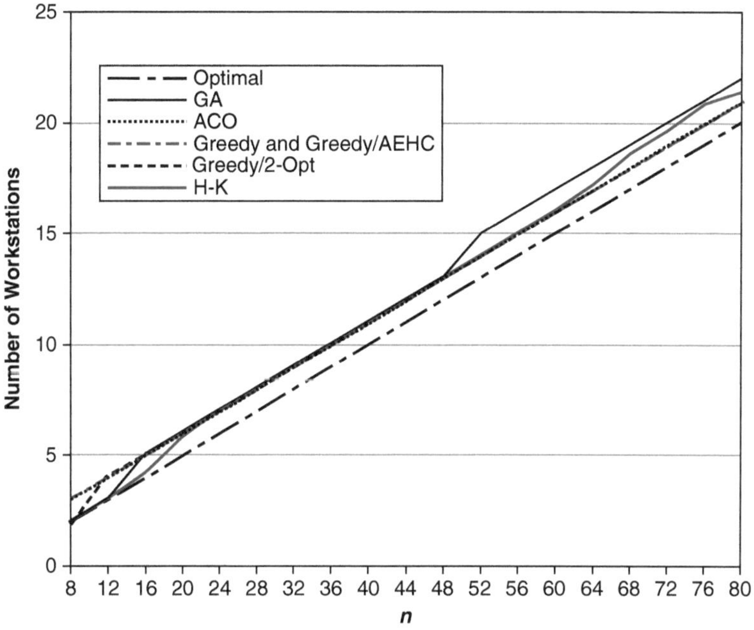

FIGURE 20.1 Workstation calculations for each DLBP combinatorial optimization method.

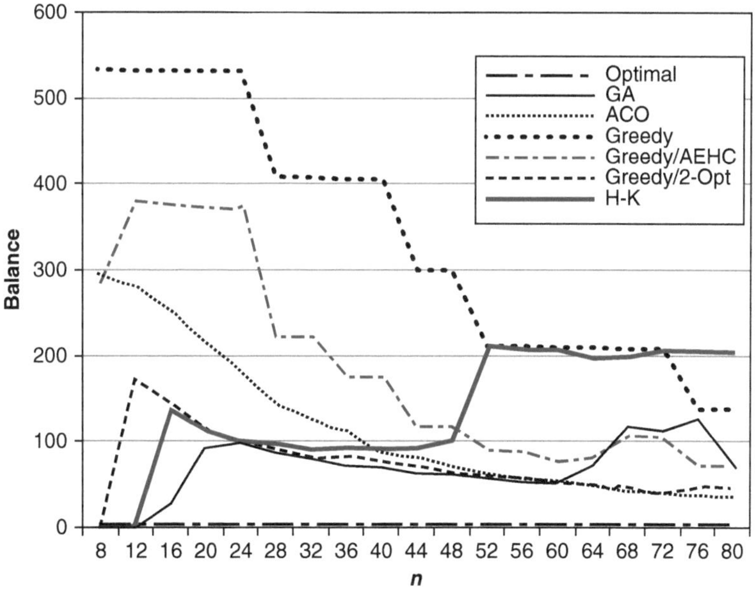

FIGURE 20.2 Detailed DLBP combinatorial optimization methods' balance performance.

general-purpose ACO since all edges in DLBP are directed edges with p, q not necessarily equivalent to q, p.

It can also be seen that the performance of the DLBP versions of ACO and the Greedy/AEHC hybrid improves with instance size, while that of DLBP H-K and GA tends to decrease and do so in a similar, step-function fashion (note, however, that even their normalized balance performance actually improves overall with instance size as a percentage of worst case as is indicated by their improved efficacy indices with instance size, as illustrated in Fig. 20.3). This is to be expected with H-K and GA. For H-K this has to do with the fact that as the instance size grows, a lower and lower percentage of the search space is investigated (assuming the skip size range is not allowed to increase with instance size, which is the case in this book). For GA, efficacy falls off with instance size because, unlike hill-climbing and k-optimal processes that continue to search until no better nearby solution can be found, GA is set up to terminate after a fixed number of generations (in addition, here DLBP GA's population does not increase with increases in instance size).

Normalized efficacy index sample means range from a low of $\overline{EI}_F = 83\%$ (Greedy) to a high of $\overline{EI}_F = 94\%$ (DLBP GA).

All of the processes perform well in hazardous part placement but none as much so as the three Greedy processes (Fig. 20.4). The

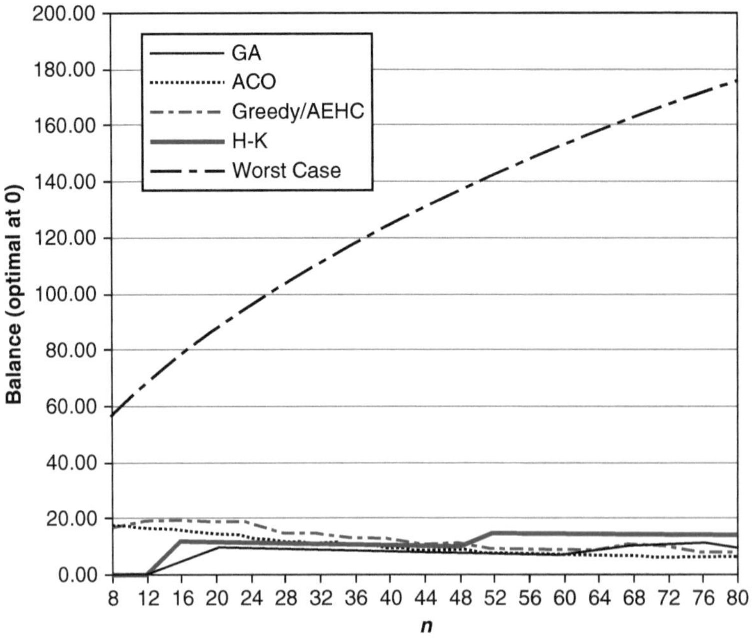

FIGURE 20.3 Normalized DLBP combinatorial optimization methods' balance performance.

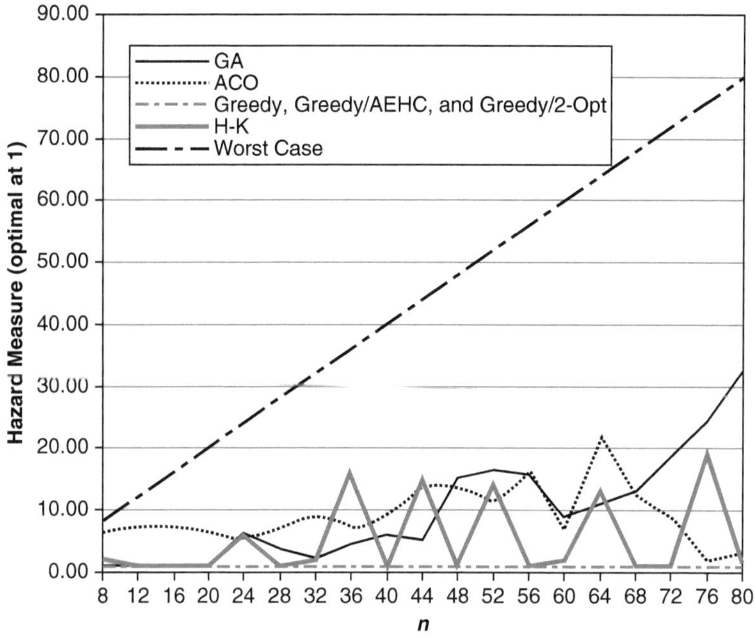

FIGURE 20.4 DLBP combinatorial optimization methods' hazard performance.

hazardous part was regularly optimally placed by these methodologies as a result of the sorting process that placed the hazardous parts at the front of the sorted list. Other techniques that did not possess this preprocessing of the data regularly, suboptimally placed the lone hazardous part. These results are expected since hazard performance is designed to be deferential to balance and affected only when a better hazard measure can be attained without adversely affecting balance. Efficacy index sample means range from a low of $\overline{EI}_H$ = 74% (DLBP ACO) to a high of $\overline{EI}_H$ = 100% (Greedy, Greedy/AEHC, Greedy/2-Opt).

All the processes also suboptimally place the high-demand part, though at a higher rate than the hazardous part. As seen in Fig. 20.5, Greedy appears to be the most predictable of the six methodologies, while DLBP H-K seems to be the least. Efficacy index sample means range from a low of $\overline{EI}_D$ = 49% (DLBP H-K) to a high of $\overline{EI}_D$ = 78% (DLBP GA).

With part removal direction structured as to be deferential to balance, hazard, and demand, all of the methodologies are seen to decrease in performance, and do so in a haphazard fashion. This decrease in performance is seen both when compared to the best case and when compared to the worst case (Fig. 20.6). Again, these results are as expected due the prioritization of the multiple objectives.

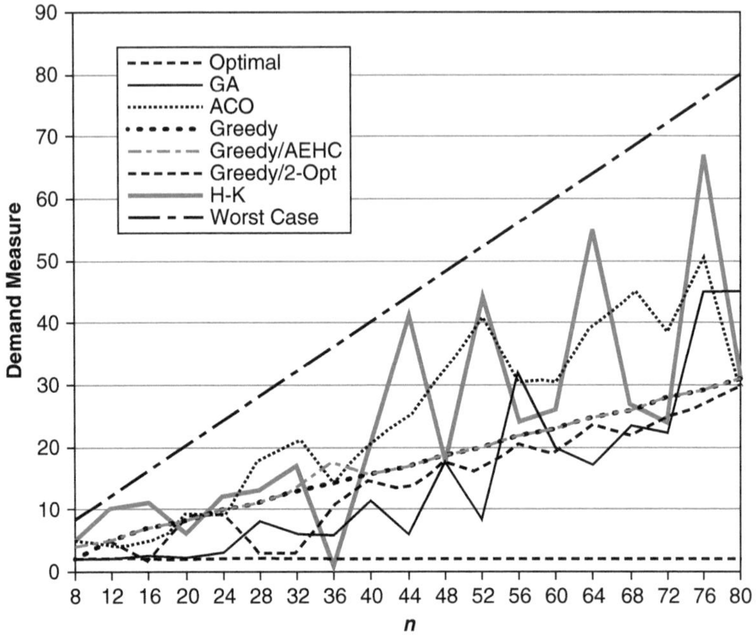

Figure 20.5 DLBP combinatorial optimization methods' demand performance.

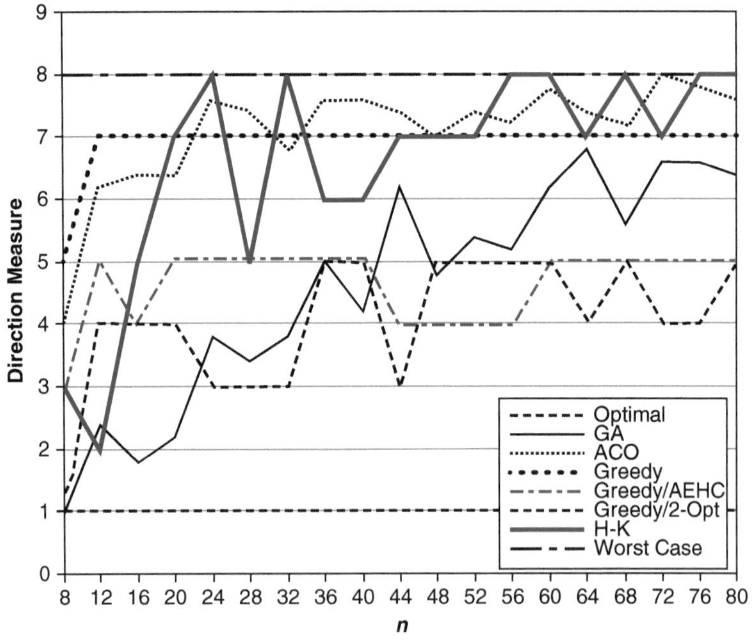

Figure 20.6 DLBP combinatorial optimization methods' part removal direction performance.

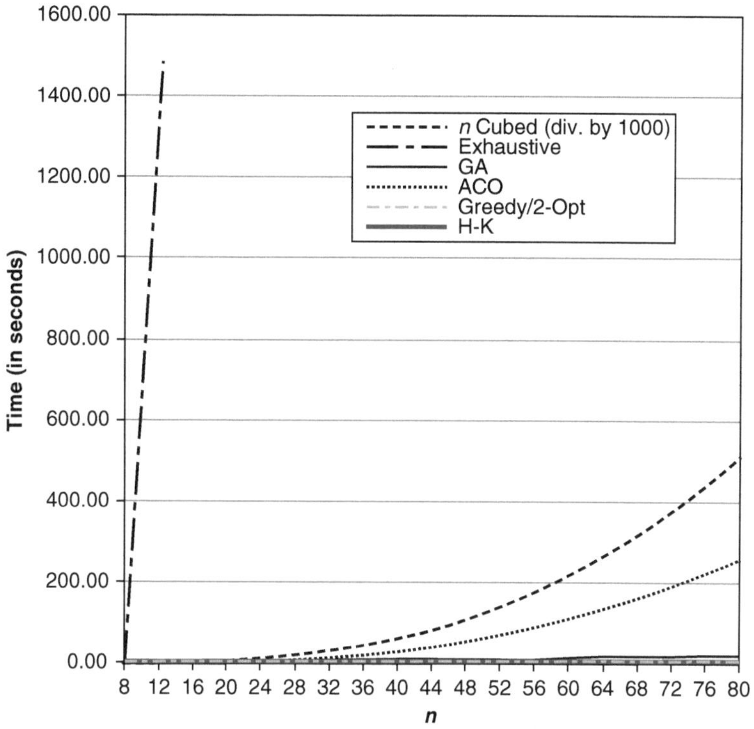

FIGURE 20.7 Time complexity of DLBP combinatorial optimization methods.

Efficacy index sample means range from a low of $\overline{EI}_R = 13\%$ (DLBP ACO) to a high of $\overline{EI}_R = 56\%$ (Greedy/2-Opt).

Finally, time complexity is examined using the A Priori data. All of the techniques are seen to be very fast (Fig. 20.7). Each is significantly faster than exhaustive search and each is seen to be as fast as or faster than third order.

The runtimes can be examined in greater detail in Fig. 20.8. Specifically designed for the DLBP, the Greedy and Greedy hybrids proved to be, as one may expect, the fastest techniques examined on the A Priori instances of $n = \{8, 12, 16, \ldots, 80\}$; however, DLBP GA was seen to be only slightly slower. Note that due to the time complexity, at some point none of these techniques can be faster than DLBP GA due to its linear growth. However, by that point DLBP GA's population or number of generations may need to be adjusted to allow the generation of adequate solutions and this will, of course, increase its runtime. DLBP H-K is also very fast, even with $\Delta\psi$ varying from 1 to 10 and with forward and reverse data runs. The H-K process grows approximately exponentially in $1/\psi$, taking, for example, from two 1/100ths of a second at $\psi = 5$ (actually $5 \leq \psi \leq 11$ and 0.02 seconds) up to just under 20 seconds at $\psi = 2$ (i.e., $2 \leq \psi \leq 11$ and 19.34 seconds)

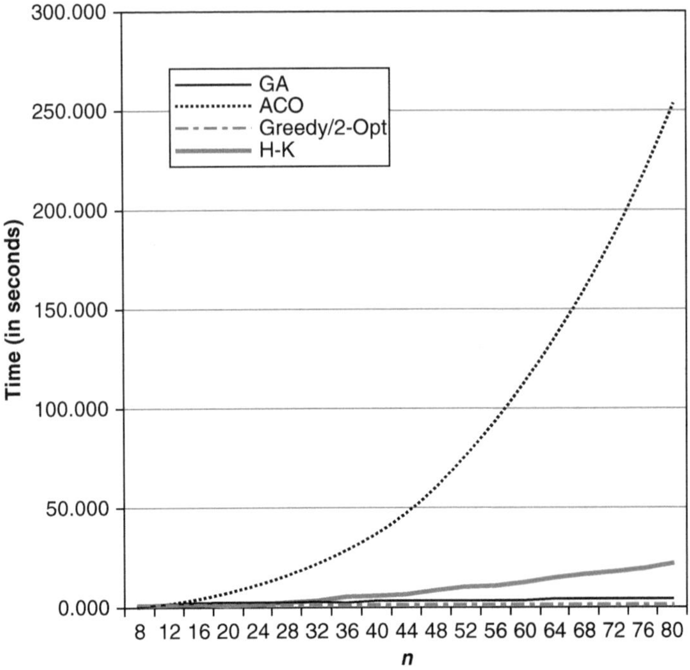

FIGURE 20.8 Detailed time complexity of DLBP combinatorial optimization methods.

with $n = 12$ and forward and reverse data, while exhaustive search was seen to take almost 25 minutes (24.61 minutes) on the same size data. The slowest is seen to be DLBP ACO but this can be partially attributed to the dynamic probability calculations, which is more a consequence of the challenges posed by DLBP than any concerns with the ACO methodology.

As previously discussed, these suboptimal results are not indicative of poor heuristic performance but are, more precisely, indicative of a successful DLBP A Priori design. Suboptimal solutions typically result when this specially designed set of instances is processed, even at seemingly small n values. For each of these methodologies, their DLBP-specific engineering element of searching exclusively for precedence-preserving solution n-tuples means that the inclusion or addition of precedence constraints will reduce the search space and increasingly move any of these methods toward the optimal solution. Larger n appears to increasingly move the DLBP GA/AEHC heuristic hybrid and DLBP ACO toward the optimal solution, while a smaller ψ or multiple runs with randomized (or other formatted) data sets will increasingly move the DLBP H-K metaheuristic toward the optimal solution.

Time complexity improvements can be gained in DLBP GA with smaller generations or smaller populations, in DLBP ACO with a smaller number of cycles or probability calculations that are not dynamic, and in DLBP H-K with increases in ψ. Each of these can be expected to be at the cost of solution performance. Due to its problem-specific design, there are no obvious ways to speed up the DLBP Greedy algorithm, the DLBP Greedy algorithm/AEHC heuristic hybrid, or the DLBP Greedy algorithm/2-Opt heuristic hybrid; however, a gain can be easily realized through replacing its notoriously slow sort routine (bubble sort) to a faster algorithm such as quick sort or merge sort.

The calculated asymptotic bounds of each of the seven methodologies are compiled into one table as seen in Table 20.1. The table contains each of the combinatorial optimization methodologies demonstrated here and lists their calculated theoretical best-case, worst-case, and (for those they exist for) tight bounds in time complexity. Shading indicates values for which precise calculations are not available; the dark shading indicates a minimum worst-case value (the actual worst case is expected to be larger), while the light shading indicates values based on experimental average-case results using the DLBP A Priori data.

Each of the collected quantitative values is then compiled into one table for reference and comparison (Table 20.2). This table contains each of the combinatorial optimization methodologies demonstrated here and lists their number of workstations, balance, hazard, demand, and part removal direction sample mean efficacy indices; regression model average-case experimentally determined time complexity; and associated asymptotic upper bound (experimentally determined average case using the DLBP A Priori data).

The shading provides a quick reference to performance, with darker shades indicating worsening performance. While Exhaustive Search is included in the table, none of its row elements are considered

	Big-omega	Big-oh	Big-theta
Exhaustive	$\Omega(n!)$	$O(n!)$	$\Theta(n!)$
GA	$\Omega(n)$	$O(n)$	$\Theta(n)$
ACO	$\Omega(n^2)$	$O(n^3)$	None
Greedy	$\Omega(n^2)$	$O(n^2)$	$\Theta(n^2)$
Greedy/AEHC	$\Omega(n^2)$	$O(n^2)$	Unlikely
Greedy/2-Opt	$\Omega(n^2)$	$O(n^2)$	Unlikely
H-K	$\Omega(n^2)$	$O(n^2)$	$\Theta(n^2)$

TABLE 20-1 Summary of Each Methodology's Asymptotic Notation

	$\overline{EI}_{NWS}$	$\overline{EI}_{F(norm)}$	$\overline{EI}_H$	$\overline{EI}_D$	$\overline{EI}_R$	$T(n)$	Big-oh
Exhaustive	100%	100%	100%	100%	100%	$1.199n!$	$O(n!)$
GA	97%	94%	84%	78%	49%	$0.033n$	$O(n)$
ACO	95%	90%	74%	51%	13%	$0.001n^3$	$O(n^3)$
Greedy	95%	83%	100%	64%	16%	$7 \times 10^{-8}n^2$	$O(n^2)$
Greedy/AEHC	95%	87%	100%	65%	48%	$3 \times 10^{-7}n^2$	$O(n^2)$
Greedy/2-Opt	96%	93%	100%	74%	56%	$5 \times 10^{-5}n^2$	$O(n^2)$
H-K	96%	92%	90%	49%	20%	$0.003n^2$	$O(n^2)$

TABLE 20-2 Summary of Quantitative Measures for all Methodologies

for shading since its purpose in the table is as the benchmark for comparison. Note that the balance measure sample mean efficacy index is based on the normalized balance. This is done in the interest of providing a more appropriate and consistent scale across the table.

The regression model column only contains the highest-order base in the polynomial (since the leading term of a polynomial determines its order) and its coefficient. This is performed in the interest of space since these are the portions of the regression model that result in the fastest runtime growth with instance size. The coefficients of determination for all of the methodologies are all very high indicating a high confidence in the accuracy of the regression curves. The coefficients of determination expressed as a percentage are calculated to be 99.2 percent for DLBP GA, 100.0 percent for DLBP ACO, 95.7 percent for Greedy, 99.1 percent for Greedy/AEHC, 92.0 percent for Greedy/2-Opt, and 99.7 percent for DLBP H-K.

Also, the "Big-oh" notation column is not considered for shading since it bears a direct correlation to the information in the regression model column, but with less resolution. In addition, each of the regression models and some of the big-oh data are understood to be average-case results when using the DLBP A Priori data sets and not necessarily actual average case, worst case, or best case and not necessarily for all instances.

Finally, the information in this table should not be taken alone and should only be used as an accompaniment to the qualitative graphs, remarks, and analyses in this chapter and in each of the chapters describing and evaluating the seven combinatorial optimization methodologies.

20.3 Conclusions

The results of seven combinatorial optimization solution methods applied to the DISASSEMBLY LINE BALANCING PROBLEM were quantitatively and qualitatively compared in this chapter. This was performed using graphical comparisons of efficacy measures with instance size, as well as through the use of efficacy indices applied to the same measures and using Exhaustive Search as a time and performance benchmark. All of the methodologies perform very well—though consistently suboptimally—with different techniques demonstrating a variety of performance subtleties that show no one technique to be ideal in all applications. While it should be noted that these measures are solely dependent on the A Priori data set instances used, this data set can be seen to have posed—even with its multiple optimum extreme points—a nontrivial challenge to all of the methodologies and in all instance sizes. This would seem to indicate that the data set meets its design expectations and does not contain any unforeseen nuances that would render it a weak benchmark for the subject problem of this text, the DISASSEMBLY LINE BALANCING PROBLEM.

CHAPTER 21

Other Disassembly-Line Balancing Research

21.1 Overview

This chapter concludes Part II and the discussion of the DISASSEMBLY LINE BALANCING PROBLEM (DLBP) by presenting other directly related areas of study. The chapter is organized as follows: Section 21.2 presents an overview of extensions to the DLBP. Section 21.3 reviews the wide range of potential solution methodologies that can be applied. The next two sections provide a background on probabilistic considerations and efforts. Specifically, Sec. 21.4 discusses the variety of approaches that have been taken to-date to solve stochastic assembly-line problems and Sec. 21.5 focuses on the potential for generation of stochastic benchmark data sets and case studies. Suggestions for future research are listed in Sec. 21.6.

21.2 Problem Extensions

Some appropriate extensions include designing and adding additional prototypes (Chap. 8), changing the multicriteria ordering of priorities (Chap. 12), evaluation under the condition of incomplete or partial disassembly, measuring the level of unbalance (Chap. 29), deleting some evaluation criteria (this is especially applicable to the case where the data set would be used in measuring the unbalance resulting from a heuristic applied to an assembly-line study rather than a disassembly line), or model evaluation using other solution methodologies.

Though the disassembly line is most suitable for automated disassembly (Güngör and Gupta, 2001a), disassembly operations can also be performed at a single workstation or in a disassembly cell. While less efficient—since they are unable to achieve the high productivity rates provided by disassembly lines—they are generally considered to be more flexible and therefore, potentially worthy of further study.

Some researchers use mathematical programming techniques to solve the DLBP. While not addressing the line-balancing aspect of the DLBP, Altekín et al. (2008) provide a mixed integer programming formulation that allows for partial disassembly. That model simultaneously determines the feasible disassembly sequence, the number of workstations, and the cycle time. Disassembly-line balancing in real time (Duta et al., 2008a) has been modeled using mixed integer quadratic programming and solved with a branch-and-cut-algorithm-based method. Balancing the disassembly line in the presence of task failures (Güngör and Gupta, 2001a) has been considered with a balancing algorithm that assigns tasks to workstations such that the effect of the defective parts on the disassembly line is minimized. Existing concepts in assembly-line balancing have been modified to demonstrate the applicability of some important factors in balancing a paced disassembly line (Güngör and Gupta, 2002). Two-phase Petri nets and simulation have been shown to maximize disassembly system throughput and revenue (Tang and Zhou, 2006; Tang et al., 2006). This was accomplished by dynamically configuring the disassembly system into multiple disassembly lines while considering line balance, different process flows, and meeting different order duedates.

21.3 Additional Solution Methodologies

While the solution methodologies demonstrated in this book provide a range of breadth and depth, many other options exist. These include the application of accepted assembly-line balancing rules and algorithms (including extensions of these for use in disassembly), as well as the use of other combinatorial optimization heuristics and metaheuristics.

Related assembly-line balancing heuristics include the Kilbridge-Wester heuristic, the Moodie-Young method, Helgeson-Birnie method (also known as the positional-weight technique), immediate-update first-fit heuristic, and the rank-and-assign heuristic. Elsayed and Boucher (1994) thoroughly detail each of these line-balancing algorithms and provide relevant examples. While all of these are not directly applicable to all aspects of the DLBP, they each have potential application to disassembly-line balancing. These algorithms are directly applicable in the case of pure, basic, complete disassembly; for example, in cases where: parts are not demanded, all end-of-life products are the same and in serviceable condition, hazardous parts are not present or are not a consideration, line efficiency in terms of part removal directions is not of interest, and so on. They are also valuable as starting points for both understanding a related heuristic and a basis for future DLBP heuristics.

Also, many other general-purpose heuristics, metaheuristics, and search algorithms exist and may be applied to the DLBP. These include: best-first search, A* search algorithm, artificial bee colony algorithm, artificial immune system, culture algorithm, differential evolution algorithm, ejection chain algorithms, extremal optimization, firefly algorithm, foraging algorithms, glowworm swarm optimization, greedy randomized adaptive search procedure, harmony search, honeybee algorithm, hyper-heuristics, local search, memetic algorithm, particle swarm optimization, path re-linking, random optimization class of algorithms, random restart hill climbing, reactive search optimization, scatter search, simulated annealing, stochastic diffusion search, strategic oscillation, tabu search, and variable neighborhood search, among others. It should be noted that, among various criticisms of metaheuristics, the "no-free-lunch theorem" says that each optimization algorithm will perform, on average, no better than any other over the set of all possible mathematical problems.

21.4 Probabilistic Disassembly-Line Balancing

Probabilistic assembly-line balancing provides a foundation for some of the issues faced, models used, and solution methods considered for application to disassembly-line balancing. Probabilistic assembly lines traditionally consider task times to be the stochastic element. In these models, the task times are described by random variables that may or may not follow a probability distribution.

Elsayed and Boucher (1994) review two models of stochastic assembly lines where (1) task times follow a Gaussian distribution with known means and standard deviations (referred to as the "known distribution" case), and (2) task time distributions are unknown, though their means and standard deviations are (referred to as the "distribution free" case).

The first model is a modification of the Moodie-Young method. The first phase of Moodie-Young is basically a first-fit-decreasing (FFD) algorithm and is referred to as the *largest candidate rule* (the second of the two phases of the deterministic version then attempts to distribute the idle times equally between workstations). This rule allocates tasks to a workstation sequentially from largest to smallest times (and accounting for precedence). In the known-distribution case, an acceptable probability of exceeding the cycle time is first determined. Tasks available for assignment to a workstation are then sequentially evaluated based on each task's probability (calculated using Gaussian tables) that the station time will not exceed the cycle time if it is added to the workstation. If the considered task's calculated probability is less than a predetermined threshold, it is added. If no task meets this criterion, a new workstation is created. This continues until all tasks are assigned to a workstation.

The second model is a modification of the first. The distribution-free case is structured the same as the known-distribution case; however, Chebyshev's inequalities—rather than Gaussian tables—are used for estimating the probability that a station time will exceed the cycle time.

Queueing theory is also commonly used to model and to provide analysis of stochastic assembly lines. This field of study is equally applicable to disassembly lines. (Chap. 29 provides an overview of queueing theory.)

Stochastic DLBP research efforts also include the use of a collaborative ant colony optimization algorithm (Agrawal and Tiwari, 2006) for balancing a stochastic, mixed-model, U-shaped disassembly line.

21.5 Probabilistic Disassembly-Line Data

While the A Priori benchmark data set has not been designed with stochastic part removal times in mind or for use with an unpaced (e.g., job shop) scheduling model, these items can be addressed in several ways.

It is recognized that the work-sampling operations required to define the standard deviation of stochastic task times on an assembly line can be excessively time consuming (Tiacci et al., 2003). Tiacci et al. studied the significance of the task times' standard deviation on the performance of the system model in order to determine if it is worth collecting enough data to calculate the standard deviation. Ultimately, they proposed a methodology for performing assembly-line balancing with a reduced set of data by using the means of the stochastic task times as the sole inputs (i.e., without their standard deviations).

The A Priori benchmark (Chap. 10) could also be used in this fashion by setting $E[x] = \mu_x = x$ and $\sigma_x^2 = 0 \ \forall x \in \{PRT_k\}$. This would be applicable in other cases, including those where use of probabilistic data is performed purely deterministically (e.g., by assigning the deterministic average-case or worst-case part removal time to the problem) or effectively deterministically (e.g., calculate the lowest probability of any candidate task exceeding the cycle time based upon each candidate's means and variances then assign the task with the best discrete mean value; Elsayed and Boucher, 1994).

Alternatively, the stochastic part removal time could be represented deterministically as being within some multiple of σ; for example, if a part's removal time were given by a normal distribution, this time could be represented deterministically as being within 1σ, 2σ, 3σ (68.27, 95.45, or 99.73 percent, respectively) or some other tolerance of interest.

21.6 Future Research Directions

While much has been accomplished in the DLBP area, a great deal still remains. Many concepts for future research efforts have been suggested throughout Part II and some additional ones are described here.

Though the objectives used in Part II allowed for defining the DLBP and provided different methodologies with the ability to solve the DLBP quantitatively, these newly-defined criteria and their associated formulae are not all encompassing; other evaluation criteria, including tool-type required to remove a part, could be researched and defined.

It may be of interest to vary the multicriteria ordering of the objectives; two possibilities include a reordering of the objectives based on expert domain knowledge, and a comparison of all permutations of the objectives in search of patterns or unexpected performance improvements or decreases.

While the multicriteria decision-making approach used here made use of preemptive goal programming, many other methodologies are available and should be considered to decrease processing time, for an overall or selective efficacy improvement, or to examine other promising methods including weighting schemes.

Per Sec. 21.3, other types of combinatorial optimization methodologies might show interesting results when applied to the DLBP; techniques such as tabu search and particle swarm optimization could be used on the included DLBP instances.

The four-problem instances provide a variety of data sets differing in size, number, and type of precedence constraints; product application; and intended use; however, consideration should be given to developing other DLBP-specific data sets including those from actual problems as well as specially designed experimental benchmarks.

While this book made use of a variety of heuristics and metaheuristics, subsequent research could consider optimal solution-generating methodologies including integer programming, branch-and-bound, and dynamic programming.

All of the part removal times in Part II of this book were deterministic; although use of probabilistic data is often performed purely deterministically (e.g., by assigning the deterministic average-case or worst-case part removal time to the problem) or effectively deterministically (e.g., calculate the lowest probability of any candidate task exceeding the cycle time based upon each candidate's means and variances then assign the task with the best discrete mean value, Elsayed and Boucher, 1994), actual part removal times on a disassembly line may be more accurately represented probabilistically so a study making use of this type of data is warranted (with an understanding of the comments in Secs. 4.3 and 21.4, and the work of Tiacci et al., 2003).

Per the previous recommendation, development of data sets with probabilistic part removal times is necessary (Sec. 21.5).

Many of the combinatorial optimization methodologies can be run separately on stand-alone machines, searching different areas of an instance's search space; this is especially timely with the advent of distributed computing and as such, a study of the use of distributed computing (or some similar hardware and software technology) using some of the applicable methodologies could be of value.

It may be of interest to make use of the promise of H-K in generating uninformed solutions from throughout the search space through the use of H-K as a first phase to adjacent element hill-climbing or to hot start a genetic algorithm.

Throughout this part of the book, complete disassembly was typically assumed; while this is of great theoretical interest (since it allows for a worst-case study in terms of problem complexity and it provides consistency across the problem instances and methodologies), in practical applications it is not necessarily desired, required, practical, or efficient—additional studies could consider allowing for incomplete or partial disassembly.

Per the prior recommendation and an earlier one, different multicriteria ordering of the objectives could simplify the determination of the optimal level of incomplete disassembly; for example, if the main objective is to remove several demanded parts of a product containing hundreds of parts, by making the demand measure a higherpriority objective than the balance, it may be found that these parts can be removed relatively early on in the disassembly process thereby allowing the process to terminate significantly earlier in the case of partial disassembly.

Finally, application of this book's research results and comparison to the performance on an actual disassembly line may be of interest in a more applied study.

This concludes some suggestions for future research. In addition to the items listed above, any further developments and applications of recent methodologies to the DISASSEMBLY LINE BALANCING PROBLEM that will help to extend the focus area of the research may be appropriate.

PART III

Further Disassembly-Line Considerations

Chapter 22
Overview of Additional Disassembly-Line Related Problems

Chapter 23
Disassembly-Line Product Planning

Chapter 24
Disassembly-Line Design

Chapter 25
Disassembly-Line Sequencing and Scheduling

Chapter 26
Disassembly-Line Inventory

Chapter 27
Disassembly Line Just-in-Time

Chapter 28
Disassembly-Line Revenue

Chapter 29
Unbalanced Disassembly Lines

CHAPTER 22

Overview of Additional Disassembly-Line Related Problems

22.1 Introduction

Part III of the text provides a review of problems that are directly related to the disassembly line but do not involve balancing the line. Some of these problem areas are quite closely related to line balancing, such as the determination of part removal sequences and the evaluation of lines that are—intentionally or otherwise—unbalanced. Others are more peripheral, such as product design and inventory.

This chapter presents an overview of disassembly-line research that is not directly related to balancing the line (which is covered in Part II). Section 22.2 introduces the mathematical models used in these additional disassembly-line problems. Section 22.3 considers the computational complexity of disassembly-line problems, which dramatically affects the ability to solve these problems optimally as well as the choice of methods available. Section 22.4 lists some available case studies, while Sec. 22.5 reviews the variety of solution methodologies used in these problems.

Part III of this book covers the following: designing new products with end-of-life processing in mind (Chap. 23), the design of the disassembly facility and the line itself (Chap. 24), part removal sequencing and scheduling (Chap. 25), demanded-part inventory (Chap. 26), just-in-time (Chap. 27), revenue (Chap. 28), and unbalanced disassembly lines (Chap. 29).

22.2 Mathematical Models

The variety of additional disassembly problems requires a wide range of modeling techniques. These models encompass operations research and management science, closed-form expressions (i.e., not requiring numerical methods or computer simulation to solve), heuristics, linear programming, queueing theory, inventory theory, and simulation. In addition, some are direct applications without modification, while others are extensions having varying degrees of complexity, and still others are almost completely original techniques.

Designing new products for disassembly (Chap. 23) is addressed using measures of the ease of disassembly. Designing products for assembly has been an obvious attribute of all items of any manufacturing significance since the earliest tools and weapons. Including disassembly has truly only been considered in the cases of items with long lifespans undergoing remanufacturing; primarily in the aerospace and defense sector. Even here, however, designing for remanufacturing is not necessarily the same as designing products for disassembly and disposal at their end of life, especially with hazardous part and demanded part considerations. Product design can be modeled using indices to quantify the desirability of various product and part characteristics.

The design of the disassembly facility and of the line itself shares many attributes with assembly facilities and details of these are provided (Chap. 24). One of disassembly's unique characteristics is the likelihood that the line will be used to disassemble more than one product (or one product that may have been modified or damaged during its useful life or in transport and hence is no longer a duplicate of the other products on the line). Facility and line design are often qualitative exercises, though equations providing evaluation metrics for line design are provided here as well.

The reviewed part removal sequencing and scheduling problems focus on the general rules that are used in this area (Chap. 25). Sequencing and scheduling problems have also been formulated as linear programming, mixed integer programming, and integer programming models. Also, if stochastic part removal times are used, the system is then modeled using queueing theory.

Demanded-part inventory models (Chap. 26) range from general inventory theory formulations (e.g., the economic order quantity model), to ordering-rules and algorithms (e.g., material requirements planning), to disassembly-unique models such as the disassembly-to-order model (formulated as an integer programming model).

The just-in-time approach to demanufacturing (Chap. 27) has been applied in several ways. One methodology uses the just-in-time equations and model structure as is. This is effective in providing a fundamental understanding of the made-to-order or "pull" system and its disassembly-related challenges, as well as allowing for some simplified analysis and just-in-time line design. A more promising technique uses multiple kanbans to address the disassembly-unique uncertainties in product arrival, part demands, inventory fluctuation, and production control mechanisms. The result, however, does not lend itself to closed-form analysis, and necessitates analysis through the use of simulation. Both models make use of problem-specific equations.

When revenue is a study's primary consideration (Chap. 28), the model can be formed using mixed integer programming.

Unbalanced disassembly lines (Chap. 29) are often used when the part removal times are stochastic. In this case, it has been found that intentionally unbalancing a line can result in a more efficient and predictable line, making it less likely that cycle times will be exceeded. The models used for this type of problem come from the field of queueing theory. As such, they allow for a detailed analysis of the line, but in general do not directly provide an optimal solution of any type.

22.3 Computational Complexity

The variety of disassembly-line models can then be viewed from a complexity perspective. The study of each problem's computational time complexity is a factor in the types of solution methodologies that are justifiable and which may be subsequently applied, as well as the likelihood of obtaining an optimal solution. Of the problems considered in Part III, many again exhibit NP-complete characteristics and associated complexities and solution approaches (Chap. 6). Other problems, however, are solved to optimality, are described by closed-form equations, or are only addressed through an analysis perspective. The complexity of each type of problem is considered individually in this section.

As a set of indices, designing new products with end-of-life processing in mind is not an NP-complete problem and can therefore be addressed in a straightforward manner through the generation of the index values.

This is not the case for many facility location problems and line layout problems. These problems are both NP-complete [many location problems are related to the TRAVELING SALESPERSON PROBLEM (TSP) and many layout problem are similar to KNAPSACK], though facility layout problems can often be efficiently addressed with heuristics due to the limited options in an actual application. In addition, the facility location and line layout examples presented in Chap. 24 do not have NP-complete complexities since they are focused on qualitative evaluation of the facility or on the calculation of line performance metrics.

Optimal sequencing and scheduling are well known as being NP-complete problems. By their nature, rules (i.e., algorithms) for sequencing and scheduling are not. In addition to rules, in Chap. 25

stochastic part removal times are discussed, as well as general sequencing and scheduling formulations. The analysis of stochastic part removal times is not NP-complete, as queueing formulations are used. General sequencing and scheduling tends to be NP-complete, with representations often being in the form of an integer programming model.

Inventory theory makes use either of rules (where generating a solution would not be an NP-complete problem) or of first- or higher-order equations that can typically take advantage of optimization using calculus (due to the continuous nature of order-size versus time). Some disassembly models are an exception to these cases; the disassembly-to-order system is structured as an integer programming model and can therefore be expected to be NP-complete.

When considering a just-in-time model, though the multiple kanban version used in disassembly is complex enough to require analysis using simulation, since basic just-in-time models are a collection of rules defined by equations, they are not NP-complete.

Since the disassembly revenue problem is formulated as a mixed integer programming model, it is therefore expected to be NP-complete.

Consideration of unbalanced lines is not an NP-complete problem since it is modeled using queueing theory.

22.4 Case Studies

Case studies and data sets are an essential component in studying the disassembly line and its associated problems. While the number and variety of data available is relatively limited, there are case studies and data sets that exist in addition to those presented in Chap. 10. For example, O'Shea et al. (1999) provide a commercial washing machine valve as a product component for disassembly. Wang et al. (2003) make use of a nine-part wheel support assembly as well as a 16-part driver assembly. Both examples include part removal direction information. Hong and Cho (1997) provide case studies including an electrical relay and an automobile alternator. Lapierre and Ruiz (2004) present a case study of an appliance. Li et al. (2010) make use of a 10-part gear reduction assembly. Other data for the disassembly of different types of products includes vehicles (Kazmierczak et al., 2005, 2007), electronics (Kuo, 2000; Scholz-Reiter et al., 1999), and consumer appliances (Kara et al., 2006; Uhlmann et al., 2004).

Other texts can also provide various data for use in different disassembly studies. Lambert and Gupta (2005) provide many other product examples, including the box with lid, strongly connected four-component product, intermediate state and nonmonotone products (six examples), toaster, noncontact coherent product, topologically unconstrained/geometrically constrained product, ballpoint pen, and so on.

Finally, the case studies presented in Chap. 10 are easily extended to other types of disassembly problems, especially those that deal, either directly or indirectly, with sequencing and scheduling. In addition, many case studies and data sets from the area of assembly-line balancing can be used as is or extended, as can those from related problems (e.g., KNAPSACK, BIN PACKING, CUTTING STOCK, packing problems).

22.5 Analytical and Solution Methodologies

The methodologies used to address each of the problems in Part III are dependent on how they are modeled and their computational complexity.

The design of new products with end-of-life processing in mind makes use of a set of defined indices, which are calculated for different products being disassembled on an established line. These indices quantify how compelling a competing product design is when considering disassembly.

The facility problems described in this part of the text are primarily qualitative, weighing a variety of options. The line design considerations make use of simple equations to provide a metric for gauging the performance of different line options. Other line and facility layout problems are often addressed using heuristics that iteratively try different layouts in search of a desirable configuration.

Chapter 25 introduces popular rules for designing a sequence, including the shortest-processing-time-first rule and the earliest-due-date rule. Established queueing theory formulations are used to provide a quantitative analysis of disassembly lines having stochastic part removal times. Other models make use of simplex (for schedules that are set up as a linear programming model) or other techniques such as branch-and-bound (for the integer or mixed integer programming models) for obtaining a solution.

Where part demand can be modeled using traditional inventory theory—such as the economic order quantity model, material requirements planning, and so on—these equations or algorithms are applied. In some new areas, such as disassembly-to-order, their structure requires the use of branch-and-bound or some other integer programming heuristic in order to generate a solution.

Where just-in-time does not exclusively make use of associated rules and calculations, simulation is used as an analysis tool.

As is the case with other mixed integer programming models, the disassembly revenue problem makes use of some type of general-purpose heuristic (such as branch-and-bound) or a problem-specific heuristic.

Finally, analysis of unbalanced lines can be performed using established queueing theory equations.

CHAPTER 23

Disassembly-Line Product Planning

23.1 Introduction

While products are frequently designed for ease of assembly, there is growing need to design new products that are equally efficient at being disassembled later. Disassembly possesses considerations that add to its complexity when compared to a traditional assembly line, including treatment of hazardous parts and a used-part demand that varies between components. Section 23.2 provides an overview of research directed at designing products for disassembly. Section 23.3 uses the DISASSEMBLY LINE BALANCING PROBLEM (DLBP) measures as metrics for quantitatively comparing competing new-product designs for end-of-life disassembly on a reverse-production line. It also includes a case study consisting of three design alternatives of a consumer electronics product that is analyzed to illustrate application of the metrics. Section 23.4 provides an overview of some additional research, while Sec. 23.5 summarizes the chapter.

23.2 Sustainable Product Design Overview

Graedel and Allenby (1995) suggest that a product's design has the highest influence on the product's life cycle and list design as being the first priority toward the greening of products. In order to assess the environmental impact of a product, *green design* must take into consideration each stage of that product's life cycle. To connect both ends of a product's life cycle, its design must not only satisfy functional specifications and be easy to assemble, but should also lend itself to disassembly as well as possessing a host of other end-of-life attributes. This has led to the emergence of concepts such as design for environment, design for disassembly, planning for disassembly, and design for disassembly.

Numerous analysis tools have been developed to assist or evaluate different aspects of product design. Ishii et al. (1994) developed a methodology to design a product for retirement using a hierarchical semantic network that consists of components and subassemblies. Navin-Chandra (1994) presented an evaluation methodology for design for disassembly and then developed software that optimizes the component-recovery plan. Subramani and Dewhurst (1991) investigated procedures to assess service difficulties and their associated costs at the product design stage. They use a serviceability metric to provide an overall rating of a product's design as compared with its expected lifetime servicing costs. Isaacs and Gupta (1997) developed an evaluation methodology that enables an automobile designer to measure disassembly and recycling potential for different automobile designs using goal programming to analyze the trade-off between the profit levels of disassembling versus shredding. Johnson and Wang (1995) used a disassembly tree for designing products that enhance material-recovery opportunities. Their technique employs selective disassembly of a product by performing a profit/loss analysis on various combinations of components. Vujosevic et al. (1995) have studied the design of products that can be easily disassembled for maintenance. Torres et al. (2004) reported a study for nondestructive automatic disassembly of personal computers.

Quantifying the merits of different product designs allows manufacturers to intelligently plan for a wide variety of potential future contingencies. In this chapter a method to quantitatively evaluate product design alternatives with respect to the disassembly process is demonstrated. A product design can make a significant difference in the product's retirement strategy. As it is not uncommon for a designer to be faced with the dilemma of choosing among two or more competing design alternatives, a product designer may wish to place equal importance on designing products that accommodate disassembly, reuse, and recycling, in addition to a product's appeal and functionality.

23.3 New-Product Design Metrics for Demanufacturing

23.3.1 Design Metrics for End-of-Life Processing

The measures from Chap. 8 can be used as design-for-disassembly metrics. Their values then provide quantitative end-of-life product disassembly-related performance measures that allow for comparing multiple, competing product designs. Also, here, as in Chap. 8, the H, D, and R metrics form the three basic prototypes of any additional end-of-life processing design evaluation criteria; these three different models are then the basis for developing differing or additional metrics. In addition, the efficacy index is also applied. For this problem, however, it is slightly modified to optionally account for the best-case and worst-case design options. That is, the efficacy index is now more broadly defined as the ratio of the difference between a calculated measure x and its worst-case measure x_{worst} to the difference between the best-case measure x_{best} and the worst-case measure as given by

$$\mathrm{EI}_x = \frac{100 \cdot |x_{\text{worst}} - x|}{|x_{\text{worst}} - x_{\text{best}}|} \tag{23.1}$$

This generates a value between 0 and 100 percent, indicating the percentage of optimum for any given measure and any given design being evaluated. This new formulation allows for calculation using best-case and worst-case design options instead of the theoretical bounds given in Chap. 8.

23.3.2 Case Study

Product data consists of the personal computer instance from Sec. 11.2. A simple extension to this data clearly illustrates the methodology. Using, for example, the assumption that parts 3 and 7, and parts 5 and 6 have the same footprints and are completely interchangeable (as a result, only the precedence is ultimately affected in this example), three design versions of this product are considered: A, B, and C. Design A is reflected in Table 11.1, design B swaps parts 3 and 7 in Fig 11.1, while design C swaps parts 5 and 6. The Greedy/AEHC hybrid from Chap. 17 is used to generate the numerical results.

23.3.3 Results and Analysis

The Greedy/AEHC hybrid generated the sequence $\langle 1,5,3,6,2,8,7,4 \rangle$ for design A, $\langle 1,7,5,6,2,8,3,4 \rangle$ for design B, and $\langle 1,6,2,5,3,8,7,4 \rangle$ for design C with the metrics shown in Table 23.1 (best-case values are bold).

The bound formulations are used to calculate the case studies upper and lower theoretical bounds as seen in Table 23.2. These values are used along with the values in Table 23.1 to provide the

Case	NWS	$\sqrt{F}$	H	D	R
A	**4**	**5.74**	7	19,025	6
B	5	31.45	**2**	20,570	**5**
C	**4**	6.71	7	**18,845**	6

TABLE 23.1 Personal Computer Instance Metrics for the Three Design Alternatives

Bound	NWS	$\sqrt{F}$	H	D	R
Upper	8	64.23	8	21,915	7
Lower	4	5.50	1	16,470	3

TABLE 23.2 Upper and Lower Metric Bounds for the Personal Computer Instance

Case	E_{NWS}	$E_{\sqrt{F}}$	E_H	E_D	E_R	E_{NWS}	$E_{\sqrt{F}}$	E_H	E_D	E_R
A	100%	100%	0%	89.6%	0%	100%	99.6%	14.3%	53.1%	25%
B	0%	0%	100%	0%	100%	75%	55.8%	85.7%	24.7%	50%
C	100%	96.3%	0%	100%	0%	100%	97.9%	14.3%	56.4%	25%

TABLE 23.3 Efficacy Index Metrics Calculated Using the Best and Worst of the Three Alternatives (Center Column) and the Upper and Lower Theoretical Bounds (Right Column)

efficacy indices (Table 23.3) as compared to the best and worst values provided by the alternative designs as well as when compared to the theoretical values.

If minimizing the number of workstations is the priority, then designs A and C are equally acceptable. If balancing the workstations is the goal, design A is the preferred option. In the case of removing the hazardous part as quickly as possible, design B is the preference since the sole hazardous part is removed in the first workstation as opposed to the last workstation in the other two designs. When considering the rapid removal of demanded parts, all are quite close, with the preference given to design C. Finally, design B provides a marginal improvement in minimizing the number of part removal direction changes encountered.

Also, due to the size of the case-study product and the short 40-second cycle time, if all metrics are considered by the designer, design A or B would most likely be chosen since an equitable balance and the savings of an entire workstation (20 percent less than the alternative) may outweigh any desire to remove hazardous or demanded parts earlier in the short 2-minute and 40-second disassembly sequence.

While this small example with minimal alternatives (i.e., eight parts and only differing in the precedence of two parts) is used here to clearly illustrate application of the metrics, products with a greater number of parts, the use of additional metrics (i.e., the prototypes), or a larger number of design options—all of which would be expected in real-world applications—provides a wider range of metric values, enabling designers to quantitatively measure a variety of end-of-life parameters prior to committing to a final new-product design.

23.4 Additional Design-for-Disassembly Studies

A technique for analyzing the design efficiency of electronic products is documented in Veerakamolmal and Gupta (1999). Design efficiency is measured using an index which is based on the product's disassembly tree. The cost considerations used by this analysis include disposal and disassembly costs, while the benefit is derived from the sales of recovered components and materials for reuse and recycling revenue. In order to judge if one design is better than another, the benefit from retrieving a set of components can be weighed against the cost of disposing of the remains (e.g., a chosen set of reusable computer components may include the motherboard and hard drive, with the rest of the components slated for recycling or disposal). To compare the merits of two or more designs, not only is it desirable to evaluate the feasible permutations of parts, but it may also be of value to find the permutation with the highest cost-benefit ratio (this requires that all permutations be listed and enumerated with respect to the cost-benefit functions). Veerakamolmal and Gupta's proposed cost-benefit objective function (in terms of revenue) to be maximized consists of the sum of four terms: total resale revenue, total recycling revenue, total processing cost, and total disposal cost. Veerakamolmal and Gupta (1998) also present an earlier mathematical programming model for product disassembly and recycling, while Lambert and Gupta (2002) build on this research. Arola et al. (1999) discuss the impact of plastic selection in product design and provides an overview of three different plastic-sorting methods and technologies applied to an economic case study of recovering plastic from demanufactured consumer electronics.

23.5 Summary

Disassembly of end-of-life products is becoming more prevalent for various combinations of legal, public relations, and financial reasons. Designing products with the expectation of end-of-life disassembly can lead to efficiencies that can minimize future costs and potentially increase future profits. Rather than take an intuitive or qualitative approach to design-for-disassembly, metrics can provide compelling data for the selection of one design over another. The metrics demonstrated here also provide a measure of goodness, showing not only that one design is more efficient during disassembly than another, but in what areas of interest and by how much. This allows a design decision maker to make trade-offs where one design may be quantitatively preferable, but not by a significant enough margin to justify some other trade-off.

CHAPTER 24
Disassembly-Line Design

24.1 Introduction

Disassembly-line design is addressed in this chapter in two ways—location of the facility and the internal layout of the facility. Section 24.2 provides a background in the broad field of facilities engineering, including much of the commonly used terminology. Section 24.3 provides a qualitative overview of the wide range of considerations, options, and algorithms and associated software used in locating and laying out a facility. Section 24.4 provides a design for a mixed-model disassembly line. The formulations from Part II are used in Sec. 24.5 as metrics for comparing different line designs.

24.2 Background

The discipline of *facilities engineering* encompasses a variety of functions that focus on the design, construction, and maintenance of facilities (while summarized here, see Defense Acquisition University, 2009a for a detailed overview). Facilities provide the necessary shelter and work spaces to support the assigned production (or end-of-life) tasks.

Facilities engineering is a multidisciplinary process that involves all facets of life cycle management, from planning through disposal. A first step is determining the facility requirements. Specific requirements differ depending on production, storage, and support needs. Also, the product must be compatible with the facilities and vice versa.

The facilities engineering tasks are typically performed by facilities engineers. Facilities engineers perform various functions related to facilities engineering planning, real estate, engineering and construction, environmental engineering, and general support.

24.2.1 Facilities Engineering Planning

Facilities engineering planning requires knowledge in environmental planning, functional and operational planning, encroachment planning, facilities planning for production systems, and traditional planning. Traditional facilities planning includes scenario planning, land use planning, requirements generation, facilities planning (also known as small-area planning), transportation planning, demographic analysis, and constraints analysis, as well as acting as a political and community liaison.

24.2.2 Real Estate

The facility real estate professionals are accountable for real and personal property. Real estate professionals plan, organize, monitor, and manage real estate activities. They also issue, execute, manage, renew, supplement, oversee, or revoke real estate documents. Real estate actions can include appraisals and cartography, acquisition of facilities and lands, use of property by other entities, management of title, utilization of space, and disposal of property.

24.2.3 Engineering and Construction

Engineering and construction includes identifying, establishing, organizing, or implementing facility-acquisition objectives and policies. It also encompasses developing specifications, performing cost analysis, establishing project budgets, accomplishing design and engineering services, administering contracts, and managing the construction. Engineering and construction personnel are involved with all planning, organizing, directing, monitoring, management, and oversight during the facility's life cycle. They may also take an active role in acquisition strategy development as well as having technical responsibilities during the initial planning process.

24.2.4 Environmental Engineering

There are two components of environmental engineering that are relevant to facilities; these two distinct aspects are facilities environmental engineering and product environmental engineering. These sensitive areas can involve legal requirements as well as the potential for criminal liability.

24.2.5 Support

Support personnel manage the workers, materials, equipment, and contracts necessary to sustain the facility. This requires expertise in planning, design, scheduling, procuring, real property maintenance and repair, minor construction, and uninterrupted operation and distribution of utility systems. Support efforts may also include providing guidance, counsel, and direction to managers and technicians.

Personnel should be knowledgeable in budget formulation and execution, cost and performance metrics, material acquisition and management, and contract administration.

24.2.6 Life Cycle Considerations

The primary phases in the life cycle of a facility are planning, design, construction, sustainment, and disposal.

The planning phase includes master planning, real estate acquisition, project scoping, criteria and standards identification and development, environmental assessment, and documentation.

The design phase of facilities engineering includes an integrated architectural and engineering concept that supports the functional requirements and needs of the user. The two primary methods for producing the facility are design-bid-build and design-build. Design-bid-build is more traditional and employs a solicitation for construction bids (or proposals) based upon fully designed plans and specifications. The design-bid-build process consists of three phases. The design phase involves the production of contract documents in the form of plans (i.e., blueprints) and specifications. The plans and specifications may be prepared by in-house engineers and architects, or by private architectural and engineering firms. The design phase is completed with the production of the contract plans and specifications. In the bid phase, construction contractors compete for the construction of the facility by submitting bids. The construction contractors base their bids on the contract plans and specifications developed in the design phase. The bid phase is completed with the award of a construction contract. The build phase involves the actual construction of the facility. Design-build is gaining more use and utilizes a solicitation for the design and the construction in one contract based upon requirements identified in the planning phase. This process also involves the production of a contract document, but that contract document is in the form of a request for proposal (RFP). The RFP can be developed by in-house engineers and architects, or by a private architectural and engineering firm. Once the RFP is completed, design-build firms compete for the contract by submitting proposals. The design-build contract is awarded to the firm whose proposal best meets the requirements of the RFP, and offers the "best value" facility. An important consideration in determining the best value proposal is the overall life cycle cost of the facility after careful evaluation of the proposed design.

The construction phase is overseen by facilities engineering and may include construction, modernization/renovation, and major repairs. The role of facility engineering in construction falls primarily into either contract administration or quality management. Construction contract administration comprises the activities performed to monitor and enforce contract compliance. It includes managing contract funds, performing labor-related activities, preparing and processing contract modifications, resolving disputes, enforcing safety and warranty requirements, managing contract schedules, preparing contractor performance evaluations, and conducting preconstruction, prework, and safety conferences. Construction quality management is defined as the quality control and assurance activities instituted to achieve the quality levels established by the contract requirements. Quality control is the construction contractor's system for managing, controlling, and documenting the activities of the contractor, supplier, and subcontractor in order to comply with the contract requirements. Quality assurance is defined as the company's system that is in place to monitor the quality control efforts of the construction contractor.

In the total life of a facility, sustainment is the longest phase due to the life expectancy of buildings, which can be over 100 years. Often, facilities will have a different use at some point than originally intended. The main focus of this phase is the operation, maintenance, restoration, and modernization of the facility, including land, buildings, structures, pavements, and utilities. Decisions made during planning, design, and construction can enhance or hinder the ability to sustain a facility. The facility owner must document how the individual facilities fit with the organization's priorities, determine methodologies for provision of services, provide engineering services, provide reliable and efficient utilities, establish effective energy management, and meet environmental, safety, and occupational health requirements such as solid and hazardous waste management and disposal, ventilation, medical surveillance, industrial hygiene monitoring, or fire suppression.

The disposal phase occurs when a facility is no longer needed or is no longer fit for its intended purpose. The structure may be reconfigured as a different facility, transferred to another organization, sold, or demolished. These decisions are normally made as part of a recurring planning phase.

24.2.7 Equipment Planning

Facilities engineering includes specific considerations such as equipment installation. The construction of a facility includes the installation of built-in equipment that is not intended to be moveable and is required for the operation of the facility. Loose, portable, and easily moved equipment is not included in the construction costs.

If the equipment is being installed in new construction, the construction should be complete in order to allow the facility to receive the equipment; if the equipment is being installed in a facility that is not new, the equipment installation costs include both the labor and material to install that equipment as well as the items to support the equipment. Also, a facility's pollution prevention programs may procure and install green equipment. Green initiatives use finances to improve industrial processes by reducing or eliminating the generation of hazardous wastes and by providing for improved worker safety.

The cost of equipment procurement includes the cost of the equipment, transportation, unpacking, assembly, and testing.

24.2.8 Scope and Funding Considerations

Issues that affect the scope and funding for a facility include regional planning, utilities, environmental laws, safety and occupational health laws, and natural and cultural resources. These issues can influence the timeliness and suitability of facilities infrastructure support and are a factor in all phases of the life cycle of a facility.

Regional planning has become important with the arrival of a focus on creating a leaner, more efficient, and effective infrastructure that is in proper balance with an organization or an industry. In order to maintain this balance between a company's mission and its supporting facilities, an organization often must respond by changing its focus to a broader regional perspective and a more comprehensive regional planning approach. A well-thought-of regional planning process helps to optimize resources and opportunities by resolving facility development issues across a region. Regional planning looks for broader methods of satisfying infrastructure requirements including outsourcing, privatization, and leasing of facilities. The regional planning process may also include identifying potential partnerships for joint or shared facility use by other parts of an organization, different levels of a nation's government, or academia in order to meet future infrastructure requirements.

Utilities, water, and power are essential infrastructure systems. The infrastructure systems must provide the required utility services with acceptable levels of quality, quantity, and reliability. The utility source may be a company-owned plant or an outside utility supplier. Distribution systems, such as pipelines and power lines, bring the service from the source to individual buildings. Where the company does not own the plant, they must purchase utility services via service contracts negotiated with utility companies. A utility service contract typically has two main components: the commodity itself (e.g., water, gas) and a distribution/delivery fee (for transportation, pipelines, wires, etc.). When proposing changes to a facility's use, consideration must be given to the impact on utility requirements. Increased requirements for capacity, quality, or reliability may require an upgrade or extension of the utility system, or additional back-up systems. Construction projects may be required to upgrade current systems or provide back-up systems. If a utility company owns the system, negotiations with the utility company may be required to obtain the necessary service improvements with the improvements reflected in the service contract. These infrastructure systems have close ties to the environmental requirements. There are normally air, water, and solid waste emissions from these systems, therefore there is an opportunity to reduce or increase these emissions with changes in the facility use and the corresponding infrastructure changes.

The body of law relevant to the environmental area is complex and growing, and the corresponding regulations are gradually becoming more stringent. A few of the more significant laws in the United States include the Clean Air Act; Clean Water Act; Resource Conservation and Recovery Act; Solid Waste Disposal Act; Comprehensive Environmental Response, Compensation, and Liability Act (also known as the Superfund); Emergency Planning and Community Right-to-Know Act, SARA Title III; Safe Drinking Water Act; Pollution Prevention Act; Occupational Safety and Health Act; and National Environmental Policy Act.

Finally, few facilities occupy land entirely devoid of historical interest or natural resources. Because historic sites can have significant cultural or economic implications, facilities must comply with a wide range of laws and regulations in this area. Responsibility for such compliance, which can strongly affect facility and program schedules and costs, falls into the realm of facilities engineering. The areas governed by the numerous applicable laws and regulations include archaeological artifacts (prehistoric and historic), historic buildings and structures, indigenous peoples' sites, natural landmarks, natural resources, and endangered species.

24.3 Facility Location and Layout Overview

Facility location and layout decisions (see Finch, 2008 for additional details) are multifaceted. Deciding on a facility location is the result of three possible scenarios: a new business (the most likely case for disassembly in the early part of the twenty-first century), a business forced to relocate for some reason, and a location that is the result of expansion.

Locating a new business is often considered the most challenging of the three since there are more alternatives available and future demand has few details. The decision alternatives may start out to be global, but all eventually work down to a region of a country, then a city (town), and finally a section of the city (town). While numerous quantitative (e.g., transportation costs) and qualitative (e.g., employee quality of life) items are part of all of these decisions, two business inputs represent the primary considerations. These inputs—which may also be conflicting—are locating near inputs and locating near customers. The raw material consumed by a manufacturing business is equivalent to a final product at its end of life for a demanufacturing facility. In the disassembly case, locating near inputs equates to locating near products at the end of their useful life. This consideration is greater when products are expensive to transport (cars, aircraft, etc.

at their end of life) or impossible or impractical to transport (e.g., buildings), and less important when transport costs are lower (cell phones, computers, batteries, etc.). Another input to be considered is employees, in terms of the following: availability, sheer numbers, skills, knowledge, and wages and other labor costs. Other inputs have to do with support of the facility itself, such as power, water (for washing, cooling, etc.), and other utilities and resources. The final product consumed by the customer of a manufacturing business is equivalent to the parts and/or raw materials output by a demanufacturing facility. In the disassembly case, locating near customers equates to locating near users of raw materials or of specific reusable parts and components of end-of-life products. While location consideration is crucial for businesses such as retailers, it is less critical for manufacturing (or demanufacturing) facilities. Issues for these facilities are more cost related (the requirement for person-to-person interaction, e.g., does not exist in manufacturing/demanufacturing businesses to the extent that it does in retail).

Some quantitative location decision-making techniques include multifactor rating, decision-tree analysis, and breakeven analysis (Finch, 2008). With the *multifactor rating* method, different factors are selected (building cost, building condition, skilled worker availability, etc.) and the decision-maker assigns weights (having values between zero and one) to each, ensuring that all of the weights sum to one. Each of the locations under consideration is given a score of between 1 and 100 in each of the areas (i.e., for each factor) and each of these scores is then multiplied by its corresponding weight. The resulting values for each location are then summed, resulting in a final score between 1 and 100 for each location. The location with the highest overall score is considered the most desirable by the multifactor rating method. While this is a quantitative process, the ratings are purely subjective.

Decision-tree analysis is a decision support tool based in the tree diagrams as found in traditional statistics; it makes use of a graph known as a decision tree. A predictive model, decision-tree analysis is used in facility location decision-making by organizing decisions based upon the revenues expected to come from each alternative location. These revenues are each multiplied by their likelihood of occurrence, resulting in an expected value (also referred to as the expected *utility*) of each location. The facility location with the largest calculated utility would be the preferred site.

In considering units of production versus costs, *breakeven analysis* generates a straight line for each location. Each of these lines is defined by the fixed and variable costs associated with that location. The slope of the line is determined by the variable costs (e.g., materials, labor, and other resources), while the line's offset is determined by the site's fixed costs (e.g., purchase, site improvements, and other capital costs). If one or more of the lowest lines intersect, each of the points of intersection on the resulting lowest segments provides a breakeven point where more of the amount of product produced (disassembled here) would dictate selection of the subsequent location corresponding to the line that defines that portion of the segment. (If the lines do not intersect, the lowest line corresponds to the location with the lowest costs.)

Once the facility's location is determined, the internal layout must be considered. Facility layouts can be grouped as process oriented, product oriented, cellular, or service (Sec. 2.4). The primary objective of process-oriented layouts is to locate the groups, departments, or equipment that interact the most, close to each other, and ones that don't (or those that need to be separated) further apart or as far apart as possible. Manual approaches and automatic approaches (such as that provided in CRAFT software) can be used to design the layout. (CRAFT starts with a matrix containing department center-of-gravity locations and then swaps all department pairs until no further improvement can be made, in search of the optimal layout.)

Product-oriented layouts are familiar production lines (assembly or disassembly). While these layouts have limited flexibility, they are very efficient at generating high volumes of product at low per-unit costs. Since product-oriented layouts refer to production lines, the information and solution methodologies detailed throughout this book are generally applicable. Patterns of flow are typically classified as *straight*, *serpentine*, *U flow*, *circular*, *L flow*, and *S flow* (see Fig. 24.1).

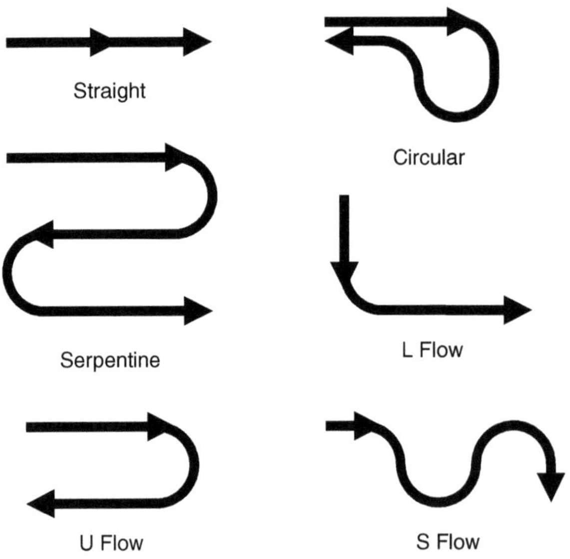

FIGURE 24.1 Product-oriented patterns of flow.

Cellular layouts are a compromise between product- and process-oriented layouts where a family of related products are produced in a cell that contains all of the tools and parts required for the entire assembly process.

Service layouts are used most often in retail settings, which depend upon customer flows and placement of popular products where they are easiest to find and purchase. The layout mechanism may be similar to that of process-oriented layouts (which allows for a great deal of flexibility) or product-oriented layouts (where an efficient flow of service, rather than product parts and components, is desirable).

Facility location and facility layout tools include the *activity relationship chart* (commonly referred to as the "rel chart"), *from-to chart*, and *isocost lines*. Facility location and facility layout problems are often mathematically described by the NP-complete QUADRATIC ASSIGNMENT PROBLEM. Solution techniques include *k*-opt and other appropriate algorithms, and the rules and associated software tools of CRAFT, COFAD, ALDEP, and CORELAP. Other models and solution methodologies include the single location, rectilinear model and methodology; the MINIMAX location formulation and methodology; and the center of gravity method. See Finch (2008) and Nahmias (2009) for detailed overviews and examples.

Ranky et al. (2003) use COMSOAL, while additional automated disassembly research can be seen in Kim et al. (2006a) and Scholz-Reiter et al. (1999).

24.4 Mixed-Model Disassembly-Line Design

One of the challenges posed by disassembly, which is not typically faced in assembly, is the frequent requirement that the facility process dissimilar items. A familiar example of this would be a disassembly line that processes consumer electronics. Consumer electronic items are popular, regularly disposed of (often before reaching the end of their functional life), varying in size, and desirous of disassembly (for both environmental and profit considerations). Opalić et al. (2004, 2010) consider a disassembly job shop that handles mixed models; specifically both smaller consumer electronics (telephones, computers, vacuum cleaners, microwave ovens, etc.) and larger consumer electronics (e.g., television sets and computer monitors). The proposed layout configuration consists of two, parallel unpaced lines; one for the smaller items and the other for the larger ones. Both lines sit on either side of a shared outgoing conveyor that carries the disassembled parts and components to a final sorting location. The small-item workstations are fed by a continuous (i.e., circular or oval) conveyor belt that allows workers to select similar items to disassemble at their workstation. The large-item workstations are fed by an overhead hoist

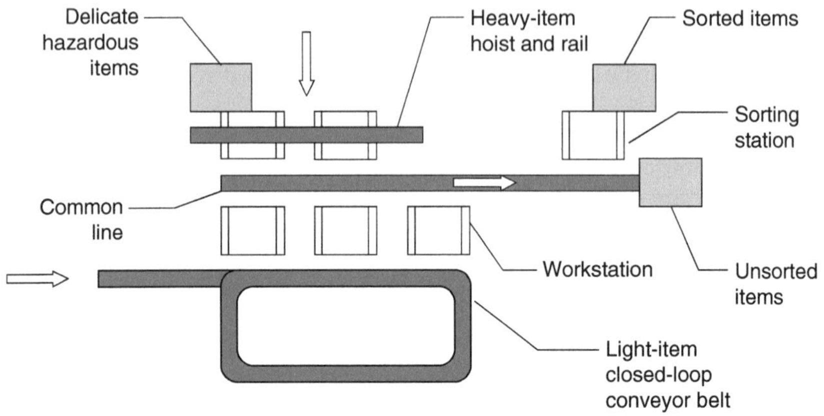

Figure 24.2 Mixed-model consumer electronics disassembly facility layout.

and rail system in order to minimize lifting of heavy electronics items. These workstations also have a separate storage container for delicate cathode-ray tubes (as opposed to use of the common conveyor belt). This disassembly-specific layout for consumer electronics (see Fig. 24.2) is claimed to increase operator speed (by allowing for workers to select items for disassembly appropriate to their workstation, tools, or experience), minimize the number of tools required at each workstation, reduce the risk of contamination by hazardous materials (both due to sorting and a reduction in breakage when using the hazardous/delicate items bins), and minimize heavy lifting with the large-item-specific features (including the overhead hoist and rail).

24.5 Disassembly-Line Design Metrics

Prior to implementing any disassembly-line design, that design can be evaluated in a variety of ways. While many decisions will be dictated by the facility's shape, size, costs, local worker skill sets and availability, desired level of automation, tool or machinery footprint/speed/cost, and type of product or products being disassembled, once these constraints are addressed there still exist line-related items that can be adjusted.

The cycle time and the sequence of machines and/or tools are two variables that can generally be tuned to best meet the needs of the line. The sequence of machines will affect the number of workstations, the balance of the line, when and where (in the facility) hazardous parts are removed, and when and where demanded parts are removed. The part removal direction can impact the required cycle time, the number of workstations, and optimal machine/tool layout; as such, a product's part removal directions are also a component in the design of a disassembly line.

All of these considerations lend themselves to using the existing DLBP formulations (Chap. 8) as metrics in evaluating different cycle times and workstation sequences. As long as the product or products to be disassembled on the line are known in advance—or can be reasonably estimated—they can be used as case studies to compare different cycle times and/or different workstation sequences.

Different cycle times can be expected to generate different part removal sequences, which will in turn potentially generate different NWS, *F*, *H*, *D*, and *R* values. These values can then be compared to allow the decision maker to select the cycle time that optimizes their selected metrics and associated trade offs.

Different workstation sequences can be used to generate different part removal sequences. These sequences can be used to then generate different NWS, *F*, *H*, *D*, and *R* values for comparison. Again, the resultant metric values can be used by the decision maker to select the sequence and type of workstation that best meets their disassembly and facility goals.

Using the formulations described in Chap. 8 provides a set of metrics for use in quantitatively evaluating the design of a new disassembly-line layout.

CHAPTER 25

Disassembly-Line Sequencing and Scheduling

25.1 Introduction

This chapter presents a discussion of sequencing and scheduling problems as related to the disassembly line, and is organized as follows: Section 25.2 presents an overview of sequencing and scheduling research as related to manufacturing. Section (25.3 provides details of sequencing and scheduling rules as applied to disassembly along with associated examples. Section 25.4 discusses the variety of approaches that have been taken to-date to address stochastic disassembly sequencing and scheduling. The next two sections provide insights into other disassembly-line sequencing and scheduling studies. Specifically, Sec. 25.5 focuses on that research which is focused on sequencing for disassembly lines and Sec. 25.6 covers research that is most directly related to disassembly-line scheduling.

25.2 Machine Sequencing and Scheduling Overview

Pinedo (2002) defines a *sequence* as a permutation of n jobs or the order in which jobs are to be processed on a given machine. A *schedule* then refers to an allocation of jobs within the setting of machines, allowing possibly for preemptions of jobs by other jobs that are released at later points in time. Sequencing and scheduling is a well-studied science with many manufacturing and nonmanufacturing applications such as service (Nahmias, 2009 provide an overview). This area includes shop-floor control and machine scheduling where efficient sequencing rules and algorithms are sought for the typically NP-complete problems associated with single or multiple machines in a manufacturing environment. Some characteristics of shop scheduling include job arrival pattern (stochastic or deterministic), number of machines (one or more); flow patterns (job sequences fixed or variable), type of machines (identical or varying), metrics (the rule selected to achieve a given goal), processing sequences (parallel or sequential machines), flow time (time from the initiation of the first job to the completion of some given job), makespan (flow time of last completed job), tardiness (positive difference between a job's flow time and due date), and lateness (positive or negative difference between a job's flow time and due date).

Sequencing and scheduling rules include FIFO or first come first served (FCFS) where jobs are scheduled in the order they arrive, shortest processing time (SPT) which schedules the remaining job with the smallest processing time next (minimizes *mean flow time*), earliest due date (EDD) which schedules jobs with the earliest due date first (minimizes *maximum lateness*), critical ratio (CR) which schedules the job with the smallest CR (the due date minus the current time, all divided by the processing time) time next, Moore's algorithm (minimizes the number of tardy jobs), and Lawler's algorithm (used when the problem involves precedence constraints).

Sequencing and scheduling on multiple machines is much more complex, though algorithms exist for n jobs on one, two, and three machines as well as for two jobs on m machines. These rules include permutation schedule (jobs are processed in the same order on all machines) which is the optimal solution for n jobs on two machines, Johnson's algorithm (for n jobs on two machines and with extensions to three), and Aker's graphical procedure for two jobs on m machines.

Other sequencing and scheduling rules include next-fit, first-fit-decreasing, and line balancing, all of which are discussed elsewhere in this text.

Scheduling is a subject commonly found in advanced computer science, where computer processor event scheduling takes advantage of many of the rules listed here (some of which originated in computer science rather than manufacturing and production). As such, there are many excellent reference texts on the subject as well as a rich body of ongoing research.

Sequencing and scheduling problems are generally considered to be NP-complete, so problems of this type—including those in disassembly—that cannot be adequately addressed by the above rules or do not fit the listed constraints for number of machines or number of jobs, necessitate the use of heuristics or metaheuristics.

25.3 Disassembly Sequencing and Scheduling

Several of the successful sequencing and scheduling rules described in the previous section are demonstrated here using much of the data provided by the personal computer instance as given in Chap. 10. This data has once again been modified, this time for use with these rules (see Table 25.1). Specifically, the hazardous part, part demands, and part removal direction information have been deleted, while due date information has been added. It should be noted that due dates could be selected as to be inversely proportionate to demand, giving high-demand parts early due dates. This is also appropriate to the hazardous parts, where a hazardous rating would translate to an earlier due date. Alternatively, a hazardous part rating (i.e., a number inversely proportionate to the level of hazard, giving a very hazardous part an early due date, or proportionate in the case where it was desirable to keep hazardous parts in the product as long as possible to avoid their potentially damaging other parts) could refine this beyond a simple binary yes/no value. Also, precedence constraints are not necessary and may not be appropriate (e.g., the goal of the schedule may be to schedule different products on a mixed-product line). Finally, for some rules it is necessary to know when a job arrives. In this case, it is assumed that the parts are numbered in the sequence at which they become accessible to the disassembler (while still considering precedence). That is, part 1 (using the data in Table 25.1) is assumed to be the first accessible part, part 2 the second, and so on; however, even though part 4 may be accessible early on, precedence constraints restrict its removal until after part 7 is removed.

These data set modifications are necessary because of the structure of sequencing and scheduling problems. It should be kept in mind that while much of the work in part 3 takes place on a disassembly line (e.g., a conveyor belt), the rules in this chapter are more applicable to a job shop (where each job has its own, predetermined route to follow among machines and not all jobs require all machines) or a flow shop with buffers (a line where jobs can wait between machines or where machines may need to wait for jobs).

Metrics that can be considered include mean flow time, average tardiness, number of tardy jobs, and the makespan. The *flow time* of

k	Part Removal Description	PRT_k	Predecessors	Due Date
1	Cover	14	—	20
2	Media drive	10	1	70
3	Hard drive	12	1	50
4	Back plane	18	7	150
5	Adaptor cards	23	1	30
6	Memory modules (2)	16	2 or 3	60
7	Power supply	20	8	140
8	Motherboard	36	2, 3, 5, 6	120

Table 25.1 Knowledge Base of the Scheduling Version of the Personal Computer Instance

job k is the time from the initiation of the first job to the completion of job k. The *mean flow time* is defined as the amount of time from the point that the job arrives at the shop to the point that the job is completed. *Tardiness* is the positive difference (or zero) between a job's flow time (same as its *completion time*) and its due date, so the *average tardiness* is the total tardiness divided by the number of jobs. (*Lateness* is similar, but can be negative rather than being set to zero.) The number of tardy jobs is self explanatory, while the *makespan* is the flow time of the last completed job. (Note that mean flow time, mean waiting time, and mean lateness are all equivalent.) SPT minimizes the mean flow time, EDD minimizes the maximum lateness, and Moore's algorithm minimizes the number of tardy jobs.

While extensions—beyond those introduced in the following sections—to these rules in order to optimize them for disassembly are either not entirely practical or do not yet exist, these tools provide the basis for both applied solutions and continued research.

25.3.1 First Come First Served

The first-come-first-served rule is ordered by the job number. Without precedence constraints (e.g., any disassembly sequence is allowable, the sequence consists of selecting between multiple products, etc.) the FCFS solution is ⟨1, 2, 3, 4, 5, 6, 7, 8⟩. For disassembly of a product, we will also consider precedence. In this case, the solution is given by ⟨1, 2, 3, 5, 6, 8, 7, 4⟩, which can be seen in Table 25.2. The tardiness is the difference between the completion time and the due date, or zero, whichever is greater.

With the sum of the completion times equal to 599 and the sum of the tardiness times equal to 44, the mean flow time is calculated to be 599/8 = 74.88, the average tardiness is calculated as 44/8 = 5.50, and there are two tardy tasks.

k	PRT_k	Due Date	Completion Time	Tardiness
1	14	20	14	0
2	10	70	24	0
3	12	50	36	0
5	23	30	59	29
6	16	60	75	15
8	36	120	111	0
7	20	140	131	0
4	18	150	149	0

TABLE 25.2 FCFS Solution (with Precedence) of the Computer Instance

k	PRT_k	Due Date	Completion Time	Tardiness
1	14	20	14	0
2	10	70	24	0
3	12	50	36	0
5	16	60	52	0
6	23	30	75	45
8	36	120	111	0
7	20	140	131	0
4	18	150	149	0

TABLE 25.3 SPT Solution (with Precedence) of the Computer Instance

25.3.2 Shortest Processing Time

The shortest-processing-time rule is ordered by the size of the task time (smaller time tasks are performed earlier). Without precedence constraints, the SPT solution is ⟨2, 3, 1, 6, 4, 7, 5, 8⟩. With precedence, the solution is given by ⟨1, 2, 3, 6, 5, 8, 7, 4⟩, which can be seen in Table 25.3.

With the sum of the completion times equal to 592 and the sum of the tardiness times equal to 45, the mean flow is equal to 74.00 (recall SPT minimizes mean flow time), the average tardiness is 5.63, and there is one tardy task.

25.3.3 Earliest Due Date

The earliest-due-date rule schedules jobs based on their due date (earlier due date tasks are performed earlier). Without precedence constraints, the EDD solution is ⟨1, 5, 3, 6, 2, 8, 7, 4⟩. With precedence, the solution is the same and can be seen in Table 25.4.

k	PRT_k	Due Date	Completion Time	Tardiness
1	14	20	14	0
5	23	30	37	7
3	12	50	49	0
6	16	60	65	5
2	10	70	75	5
8	36	120	111	0
7	20	140	131	0
4	18	150	149	0

TABLE 25.4 EDD Solution (with Precedence) of the Computer Instance

With the sum of the completion times equal to 631 and the sum of the tardiness times equal to 17, the mean flow is equal to 78.88, the average tardiness is 2.13, and there are three tardy tasks. EDD minimizes the maximum lateness (–9 here; –46 for both FCFS and SPT).

25.3.4 Critical Ratio

The critical-ratio algorithm schedules jobs based on their calculated critical ratio. The critical ratio is equal to the job's due date minus the current time, all divided by the job's processing time. This gives the priority to the jobs with longer processing times. The disadvantage is that the CR calculation needs to be performed each time a job is scheduled.

Considering only the version having precedence constraints, it can be seen that job 1 must be performed first. Also, jobs 4, 7, and 8 have to be performed in the order ⟨8, 7, 4⟩. Since the first job is known (job 1) the remaining jobs are reviewed starting at time 14 (the start time would be zero in the case of no precedence constraints or if the first job was unknown). The jobs eligible for consideration after job 1 include 2, 3, and 5. The critical ratios for all three can be seen in Table 25.5.

Since the minimum CR is job 5's at 0.70, job 5 is added to the sequence, which now stands at ⟨1, 5, -, -, -, 8, 7, 4⟩. Adding job 5 adds 30 to the time, which will now stand at 44. The eligible jobs are now 2 and 3 (see Table 25.6).

Since the minimum CR is job 3's at 0.50, job 3 is added to the sequence, which now stands at ⟨1, 5, 3, -, -, 8, 7, 4⟩. Adding job 3 adds 50 to the time, which will now stand at 94. The eligible jobs are now 2 and 6 (see Table 25.7).

This time, there are CRs that are negative. A negative CR indicates that the job is late and must be scheduled next. In this example, there is more than one negative CR. Multiple late jobs are scheduled

k	PRT_k	Due Date	Due date – Time	Critical Ratio
2	10	70	56	5.60
3	12	50	36	3.00
5	23	30	16	0.70

TABLE 25.5 CR Solution (with Precedence) of the Computer Instance at Time 14

k	PRT_k	Due Date	Due Date – Time	Critical Ratio
2	10	70	26	2.60
3	12	50	6	0.50

TABLE 25.6 CR Solution (with Precedence) of the Computer Instance at Time 44

k	PRT_k	Due Date	Due Date − Time	Critical Ratio
2	10	70	−24	−2.40
6	16	60	−34	−2.13

TABLE 25.7 CR Solution (with Precedence) of the Computer Instance at Time 94

k	PRT_k	Due Date	Completion Time	Tardiness
1	14	20	14	0
5	23	30	37	7
3	12	50	49	0
2	10	70	59	0
6	16	60	75	15
8	36	120	111	0
7	20	140	131	0
4	18	150	149	0

TABLE 25.8 CR Solution (with Precedence) of the Computer Instance

according to SPT. (If some were positive, the negative CR jobs would still take precedence.) Adding job 2 and then job 6 per SPT completes the sequence, which is ⟨1, 5, 3, 2, 6, 8, 7, 4⟩ and can be seen in Table 25.8.

With the sum of the completion times equal to 625 and the sum of the tardiness times equal to 22, the mean flow is equal to 78.13, the average tardiness is 2.75, and there are two tardy tasks.

25.4 Stochastic Sequencing and Scheduling

While this chapter considered deterministic sequencing and scheduling, stochastic models and rules exist as well. Stochastic sequencing and scheduling can be either *static* or *dynamic* (see Nahmias, 2009 for examples).

Static stochastic sequencing and scheduling treats the job times as being stochastic. This statistical treatment seeks to minimize the expected average weighted flow time and is performed on one machine or many (where the distribution of job times must be exponential). An additional option is the stochastic application of Johnson's algorithm.

Dynamic stochastic sequencing and scheduling refers to jobs arriving randomly over time. In dynamic stochastic sequencing and scheduling, analysis of the mean flow time is performed using queueing theory (reviewed in Chap. 29).

25.5 Additional Disassembly-Sequencing Studies

End-of-life processing researchers define disassembly sequencing as dealing with the problem of determining the best order of operations in the separation of a product into its constituent parts or other groupings (Dong and Arndt, 2003; Moore et al., 1998).

The use of mathematical programming techniques in disassembly sequence generation is a popular approach. Lambert (1999) presents an algorithm based on linear programming for the determination of optimal disassembly sequences and later (2006) proposes a methodology based on the iterative use of binary integer linear programming in case of sequence-dependent costs and using a disassembly precedence graph representation. Lambert (2007) applies this same methodology for problems with AND/OR representations.

Due to the combinatorial nature of disassembly sequencing, there is a continuing trend in the application of metaheuristics. Seo et al. (2001) develop a genetic-algorithm-based approach to determine the optimal disassembly sequence considering both economic and environmental aspects. Li et al. (2005) integrate a disassembly-constraint graph and a genetic algorithm to develop an object-oriented intelligent disassembly-sequence planner. Kongar and Gupta (2006b), Giudice and Fargione (2007), Duta et al. (2008b), and Hui et al. (2008) present genetic algorithms for disassembly sequencing of end-of-life products. Gonzalez and Adenso-Díaz (2006) propose a scatter search methodology in seeking the optimum disassembly sequence for complex products with sequence-dependent disassembly costs by assuming that only one component can be released at each time. Chung and Peng (2006) develop a genetic algorithm to generate a feasible selective disassembly plan considering batch disassembly and tool accessibility. Shimizu et al. (2007) apply genetic programming to derive an optimal disassembly sequence. Reveliotis (2007) presents reinforcement learning to provide near-optimal disassembly sequences. Tripathi et al. (2009) present a fuzzy disassembly-sequencing formulation by considering the uncertainty inherent in the quality of the returned products. They develop an ant-colony-optimization-based metaheuristic to determine the optimal disassembly sequence as well as the optimal depth of disassembly.

In some studies, heuristic procedures are developed. Güngör and Gupta (1997) develop a methodology to evaluate different disassembly strategies and then propose a heuristic procedure to determine the near-optimal disassembly sequences. Güngör and Gupta (1998) address the difficulties associated with uncertainty in disassembly-sequence planning. They present a methodology for disassembly-sequence planning for products with defective parts in product recovery. Kuo (2000) provides a disassembly sequence and cost-analysis study for electromechanical products during the design stage. Disassembly planning is divided into four stages: geometric assembly representation, cut-vertex search analysis, disassembly-precedence-matrix analysis, and disassembly sequences and plan generation. The disassembly cost is categorized into three types: target disassembly, full disassembly, and optimal disassembly. Güngör and Gupta (2001b) use a branch-and-bound algorithm for disassembly-sequence-plan generation. In Erdos et al. (2001), a heuristic is used to decompose the problem by discovering the subassemblies within the product structure and then a calculation is performed to determine the optimal disassembly sequence. Kang et al. (2003) propose an algorithm based on mini-max regret criterion to solve the problem of disassembly sequencing with interval profit values in the objective function. Mascle and Balasoiu (2003) propose an algorithm to select the disassembly sequence of a specific component of a product. Lambert and Gupta (2008) present a heuristic algorithm for detecting "good enough" solutions for disassembly-sequencing problems in case of sequence-dependent costs. They apply both their heuristic algorithm and the iterative binary integer linear programming method (Lambert, 2006) using the disassembly-precedence graph of a cell phone. Sarin et al. (2006) propose a precedence-constrained asymmetric TRAVELING SALESPERSON PROBLEM formulation together with a three-phase iterative solution procedure. Adenso-Díaz et al. (2008) propose a greedy randomized adaptive search procedure and path-relinking-based heuristic methodology to solve a bicriteria disassembly-planning problem. Torres et al. (2003) develop an algorithm based upon the product representation to establish a partial nondestructive disassembly sequence of a product.

Some researchers present case-based reasoning applications for disassembly sequencing. Zeid et al. (1997) apply case-based reasoning to develop a disassembly plan for a single product. Veerakamolmal and Gupta (2002) present a case-based reasoning approach for the automatic generation of disassembly process plans for multiple products. Pan and Zeid (2001) develop a knowledge base to assist users in indexing and retrieving disassembly sequences.

Additional disassembly sequencing can be found in Andrés et al. (2007), Srinivasan et al. (1999), Chung and Peng (2005), and Kim et al. (2006a).

25.6 Additional Disassembly-Scheduling Studies

Disassembly scheduling has been defined as the scheduling of the ordering and disassembly of end-of-life products to fulfill the demand for parts or components over a planning horizon (Veerakamolmal and Gupta, 1998). Using this inventory-theory-related definition, disassembly-scheduling problems can be categorized as *uncapacitated* and *capacitated*.

For the uncapacitated case, Gupta and Taleb (1994) propose a material requirements planning (MRP)-like algorithm for disassembly scheduling of a discrete, well-defined product structure. The algorithm determines the quantity and timing of the disassembly of a single product to fulfill the demand for its various parts. Taleb et al. (1997) and Taleb and Gupta (1997) extend this previous work by considering components/materials commonality and the disassembly of multiple product types. Lee and Xirouchakis (2004) suggest a two-phase heuristic algorithm with the objective of minimizing various costs related to the disassembly process. In the first phase, the algorithm of Gupta and Taleb (1994) is used to find an initial solution; the second phase improves this initial solution. Barba-Gutíerrez et al. (2008) extend the reverse MRP algorithm of Gupta and Taleb (1994) by developing a methodology which allows for lot sizing in reverse MRP. Kim et al. (2003) propose a heuristic algorithm based on linear programming relaxation for the case of multiple product types having part commonality with the aim of minimizing the sum of setup, disassembly operation, and inventory-holding costs. Kim et al. (2006c) develop a two-phase heuristic by extending the Kim et al. (2003) study. The first phase involves the construction of an initial solution using the linear programming relaxation heuristic suggested by Kim et al. (2003). The second phase improves the initial solution using dynamic programming. Lee et al. (2004) present three integer-programming models for the three cases of the uncapacitated disassembly scheduling problem; that is, a single product type without parts commonality and single- and multiple-product types with parts commonality.

For the capacitated case, Meacham et al. (1999) present an optimization algorithm by considering common components among products, and a limited inventory of products available for disassembly. Lee et al. (2002) develop an integer programming model with the objective of minimizing the sum of the disassembly operation costs and inventory-holding costs. However, the model requires large computation times to find optimal solutions for practical-sized problems. As an extension, Kim et al. (2006b) develop a Lagrangean heuristic algorithm to find an optimal solution for practically sized problems in a more reasonable amount of time, while modeling disassembly setup costs in the objective function. Kim et al. (2006d) propose an optimal algorithm for the case of a single product type without parts commonality with the objective of minimizing the number of disassembled products. In this algorithm, an initial solution is first determined using Gupta and Taleb's (1994) algorithm. The feasibility of this solution is then checked; if the solution is not feasible, it is modified to satisfy the capacity constraints. Stuart and Christina (2003) define disassembly and bulk recycling scheduling rules based on product turnover. Rios and Stuart (2004) extend Stuart and Christina's work by considering the outgoing plastics' demand as well. In both studies, scheduling rules are evaluated using simulation. Brander and Forsberg (2005) develop a cyclic lot scheduling heuristic for the disassembly processes by considering sequence-dependent setups.

CHAPTER 26

Disassembly-Line Inventory

26.1 Introduction

The application of inventory considerations to the disassembly line is reviewed in this chapter. With an introduction to the concepts of inventory theory and the detailing of several disassembly-line inventory models, the chapter is organized as follows: Section 26.2 presents an overview of traditional inventory theory. Section 26.3 introduces the disassembly-line concept of a disassembly-to-order system. In this section, two disassembly-to-order models are detailed—one with deterministic yields and the other where the yields are stochastic. Section 26.4 reviews some other disassembly-based inventory studies.

26.2 Inventory Theory Background

With similarities to scheduling (since it is focused on production schedules; i.e., when things are ordered, when they arrive, when they are depleted), inventory theory is often dated to F. W. Harris' 1913 development of the economic order quantity model. Inventory theory is also well established as well as being applicable to the disassembly line. Inventory models are concerned with economic parameters, demand, ordering cycle, delivery lead/lag time, stock replenishment, planning horizon, number of supply echelons, and number of items.

Economic parameters include setup (or ordering) costs (i.e., the fixed charges associated with the placement of an order or with the initial preparation of a production facility), purchase price (or production cost; only of interest when quantity discounts or price breaks can be applied), selling price, holding cost (or carrying cost; includes the interest on invested capitol, storage costs, handling costs, and depreciation), shortage costs (includes costs due to loss in customer's good will and potential loss in income). *Demand* can be deterministic or probabilistic, static or dynamic, and instantaneous (discrete) or over a period of time. The *ordering cycle* is the time period between two successive placements of orders and may be *continuous review* (where a record of the inventory level is updated continuously until a certain lower limit is reach, at which point a new order is placed) or *periodic review* (where orders are placed usually at equally spaced intervals of time). The *delivery lead/lag time* can be deterministic or probabilistic and includes the time between the placement of an order and its receipt). *Stock replenishment* may be instantaneous or uniform, as well as ordered from outside or produced internally. The *planning horizon* is the period over which the inventory level will be controlled.

Deterministic models include the economic lot size model (or the economic order quantity model, EOQ) with shortages allowed, EOQ without shortages (also known as EOQ with planned shortages), and EOQ with quantity discounts and no shortages. Stochastic models include the single-period model with no setup costs, the single-period model with initial stock levels (i.e., initial inventory greater than zero), single-period model with nonlinear penalty and holding costs, and the single-period model with a setup cost (the SS policy model); see Hillier and Lieberman (2005) for practical reviews of each.

Each of these considerations and models has applicability to the disassembly line, though demand uncertainties in end-of-life processing have produced the most disassembly-unique research. Other inventory-related models and algorithms that are frequently used include material requirements planning, lot-for-lot production rules, the Silver-Meal heuristic, least unit cost heuristic, part period balancing, just-in-time (see Chap. 27), Wagner-Whitier algorithm, period order quantity, and lot sizing with capacity constraints; see Nahmias (2009) for examples of each.

26.3 Disassembly to Order

The objective of disassembly-to-order (DTO) systems is the determination of the optimal lot-sizes of end-of-life products to disassemble in order to satisfy the demand of various components from a mix of different product types that have a number of components and/or modules in common (Lambert and Gupta, 2002). In a DTO system, a demand for a remanufactured product triggers the disassembly of an end-of-life product. The DTO system is tasked with determining how many products to disassemble in order to fulfill demand. Complications over conventional assembly-based inventory theory include the desire to have an alternative source of parts (and the associated holding and other costs, as well as a potential loss of economies of scale due to possible small or infrequent orders or unexpectedly high or low demand), a level of redundancy due to a supporting new-part inventory system, disposal costs, disassembly of too few or too many products, and larger than normal holding costs for parts that are not demanded as quickly, as well as all of the typical disassembly costs and challenges (e.g., unknown quality and quantity of demanded parts in an end-of-life product). The two primary areas of DTO research include studies having deterministic disassembly yields, and those having disassembly yields with uncertainty.

The first line of research involves the heuristics developed under the assumption of a deterministic disassembly yield. Lambert and Gupta (2002) develop a method called the *tree network model* by modifying the disassembly graph for use with a multiple-product demand-driven disassembly system with multiple common parts. Kongar and Gupta (2002b) propose a single-period integer goal-programming model for a DTO system to determine the best combination of multiple products to selectively disassemble in order to meet the demand for items and materials under a variety of physical, financial, and environmental constraints and goals. Kongar and Gupta (2006a) extend their previous work by using fuzzy goal programming to model the fuzzy aspiration levels of various disassembly-related goals. Langella (2007) develops a multiperiod heuristic that considers holding costs and external procurement of items. Gupta et al. (2009) use neural networks to solve the DTO problem. Kongar and Gupta (2009b) propose a linear-physical-programming-based solution methodology which can satisfy tangible or intangible financial-, environmental-, and performance-related measures of DTO systems then address the DTO problem using tabu search (Kongar and Gupta, 2009a).

The studies in the second line of research take into consideration the uncertainty related to disassembly yield. Inderfurth and Langella (2006) develop two heuristic procedures (i.e., one-to-one, one-to-many) to investigate the effect of stochastic yields on the DTO system. Imtanavanich and Gupta (2006) use these heuristic procedures to address the stochastic elements of the DTO system and then they use a goal-programming procedure to determine the number of returned products that satisfy various goals.

In this chapter, two single-period DTO models (first considering deterministic and then stochastic yields) are reviewed (both are by Inderfurth and Langella, 2008). The multiple end-of-life products used in these models all have reusable parts in common and all undergo complete disassembly. Both models allow for partial disassembly as well as for multiple time periods.

26.3.1 Single Period, Disassembly to Order (Deterministic Yields)

The single-period DTO system can be formulated as an integer programming model. Assumptions in the model include

- Products are completely disassembled.
- Demand for parts is deterministic.

- Part yields are deterministic.
- Demand for parts is completely fulfilled (through a combination of disassembly and new-part acquisition, each having an associated new-part-acquisition or excess-part-disposal cost).
- No lead times (for acquisition or for disassembly).
- No limits on the number of end-of-life products available for disassembly.

The objective function seeks to minimize the total costs. This expense is a function of the end-of-life product costs (acquisition, transportation, disassembly, part cleaning, inspection, sorting, etc.) CEP_g for product g of G products, new-part-acquisition cost CNP_k for part k of a total of na parts (where na accounts for all of parts in all of the multiple end-of-life products), and part disposal cost CPD_k. This is formulated as

$$\text{Minimize } Z = \sum_{g=1}^{G} \text{CEP}_g \cdot X1_g + \sum_{k=1}^{\text{na}} (\text{CNP}_k \cdot X2_k + \text{CPD}_k \cdot X3_k) \quad (26.1)$$

Where the decision variable $X1_g$ is the amount of end-of-life product g to acquire for disassembly, $X2_k$ is the amount of new part k to acquire in order to meet the overall demand, and $X3_k$ is the number of excess removed parts k that need to be disposed of.

The objective function is subject to two constraints. The first constraint ensures that the demand is satisfied through some combination of disassembly, and (if necessary) through the acquisition of new parts. This is modeled as

$$d_k \le X2_k - X3_k + \sum_{g=1}^{G} \text{YP}_{g,k} \cdot X1_g, \ \forall k \quad (26.2)$$

where d_k is the familiar demand of part k, and $\text{YP}_{g,k}$ is yield of part k from product g (the amount of part k obtained after disassembly of end-of-life product g). Also, Eq. (26.2) is slightly modified here from the original with the replacement of the equal sign with the less restrictive less-than-or-equal-to sign.

The final constraint forces the decision variables to be nonnegative integers and is given by

$$X1_g, X2_k, X3_k \ge 0 \text{ and integer}, \ \forall g, k \quad (26.3)$$

Integer programming models are generally considered to be NP-complete (for the decision version and NP-hard for the optimization version) and this is the case here. Also, this DTO model is similar to the NP-complete UNBOUNDED KNAPSACK PROBLEM. (While KNAPSACK asks for a selection of items from a group that will maximize the total value while taking a minimum weight, the unbounded version places no upper bound on the number of copies of each item in the group.) The structure of integer programming models often results in their complexity being addressed using branch-and-bound algorithms or the cutting-plane method (or commercial-grade software, which often obtains its solution using these algorithms exclusively or in concert with other techniques). It should be noted that the UNBOUNDED KNAPSACK PROBLEM has been shown to be solvable in pseudopolynomial time using dynamic programming (Andonov et al., 2000).

26.3.2 Single Period, Disassembly to Order (Stochastic Yields)

The quality of the parts obtained through the disassembly of end-of-life products is unknown and can be expected to be variable. As a result, the quantity of parts obtained is also unknown and potentially variable. These complications require the exploration of modeling stochastic yields. While several approaches have been used for this purpose (see Yano and Lee, 1995), Inderfurth and Langella model yield-uncertainty using the stochastically proportional yield approach. In this method, the quantity of usable parts is obtained from the number of end-of-life products disassembled multiplied by a weight (using a random fraction). This is similar to the single-period DTO with deterministic yields but making use of stochastic yields. The process uses *recourse models* (as introduced by Dantzig, 1955). Recourse models enable some degree of stochasticity when using linear programming. This is achieved by splitting decisions into two groups: proactive (made before the realization of the random variables) and reactive (or *recourse*; i.e., decisions made after). This is incorporated using *scenarios* (see Sen and Higle, 1999 for a detailed tutorial).

In a linear programming model with stochastic constraint parameters, the decision on the value for a decision variable (using typical linear programming notation) X in the objective function must be made before a realization of the constraint's random parameters of x are known (though their distribution is known). Using a recourse model allows these constraints to be violated, but at the cost of influencing the choice of X. As such, a second linear programming model then describes how each of these violations would be addressed. Recourse models have a second-stage value function and an expected value function and are formed into a deterministic equivalent. This deterministic equivalent is now a traditional linear programming model; however, it is extremely large. The number of scenarios $|SC|$ (the cardinality of the set of scenarios SC), using the DTO problem as an example, is the size of the set all possible outcomes for all random variables as calculated by

$$|SC| = w^{\sum_{g=1}^{G} n_g} \tag{26.4}$$

where w is the number of possible outcomes, and n_g represents the number of parts in end-of-life product g. Note that SC must obviously be a countable set. For example, consider two different end-of-life products, one consisting of two parts and one consisting of three parts (for a total of five parts, which may or may not be unique). If each part had two possible outcomes (yield of one or yield of zero; i.e., usable part or unusable part), Eq. (26.4) would give 2^5 or 32; with 10 parts, the number of scenarios jumps to 1024, and so on. This rapid growth ultimately limits the practicality of this method.

Using a recourse formulation, a single-period DTO system with stochastic yields can be modeled as a mixed integer programming model. The process here proceeds as

1. Proactively determine the number of each end-of-life product to be disassembled $X1_g$.
2. Realize the random yields.
3. Reactively make the new-part-acquisition $X2_{g,sc}$ and excess-part-disposal $X3_{g,sc}$ decisions (reactively since these decision variables depend on the yield rate realization).

The objective function is similar to that in the deterministic version. It still seeks to minimize the total costs, but it now contains both proactive and reactive cost decision variables. The proactive decision variable remains as $X1_g$, while the recourse decision variables (i.e., the variables dependent on the particular scenario sc) are denoted by $X2_{g,sc}$ and $X3_{g,sc}$. The objective function is formulated as

Minimize Z =

$$\sum_{g=1}^{G} \text{CEP}_g \cdot X1_g + \sum_{k=1}^{na} \sum_{sc=1}^{|SC|} \Pr(sc) \cdot (\text{CNP}_{k,sc} \cdot X2_k + \text{CPD}_k \cdot X3_{k,sc}) \tag{26.5}$$

where Pr(sc) is the probability of scenario sc occurring and

$$\sum_{sc=1}^{|SC|} \Pr(sc) = 1 \tag{26.6}$$

When considered over all scenarios and all parts, this probability provides the statistical expectation of the cost resulting from the recourse decisions.

The objective function is subject to three constraints, two of which are the nonnegativity constraints. As in the deterministic DTO model, the first constraint ensures that the demand is satisfied through some combination of disassembly, and (if necessary) through the acquisition of new parts; however, now the constraint exists for each scenario and part, not just for each part. This links yield-rate realizations with their corresponding recourse decisions in order to satisfy the demand. This is modeled as

$$d_k \leq X2_{k,sc} - X3_{k,sc} + \sum_{g=1}^{G} YP_{g,k,sc} \cdot X1_g, \forall k, sc \tag{26.7}$$

where $YP_{g,k,sc}$ is the yield of part k from product g in scenario sc. Equation (26.7) is again slightly modified from Inderfurth and Langella's model with a less than or equal to sign.

The final two constraints are the nonnegativity and integer constraints

$$X1_g \geq 0 \text{ and integer}, \forall g \tag{26.8}$$

and

$$X2_{k,sc}, X3_{k,sc} \geq 0, \forall k, sc \tag{26.9}$$

where the integer requirement falls only on the end-of-life product decision variable since the two part-decision variables now have a stochastic influence. With the scenarios possibly containing yield rates that are not necessarily integer, disposal and acquisition decisions will need to be treated as continuous in order to satisfy the integer-valued demand. If any results are applied on an actual DTO system, rounding needs to be used in order to implement the policy.

26.4 Additional Disassembly-Inventory Studies

Consideration of product return and remanufacturing options contributes two additional complexities to traditional inventory management approaches. First, uncertainty is added due to uncertainty in product returns. Second, there is a need for coordination between the remanufacturing supply chain and the traditional supply chain (Inderfurth and van der Laan, 2001). Researchers are continuing to develop various inventory models to deal with the complexities of demand and returns, while at the same time discovering and defining other inventory management issues affected by disassembly including costs, valuation, lead time effects, and spare part inventories. Contemporary disassembly-inventory research can be considered to encompass two types of inventory models: deterministic and stochastic.

In deterministic models, both stationary and dynamic demands are considered. Deterministic models are based on the assumption that demand and return quantities are known for entire planning horizon. They attempt to find an optimal balance between fixed setup costs and variable inventory holding costs.

With stochastic models, stochastic processes are employed to model demand and returns. Continuous and periodic review policies are two common approaches used in stochastic models. Continuous review models operate using continuous time and seek to determine the optimal static control policies in order to minimize the long-run average costs per unit of time (Fleischmann et al., 1997). Heyman (1977) presents the first study in this area by considering a continuous review strategy for a single-item inventory system with remanufacturing and disposal; an optimum disposal level is derived by assuming no fixed ordering costs and instantaneous outside procurement. Periodic review models search for optimal policies based on the minimization of expected costs over a finite planning horizon (Fleischmann et al., 1997).

CHAPTER 27

Disassembly-Line Just-in-Time

27.1 Introduction

While directly related to inventory theory, just-in-time is significant enough (and in the case of the disassembly application, complex enough) to warrant its own chapter. Section 27.2 presents an overview of just-in-time then Sec. 27.3 provides an introduction to disassembly just-in-time research. Section 27.4 reviews the multikanban system designed for use with a disassembly line.

27.2 Just-in-Time Background

Originally motivated by high land costs in Japan, *just-in-time* (JIT) inventory theory is based primarily on the Toyota *kanban* (Japanese for "card" or "ticket") system of material flow and on the concept of single minute exchange of dies (SMED) (Nahmias, 2009). SMED is a foundational element of JIT and its typical small production lot sizes. Credited to Shigeo Shingo, the success of the SMED program at Toyota led directly to JIT.

Die-changing operations are found on automobile and other manufacturing lines. In the original case of the Toyota car body stamping lines, die changing could take from 12 hours to almost 3 days due to the labor intense and precise process of exchanging the multi-ton dies and aligning them within less than a millimeter. Process improvements reduced this to less than 10 minutes by moving as many die-changing steps as possible off-line while the line continues to operate with the previous die (the Toyota case also required the integration of precision measuring devices for alignment). A component of lean production, die changes should take less than 10 minutes to be considered SMED.

Part of the Toyota Production System, kanban is a manual information system for implementing JIT (JIT, lean, and kanban are all closely related and exhibit a fair amount of overlap). Fundamental JIT concepts include

- **Work-in-progress (WIP) inventory reduction**: The WIP inventory is indicative of how tightly a JIT system is tuned.
- **Pull system**: Production at each stage is initiated only when requested.
- **Supplier involvement**: Suppliers need to be agreeable to meeting deliveries as needed; close proximity can become an important consideration.
- **An increase in plant efficiency**: Attributed to reductions in clutter, partially finished goods, rework, inspection, and so on, in addition to inventory savings.

A complete JIT system consists of

- **Production centers**: Workstations where production tasks are accomplished; these are sequential and ultimately result in a final product.
- **Holding area/store**: Intermediate storage location for batches.
- **Batches**: Groups of finished subassemblies.
- **Kanban physical tickets**: Containing part numbers; the two types are *production ordering kanbans* and *withdrawal kanbans.*
- **Posts/canisters/containers**: Holding place for kanbans; the three types are *production ordering kanban posts, withdrawal kanban posts,* and *store containers/canisters.*

An essential component of the process is the Toyota kanban calculation as given by

$$K = \frac{D_e L + W}{a} \tag{27. 1}$$

where K is the number of kanbans, D_e is the expected demand per unit time (e.g., per day), L is the production lead time (equal to the production time plus the waiting time plus the movement time), W is the buffer stock (equal to zero ideally; often 10 percent of D_eL is used), and a is the container capacity (often less than or equal to 10 percent of D_e). The maximum inventory calculation is given by

$$aK = D_e L + W \tag{27.2}$$

JIT is an alternative to other manufacturing environments including constant work in progress (CONWIP; system designed to force a desired level of WIP), reorder point, hierarchical production planning, and material requirements planning (MRP). Note also, that the existence of buffers in the JIT process distinguishes it from the constant-speed flow shop as required by disassembly-line balancing. Other considerations with JIT include

- A requirement for a constant production rate, which is not amenable to fluctuating demands.
- Unlike MRP, information is not made available at all levels, only at adjacent levels.
- Worker idle time in the case of a breakdown elsewhere on the line (often acceptable since it is claimed to be a Japanese philosophy that all workers are familiar with more than one portion of the production process).
- Small batches invite 100 percent inspection, which is good for quality control and total quality management processes.
- Process can result in increased worker idle time.
- SMED may be impossible or impractical in some environments.
- If set-up costs are high, JIT may cost more than maintaining inventory.

27.3 Just-in-Time Research

The applicability of the traditional kanban system to sudden or large variations in demand is very limited. As a result, researchers have developed various modified kanban methodologies to address demand variability while preserving the advantages of a traditional kanban system. Gupta and Al-Turki (1997) propose an algorithm for adjusting the number of kanbans in response to an unstable demand. Extending this research, Gupta et al. (1999) presented a novel system called the *flexible kanban system* that adjusts the number of kanbans in response to variations in demand and lead times. Gupta and Al-Turki (1998a, 1998b) show that the flexible kanban system is superior to a traditional kanban system when considering the interruptions caused by the preventive maintenance or breakdown of the material handling system. Tardif and Maaseidvaag (2001) suggest an *adaptive kanban system* that adjusts the number of kanbans using inventory and backorder levels. A *reactive kanban system* was presented by Takahashi and Nakamura (1999, 2000, 2002). The proposed system demonstrated adjusting of the number of kanbans and buffer sizes based upon detected changes in demand. Modification of a traditional kanban system considering the disassembly environment is another active research area. Kizilkaya and Gupta (2004, 2005a, 2005b) use discrete event simulation to test the performance of a novel pull-type disassembly-line control mechanism called *dynamic kanban system for disassembly line* which is based on the flexible kanban system originally developed by Gupta and Al-Turki (1997) for the production environment. The system dynamically changes the number of kanbans with respect to the demand and the capacity of the system. Udomsawat et al. (2003a, 2003b) and Udomsawat and Gupta (2005a, 2005b) analyze the applicability of a multikanban system for the case of multiple precedence relationships and component-discriminating demands. The effect of server breakdowns on the performance of a disassembly system controlled by a multikanban system is discussed by Udomsawat and Gupta (2005c, 2006). Takahashi et al. (2004) propose a reactive just-in-time ordering system for multistage production systems with changes in the mean and variance of demands. Improvements in disassembly-line pull systems, as provided through the use of embedded sensors present in end-of-life products, are considered by Ilgin and Gupta (in press a, in press b, in press c, in press d).

27.4 Multikanban System for End-of-Life Products

Implementing a kanban mechanism in a disassembly-line environment adds inherent complications including: product arrival, demand arrival, inventory fluctuation, and production control mechanisms. Udomsawat and Gupta (2008) propose addressing such complications by implementing a *multikanban system*. A modified inventory control mechanism, the multikanban system is designed in such a way that its implementation helps to reduce the inventory build up (which can be expected in a disassembly-line setting when a push system is implemented) in the system due to extensive supply and demand fluctuations. The multikanban system is able to ensure smooth operation of the disassembly line where multiple types of end-of-life products arrive for processing. This modified kanban system relies on the dynamic routing of kanbans, performed according to the state of the system. The effectiveness of this, or any kanban system, can be investigated using simulation.

Udomsawat and Gupta (2008) allow end-of-life products to enter the disassembly line at any workstation, depending upon the type of product being disassembled on the mixed-model line and its make-up of parts (as a result of parts being added or deleted during the product's lifespan). Similarly, the demand is allowed to occur at any workstation. The arrival points, the configuration and structural state of end-of-life products, and the variety of demanded items are the primary factors that make a disassembly line difficult to manage. These factors also lead to substantial inventory fluctuations.

27.4.1 Arrival Pattern

End-of-life products arriving at a disassembly line may consist of different combinations of a given set of parts. In this situation, the total number of possible combinations of a set of n parts $C(n)$ is given by

$$C(n) = 2^n - n - 1 \tag{27.3}$$

For example, a set of four parts {A, B, C, D} can produce up to 11 possible product or subassembly combinations: A-B, A-B-C, A-B-C-D, A-B-D, A-C, A-C-D, A-D, B-C, B-C-D, B-D, and C-D. By adding one more part to the set, the number of possible combinations increases from 11 to 26; the number of combinations increases exponentially with the increase in the number of parts. Fortunately, in most practical applications not all combinations are feasible. In some end-of-life products, such as a personal computer, parts are modular, allowing for different parts that have the same footprint able to be fitted in the same location. It is not unusual for these products to have been modified prior to disposal. Household appliances commonly arrive at a disassembly facility in fewer combinations because consumers rarely modify such products. Nevertheless, the workstation at which an end-of-life product enters the disassembly line is determined by the type of parts in the product and the combination of these parts that make up the product. Consider a disassembly line with three workstations with workstation 1 designated to disassemble part A, workstation 2 for part B, and workstation 3 for parts C and D. In this example, a product consisting of parts B, C, and D arriving at the disassembly line does not have to go to workstation 1 at all. This product can enter the disassembly line directly at workstation 2. Alternatively, an arriving product consisting of parts A, C, and D would have to enter workstation 1; however, after getting processed at workstation 1, it could skip workstation 2 entirely and go to workstation 3. Furthermore, end-of-life products with different precedence relationships must be processed through workstations in different sequences. These situations destabilize the disassembly line by causing an excess of materials at some workstations and a lack of materials at others, leading to undesirable fluctuations of inventory in the system. It is, therefore, crucial to manage the materials flow of the line.

27.4.2 Demand Fluctuation and Inventory Management

Although a disassembly line has many unique characteristics, in this chapter's line model the multilevel arrival of demand is the primary characteristic that makes the disassembly line much more complicated than a typical assembly line since demand can occur at any station on the line (in most assembly lines, demand arrives only at the last workstation). This arrival of external demand at workstations other than the last workstation creates a disparity between the number of demanded parts and the number of partially disassembled products. Thus, if the system responds to every request for parts, it would end up with a significant amount of extra inventory of parts that are in low demand. Because it is desired that the service level be maximized, it is essential to develop a sound methodology to control the system and manage the extra inventory produced.

27.4.3 Disassembly-Line Multikanban Model Overview

The multikanban model for the disassembly line promises three advantages. First, this model implements the pull concept; it carries minimal inventory and permits initiation of disassembly only when there is an actual demand for parts or materials. As a result, the facility does not have to carry large inventories (saving on both inventory-carrying costs and the cost of disassembling unwanted parts). Second, the multikanban pull system holds promise for assisting in tracking inventory. Third, the multikanban mechanism relies on routing rules that direct kanbans to the workstations that supply the most needed parts or subassemblies. In this way, customer satisfaction is retained while the kanban routing takes care of fluctuations in the demand for various parts. Design considerations in the multikanban model include material types, kanban types, kanban routing mechanisms, product selection, and kanban level.

27.4.4 Material Types

Two basic material types are considered in the disassembly-line multikanban system, parts, and subassemblies. In the multikanban system, a part is considered to be a single item that cannot be further disassembled. Parts are placed in the part buffer and await retrieval via customer demand. In contrast, a subassembly can be further disassembled and is composed of at least two parts. Both types of materials can be further distinguished as regular or overflow items. *Regular items* are items that are demanded by customers or downstream workstations. In order to fulfill this demand, a server must disassemble the demanded part or subassembly. The remaining item from this disassembly process that does not fulfill any request is referred to as an *overflow item*. Because the disassembly process is initiated by a single kanban, the overflow item will not have a kanban attached to it. However, the overflow item is routed in the same way as the regular item. The only difference between these two items is that the overflow item is given the priority of being retrieved after it arrives at its buffer. It should be noted that as long as there is an overflow item in the buffer, its demand will not initiate any further disassembly. This helps the system to eliminate extra inventory that is caused by any unbalanced demand.

27.4.5 Kanban Types

Corresponding to material types, there are two basic types of kanbans in the system, part kanbans and subassembly kanbans. A *part kanban* is attached to a disassembled part that is placed in the part buffer of the workstation where it was disassembled. Similarly, a *subassembly kanban* is attached to a subassembly that is placed in the subassembly buffer of the workstation where one of its parts was removed. A part placed in a part buffer can be retrieved via an external demand. When authorized, a subassembly placed in the subassembly buffer is routed for disassembly to the next workstation based on its disassembly precedence constraints.

At the first workstation, products arrive only from outside sources; at any other workstation x (where $1 < x \leq n - 1$; therefore, for a product with n parts the multikanban system uses $n - 1$ workstations), there are two possible types of arrivals. The first type is a subassembly that arrives from an earlier workstation and is referred to as an *internal subassembly*. There is always a subassembly kanban attached to an internal subassembly. The second type is a product (or subassembly) that arrives from outside sources and is referred to as an *external subassembly*. No kanban is attached to an external subassembly. This aspect is also true for products arriving from external sources to the first workstation. As long as there is an external product or subassembly available at an input buffer, the system will process it first before processing any available internal subassembly. This avoids unnecessary pulling of an internal subassembly from an upstream workstation. Thus, the number of kanbans attached to the internal subassemblies will remain constant throughout the process.

27.4.6 Kanban-Routing Mechanism

Given a product with n parts, consider a workstation y where $1 \leq y \leq n - 1$. When a demand for part y arrives at the part buffer of workstation y, one unit of part y is retrieved and the part kanban y attached to it is routed to the most desirable workstation. The procedure for determining the most desirable workstation to route part kanban y is as follows (this procedure is not applicable to part kanbans $n - 1$ and n; in both cases these kanbans are routed to the input buffer of the final workstation).

A part kanban originating from workstation y will be routed to a workstation x (where $1 \leq x < y$) or workstation y depending on the availability and the desirability of the subassembly that contains part y. Routing part kanban y to workstation x where $1 \leq x \leq (y - 1)$, will result in an immediate separation of part y from part x. Thus, the only subassembly located at the input buffer of workstation x that would be useful is a subassembly that contains only parts x and y. If this type of subassembly exists in the input buffer of workstation x, then workstation x is qualified. Similarly, if there is at least one subassembly in the input buffer of workstation y, then workstation y is qualified.

Next, a selection of the most desirable workstation to route part kanban y (from among the qualified workstations such that, if chosen, it will cause the least amount of extra inventory in the system) needs to be made. Choosing workstation x will increase the inventory level of part x by an additional unit. Thus, the best workstation x is the one that has the greatest need for this part. This workstation can be determined by checking the backorder levels for demand x; in the case of a tie the most downstream workstation is chosen. Choosing workstation y creates an extra subassembly that will be further disassembled at downstream workstations. If workstation y is selected, then an appropriate subassembly to disassemble must also be selected. For example, if a backorder exists at the part buffer of workstation z (where $y < z \leq n - 1$), then one option would be to disassemble a subassembly that contains only parts x and z. If more than one workstation qualifies as having low inventory, then the one with the greatest need is selected. In the case of a tie, the most downstream workstation is selected.

Finally, a comparison is made between the inventory needs of workstations x and y. If the need of workstation x is greater than or equal to that of workstation y, then part kanban y will be routed to workstation x; otherwise it will be routed to workstation y. Note that whenever an external subassembly is available, it is always chosen first. Internal subassemblies are only to be used when no external subassembly of the desired kind are available. Subassembly kanbans are routed in a fashion similar to part kanbans.

27.4.7 Selection of Products

Since multiple combinations of products are permitted, the worker may have several options when selecting a product for disassembly. If the authorization for disassembly is initiated by the subassembly kanban y_a, which can occur only at workstation x (where $1 \leq x < y$), the workers will have no option but to select the subassembly that results in immediate separation of subassembly y_a, namely subassembly x-y_a. (i.e., the subassembly consisting of part x and subassembly y_a). If instead of the subassembly kanban, the authorization of disassembly is initiated by part kanban y at workstation x (where $1 \leq x < y$), the worker will have no option but to remove subassembly x-y from the product buffer because the only subassembly that results in immediate separation of part y is the subassembly x-y. However, if the part kanban y arrives at workstation y, there are multiple options because every subassembly located in the product buffer contains part y and always results in the immediate separation of part y. In this case, it needs to be determined whether or not the part or subassembly that is created by the disassembly will result in an excess of inventory.

Subassembly y_z is then chosen such that z has the most desirable part/subassembly ranking based on either the request of subassembly kanban z at workstation y (existing kanban z at the workstation y) or on the current inventory level of subassembly (or part) z.

27.4.8 Determining Kanban Level

The kanban level plays an important role in the multikanban mechanism, as it maintains a proper flow of parts and subassemblies at a desired level throughout the system. The kanban level can be determined by considering product arrival rates, part- and subassembly-demand arrival rates, and part removal rates (using disassembly times). The number of part kanbans NPK_x for part x or the number of subassembly kanbans NSK_x for subassembly x can be computed at any point on the disassembly line using

$$\mathrm{NPK}_x = \max\left(1, \frac{\mathrm{RRP}_x}{\mathrm{FRP}_x}\right) \tag{27.4}$$

and

$$\mathrm{NSK}_x = \max\left(1, \frac{\mathrm{RRS}_x}{\mathrm{FRS}_x}\right) \tag{27.5}$$

where RRP_x is the request rate of part x, FRP_x the furnish rate of part x, RRS_x the request rate of subassembly x, FRS_x the furnish rate of subassembly x, and the max function returns the larger of the two values in parenthesis. These request rates and furnish rates can be calculated as

$$\mathrm{RRP}_x = \mathrm{DRP}_x \text{ for } 1 \le x \le n \tag{27.6}$$

where DRP_x is the demand rate of part x and n is the maximum number of parts ($n - 1$ is the maximum number of workstations);

$$\mathrm{FRP}_x = \sum_{z=1}^{x} \mathrm{LRP}_{x,z} \text{ for } 1 \le x \le n \tag{27.7}$$

where $\mathrm{LRP}_{x,z}$ is the removal rate of part x at workstation z;

$$\mathrm{RRS}_x = \mathrm{LRS}_y \tag{27.8}$$

where LRS_y is the removal rate of subassembly y (and y is the next part to be removed in the disassembly sequence); and finally

$$\mathrm{FRS}_x = \mathrm{ARS}_x + \sum_{z=1}^{j-1} \mathrm{LRP}_{y,z} \tag{27.9}$$

where y is the part most recently removed, ARS_x is the arrival rate of subassembly x (which is coming from external sources), and j is the identification of the current workstation.

In the case of a part kanban, which is requested only from a single source, the request rate is equal to the customer demand arrival rate; however, because the part kanban arrives from several sources in the system, the furnish rate is the summation of the arrival rates from all possible sources. In the case of the subassembly kanban, the furnish rate is influenced by both the time required for disassembly and by the subassembly arrival rate from external sources. Thus, all external and internal arrival rates of subassemblies at the buffer are taken into account. Similarly, the two requesting sources—the demand for the target part and the demand for a residual subassembly—affect the request rate.

The numbers of kanbans are determined at the beginning of the disassembly process. It is clear that demand, supply, disassembly time, and product structure affect the computation of the number of kanbans.

27.4.9 Simulation Results

Simulation experiments conducted in comparison with a traditional push system (all arriving products processed continuously in the order of their arrival) demonstrate the multikanban mechanism's ability to significantly reduce the average inventory while maintaining the parts' demand-fulfillment rates (Udomsawat and Gupta, 2008). Neither system offers a significant difference in service level over the same time period. For each part, the multikanban system does not sacrifice the level of service in exchange for customer demand; instead the system seeks to service the customer demand by taking the number of waiting customer demands into consideration prior to routing the kanban to the selected workstation. Adhering to the product selection rules, only the product that is the best choice for disassembly is selected. Disassembly is prioritized only to where and when there is a need. In the push environment, the system builds up inventory in order to fulfill the demand of customers. This large amount of inventory effectively addresses any fluctuation in demand; however, this large amount of inventory results in higher carrying costs and a higher cost of operation. These drawbacks become severe when disassembling products that contain parts that are low in demand, have short shelf lives, or incur high costs of disassembly. The multikanban mechanism addresses fluctuations in demand by routing the kanbans to the most suitable workstation. Studies indicate that the multikanban system is able to reduce the inventory level by an average of 33 percent, while keeping the demand of customers at levels comparable to a push system (Udomsawat and Gupta, 2008).

A customer is considered to be waiting if there is no part stored in the outbound buffer where it arrives. In the multikanban system, the average waiting time as indicated by simulation is noticeably lower than that in the push system. This is a result of the effective routing rules that tend to favor higherdemand parts. Also, the multikanban system responds to backorders faster than the push system does. This is because a free kanban is routed to the workstation where disassembly of a subassembly or the end-of-life product provides both the target part and the remaining parts that are required the most.

Despite serious complications in a disassembly-line environment, the Toyota Production System can be adapted to perform effectively. The multikanban mechanism enables the disassembly line to meet the demands of customers while stabilizing fluctuations in inventory through the use of real-time routing adjustments of kanbans rather than the traditional technique of adjusting the number of kanbans.

CHAPTER 28
Disassembly-Line Revenue

28.1 Introduction

Disassembly has been shown to have the potential for being a component of a profit center. Disassembly-line revenue has been considered by a variety of researchers, with one of these applications featured in this chapter. Section 28.2 provides a background in the concept of a profit-oriented disassembly line while Sec. 28.3 reviews a mixed integer programming model of this type of line.

28.2 Disassembly-Line Revenue Background

The potential for profit in end-of-life processing can be a significant—or in some cases, the only—disassembly motivator for some organizations. It then becomes important to understand what level of disassembly should be performed, for what products, and to what end: remanufacturing, refurbishing, or recycling. These three sustainable end states are referred by Pearce (2009) as *product reconstruction*. Remanufacturing is listed by Pearce as the process of completely disassembling a used product, repairing or replacing worn and obsolete components, and potentially adding upgrades to correspond with products currently on the market. *Refurbishing* is described as the process where a product is reconditioned; bringing it back to its original configuration and condition (a typical example of this type of product would be laser printer cartridges). Finally, recycling is given as being the process by which a product is disassembled down to some combination of parts (to be used in other products) and/or materials (such as aluminum, glass, paper, etc.). Claiming that production reconstruction can offer attractive consumer prices (50 to 75 percent lower than those of new products), high-quality goods, and higher profit margins

(20 percent versus 3 to 8 percent for new products), Pearce's research provides a management-oriented quantitative- and case-based study to determine where remanufacturing, refurbishing, or recycling makes the most sense. Six types of customers are identified as being appropriate for profitable product reconstruction:

1. Customers who need to retain a specific product due to its technically defined role in their current process (e.g., manufacturing equipment on a line with a defined speed, form, footprint, power, and material inputs and outputs).
2. Customers who wish to avoid any corporate, regulatory, or guideline- or process-oriented requirement to specify, approve, or certify a completely new product.
3. Customers with low utilization of new equipment (i.e., a refurbished or remanufactured product may be financially feasible where a new item is not).
4. Customers who wish to continue using a product that has been discontinued by the original manufacturer.
5. Customers who wish to extend the service life of used products, whether or not they have been discontinued.
6. Customers who are interested in environmentally friendly products.

28.3 Disassembly-Line Revenue Modeling

Part removal sequencing is modeled by Altekín et al. (2008) for optimizing disassembly-line profit. This study provides an example of a mixed integer programming model of the disassembly line, as well as consideration of incomplete disassembly, and a demonstration of a differentiation between tasks (work to be performed) and parts (individual, tangible components). (Where k was used to represent any of the n parts/tasks in Part II, kt is used to represent any of the nt tasks and kp to represent any of the np parts in Part III.) This revenue model is also used here to demonstrate treatment of artificial tasks (which have part removal times and costs of zero, and which do not remove any parts) when the graphical model from Chap. 5 is used.

The objective function seeks to maximize the revenue earned (as calculated by the product of the revenue per unit demanded REV_{kp} and the number of those parts that are removed q_{kp}), which is reduced by the cost of removing all released parts (the product of the cost of each disassembly task undertaken C_{kt} and the binary variable $x_{\text{kt},j}$ which is equal to one if task kt is assigned to workstation j) and the fixed costs associated with the workstations (the product of the cost per unit time of keeping one station open S, and the total of all open workstations according to the decision variable u_j as defined by Eqs. (28.3) through (28.5) and multiplied by the cycle time CT). This is modeled as

$$\text{Maximize } Z = \sum_{\text{kp}=1}^{\text{np}} \text{REV}_{\text{kp}} \cdot q_{\text{kp}} - \sum_{\text{kt}=1}^{T_{\text{AL}}} \sum_{j=1}^{\text{nt}} C_{\text{kt}} \cdot X_{\text{kt},j} - S\sum_{j=1}^{\text{nt}} j \cdot u_j \tag{28.1}$$

where

$$X_{\text{kt},j} = \begin{cases} 1, & \text{if task kt is assigned to workstation } j \\ 0, & \text{otherwise} \end{cases} \tag{28.2}$$

with, per the Chap. 5 representation, T_{AL} identifying the last of AL artificial tasks (and representing the total number of possible tasks; i.e., the sum of all nt tasks and all AL artificial tasks giving the set $\{1, 2, 3, \ldots, n, T_1, T_2, T_3, \ldots, T_{\text{AL}}\}$ when a counter starts prior to T_1). The decision variable u_j is defined as

$$u_j = \begin{cases} \text{CT}, & \text{if } \quad X_{T_{\text{AL}},j} = 1 \\ 0, & \text{otherwise} \end{cases} \tag{28.3}$$

(note that only one workstation, the one containing the final artificial task T_{AL}, will have a nonzero value),

$$u_j \le X_{T_{\text{AL}},j} \sum_{\text{kt}=1}^{T_{\text{AL}}} \text{PRT}_{\text{kt}}, \ \forall j \tag{28.4}$$

and

$$\text{CT} \le \sum_{j=1}^{\text{nt}} u_j \tag{28.5}$$

Other constraints include the requirement that the station times must remain within the cycle time CT as given by

$$\sum_{\text{kt}=1}^{T_{\text{AL}}} \text{PRT}_{\text{kt}} \cdot X_{\text{kt},j} \le \text{CT}, \ \forall j \tag{28.6}$$

and the cycle time CT must be less than or equal to its upper bound CT_{upper} (which would need to be provided by the decision maker) per

$$\text{CT} \le \text{CT}_{\text{upper}} \tag{28.7}$$

Since a product may contain multiple copies of the same part, the number of units of a positive-revenue-generating part (where P^+ represents the set of parts having a positive revenue) is determined as the minimum of its demand d_{kp} and the total count of each part removed by a given task $\mathrm{nr}_{\mathrm{kt,kp}}$ as seen in Eq. (28.8) where

$$q_{\mathrm{kp}} \le d_{\mathrm{kp}}, \forall \mathrm{kp} \in P^{+} \tag{28.8}$$

and in Eq. (28.9) where

$$q_{\mathrm{kp}} \le \sum_{\mathrm{kt}=1}^{T_{\mathrm{AL}}} \sum_{j=1}^{\mathrm{nt}} \mathrm{nr}_{\mathrm{kt,kp}} \cdot X_{\mathrm{kt},j}, \forall \mathrm{kp} \in P^{+} \tag{28.9}$$

The removed parts are further accounted for using

$$q_{\mathrm{kp}} \ge \sum_{\mathrm{kt}=1}^{T_{\mathrm{AL}}} \sum_{j=1}^{\mathrm{nt}} \mathrm{nr}_{\mathrm{kt,kp}} \cdot X_{\mathrm{kt},j}, \forall \mathrm{kp} \in P^{-} \tag{28.10}$$

where P^{-} represents the set of parts having negative revenue.

Allowing each task to be assigned to at most one workstation gives

$$\sum_{j=1}^{\mathrm{nt}} X_{\mathrm{kt},j} \le 1, \forall \mathrm{kt} \tag{28.11}$$

which may be changed to an equality for tasks that must be performed.

Allowing each task to be assigned to at most one workstation gives

$$\sum_{j=1}^{\mathrm{nt}} X_{\mathrm{kt},j} \le 1, \forall \mathrm{kt} \tag{28.12}$$

AND and OR precedence relations, and OR successor relations are constrained by

$$X_{\mathrm{kt},j} \le \sum_{y=1}^{j} X_{x,y}, \forall \mathrm{kt}, j, x \in \mathrm{AP}_{\mathrm{kt}} \tag{28.13}$$

$$X_{\mathrm{kt},j} \le \sum_{y=1}^{j} \sum_{x \in \mathrm{OP}_{\mathrm{kt}}} X_{x,y}, \forall j, \mathrm{kt} \tag{28.14}$$

$$\sum_{j=1}^{\mathrm{nt}} \sum_{x \in \mathrm{OS}_{\mathrm{kt}}} X_{x,j} \le \sum_{j=1}^{\mathrm{nt}} X_{\mathrm{kt},j}, \forall \mathrm{kt} \tag{28.15}$$

and

$$\sum_{x \in \mathrm{OS}_{\mathrm{kt}}} X_{x,j} \le \sum_{y=1}^{j} X_{\mathrm{kt},y}, \forall j, \mathrm{kt} \tag{28.16}$$

with the sets of AND predecessors, OR predecessors, and OR successors of task kt listed as $\mathrm{AP}_{\mathrm{kt}}$, $\mathrm{OP}_{\mathrm{kt}}$, and $\mathrm{OS}_{\mathrm{kt}}$, respectively.

It is desired that no task be assigned to any workstations that comes after the last artificial task T_{AL} (this also determines the final NWS count), so

$$\sum_{j=1}^{\text{nt}} j \cdot X_{\text{kt},j} \leq \sum_{j=1}^{\text{nt}} j \cdot X_{T_{\text{AL}},j}, \ \forall \text{kt} \tag{28.17}$$

There is also a need for artificial tasks to only be assigned to workstations that contain tasks that are not artificial tasks (otherwise, no real workstation is required or desired to be accounted for) which is constrained by

$$X_{\text{kt},j} \leq \sum_{\text{kt}=1}^{T_{\text{AL}}} \text{PRT}_{\text{kt}} \cdot X_{\text{kt},j}, \ \forall j, \text{kt} \in \{T_1, T_2, T_3, \ldots, T_{\text{AL}}\} \tag{28.18}$$

Finally, binary and non-negativity considerations give

$$\text{CT}, u_j, q_{\text{kp}} \geq 0, \forall j, \text{kp} \tag{28.19}$$

and

$$X_{\text{kt},j} \in \{0, 1\}, \quad \forall \text{kt}, j \tag{28.20}$$

Mixed integer programming models are usually seen to be NP-complete (for the decision version and NP-hard for the optimization version) and this would be expected to be the case here. The structure of the mixed integer programming disassembly-line revenue model can be addressed either with commercial math-programming software or problem-specific heuristics, both of which are used by Altekín et al. (2008).

28.4 Additional Disassembly-Line Revenue Studies

Arola et al. (1999) provide an economic case study of recovering plastic streams from demanufactured consumer electronics and contribute a wide variety of cost and other quantitative revenue-related data.

Goksoy (2010) considers a disassembly line with a fixed number of workstations and a limited supply of differing products to be disassembled. The demonstrated objective is to maximize the total value of the recovered parts. The study includes the determination of several upper bounds and one lower bound.

Willems et al. (2004) provide two case studies (a washing machine and an electric kitchen pot) and a linear programming model to determine the economic feasibility of different levels of demanufacturing different products. These levels range from complete disassembly and reuse, to no disassembly followed by shredding and recycling.

CHAPTER 29

Unbalanced Disassembly Lines

29.1 Introduction

While Part II dealt with balancing the disassembly line, in this chapter the concept of an unbalanced disassembly line is reviewed. Unbalancing may be unintentional or intentional (in the case of stochastic part removal times and just-in-time-style buffers between workstations). This chapter presents an overview of unbalanced lines, a necessary queueing theory introduction, and consideration of the benchmark data set from Part II, and is organized as follows: Section 29.2 provides a background of the concept of an unbalanced line while Sec. 29.3 further details stochastic line modeling. Section 29.4 reviews the field of queueing theory which is a core concept in understanding intentionally unbalanced lines. Section 29.5 reviews the Chap. 10 benchmark data set for use in quantifying the level of unbalance on intentionally or unintentionally unbalanced lines.

29.2 Background

Disassembly lines are inherently multicriteria, with its balance having the possibility of being one of the lower priorities. This is due to other criteria—for example, removing valuable or hazardous materials early on in the process—possibly being of more importance than obtaining an optimal balance. In addition, complete disassembly may not be desired, required, or even possible, resulting in only partial disassembly being conducted. The result may be a disassembly sequence that readily satisfies a decision-maker's primary requirements, but at the expense of an unbalanced line. However, obtaining the decision-maker's primary requirements may not sufficiently justify exceptionally poor balance. Alternatively, the difference between a balanced disassembly line and one that is unbalanced—but at the expense of other desirable criteria—may turn out to be so insignificant that a focus purely on balance may obfuscate the benefits of considering other criteria. Therefore, metrics and benchmark data are needed to quantitatively evaluate the merits of all considered criteria, including the level of balance—or unbalance (McGovern and Gupta, in press). The multicriteria benchmark data set and associated metrics developed in Part II can be used in quantitatively evaluating an unbalanced paced disassembly line.

29.3 Stochastic Line Modeling and Unbalanced Lines

Unbalancing an unpaced production line has been shown to potentially increase production rates when task times are stochastic and appropriate buffers are provided between workstations (Hillier and Boling, 1966, 1979). Unbalancing can also be an unintended consequence of disassembly lines, even when considering a paced line and deterministic task times.

Unbalancing a production line was first proposed by Hillier and Boling (1966) in recognition of studies conducted that modeled flow shops using queueing theory. Modeling was accomplished using independent queues in series, with an infinite queue before the first server and a finite buffer of capacity B for all others. Each workstation was modeled individually as an $M/E_K/1/B+1$ queue ($M/E_K/1$ for the first workstation, where M represents arrivals according to a Poisson distribution and E_K represents service according to an Erlang distribution having shape parameter K) and having probabilistic *arrival rates* (in agreement with the nature of an unpaced line; i.e., a flow shop with buffers) and probabilistic *service rates* (conceptually equivalent to probabilistic task times). It was found not only that balancing the line was detrimental to its production rate, but that unbalancing should be performed in such a way that the largest amounts of work should be allocated at the end workstations and the least amounts allocated to the ones in the middle (Hillier and Boling, 1966, 1979; Stecke and Solberg, 1985).

29.4 Queueing Theory Background

Queueing theory provides descriptive (i.e., describes a situation to allow for analysis but does not provide a solution) rather than prescriptive solution (i.e., prescribes a solution; e.g., linear programming) models. A queueing (i.e., waiting line) theory model consists of a calling population and the queuing system itself (see Fig. 29.1).

The *calling population* characteristics include its size (finite or infinite) and generation pattern (or arrival rate; generally interarrival rates are assumed to be exponential, i.e., following a Poisson distribution). The *queueing system* is the portion of the model that deals with

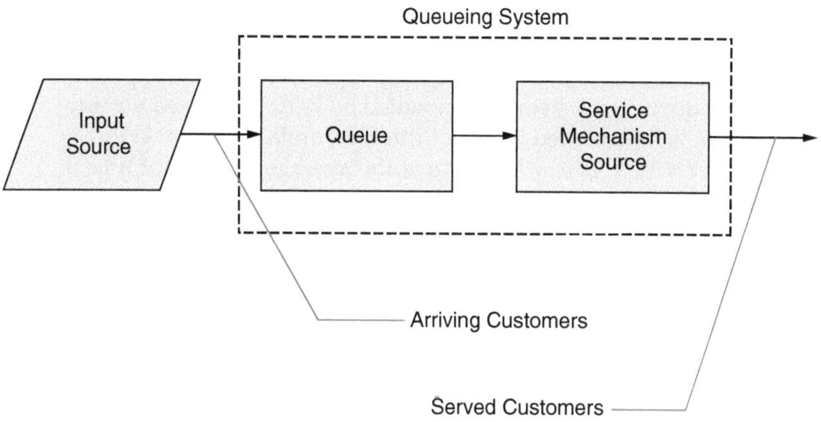

FIGURE 29.1 The basic queueing process.

the queue and the service facility. The *queue* can be finite or infinite; the *service discipline* may be first in first out, last in first out, or random; the *service facility* is defined by the number of servers, their arrangement (series or parallel), and the service pattern (generally assumed to be exponential); exits from the system occur due to completion of service, balking, or reneging (leaving the system after joining). Model assumptions include (except for some specific variations): only one event can occur at a time, arrivals occur randomly and independently of other arrivals and according to a Poisson distribution, service times vary according to an exponential distribution, and that the system is past its *transient period* and has entered steady-state operation (i.e., it is independent of its initial state). Where arrivals follow a Poisson distribution, the probability of x arrivals in a specific time period is given by

$$\Pr(x) = \frac{\lambda^x \cdot e^{-\lambda}}{x!} \qquad \text{for } x = 0, 1, 2, \ldots \tag{29. 1}$$

where λ is the mean number of arrivals per time period and e is Euler's number (approximately 2.718). Where service rates are exponential, the probability that the service time will be less than or equal to a time of length t is given by

$$\Pr(\text{service time} \le t) = 1 - e^{-\mu t} \tag{29.2}$$

where the service rate μ is the mean number of units that can be served per time period.

Several parameters describe the system. The *state* of the system k is equivalent to the number of customers in the waiting line plus the service facility. Maximum number of customers in the system is n. The probability of being in state k (i.e., the probability of exactly k

customers in the system) is $\Pr_k$. The length of the queueing system is given by L (the expected number of customers in the queueing system), while the length of the queue itself is given by L_q. The expected time in the queueing system is represented by W (the expected waiting time in line is represented by W_q). Little's formula gives $L = \lambda W$, $L_q = \lambda W_q$, and $W = W_q + 1/\mu$, where $1/\mu$ is the average amount of time a customer spends in a service facility.

Kendall notation is used to depict a chosen model's characteristics. In the first portion of the six-tuple A/B/C : D/E/F, A describes the arrival pattern (Markovian M, general G, degenerate D, Erlang E_K, etc.), B describes the service pattern (M, G, D, or E_K), and C describes the number of servers s in parallel. In the second portion, D describes the service discipline (first come first served FCFS, last come last served LCLS, rotation, random SIRO, priority PRI, or any other GD), E describes the storage capacity (the maximum number of customers allowed in the system: limited or unlimited), and F describes the calling population (finite or infinite). (Note that the default for the second portion is FCFS/∞/∞; if this describes the model, it is often not included in its Kendall notation; also, it is not uncommon to see the D/E/F portion of the six-tuple given in the order E/F/D.) For example, $M/M/1$: FCFS/n/∞ (or $M/M/1/n$) is the Kendall notation for a single server, finite waiting line queueing system with Poisson arrivals and exponential service.

When queueing models are based on the birth-and-death process (in the context of queueing theory the term *birth* refers to an input or arrival and the term *death* refers to a departure or a completed service), the system can be represented by a *rate diagram*. Set up as a Markov chain, a rate diagram is a conceptual model that describes the possible states of the model and the transitions from one state to another. The "rate in = rate out" principle can then be used to generate *balance equations*. Finally, since there are $n + 2$ equations (given by the balance equations) and $n + 1$ unknowns (given by the possible states of the rate diagram) the balance equations can be solved (where n is set to infinity, inferences can be made about the progression of the equations and their eventual convergence). These solutions, along with Little's formula, are then used to provide closed formulations of all parameters of interest. Additional background and further details are provided by Hillier and Lieberman (2005).

As an example, $M/M/1$ models are commonly used to model many capacities such as those of airports. A prototypical airport example by Hillier and Lieberman (2005) considers the entire queueing system to include holding aircraft (the holding pattern is the queue), approaching aircraft as arriving customers which are cleared one at a time on the approach (i.e., only one aircraft is on approach, landing, then taxiing clear at any time—once that aircraft is clear of the runway, a holding aircraft can commence its approach), the runway as the server, a given arrival rate, and a service rate related to the time required for the aircraft to clear the runway. The goal of the example is to determine the average number of aircraft holding and average time in holding, with further study related to the effects of increasing the arrival rate and increasing the number of runways (i.e., $M/M/2$, and $M/M/3$). Alternatively, a nonexponential distribution could be used as an extension to the example by Hillier and Lieberman; for example, an *M/G/*1-based model (specifically an *M/D/*1 model) in order to enable mimicking the separation provided by the air traffic controller. Once an aircraft has traveled a distance equivalent to the allowable separation interval (and well before reaching the runway, landing, and taxiing clear), the next aircraft is allowed out of the queue and into the service area (the approach). While both models consider that only landings are taking place on a given runway (i.e., no interfering take offs) and the queue is the holding pattern, the *M/G/*1 model considers the service mechanism to be degenerate (i.e., deterministic) and allows for multiple aircraft on approach (since an aircraft is considered serviced and leaves the queueing system once it is a safe distance ahead of the following aircraft). As long as the mean $1/\mu$ (recalling that μ represents the service rate in queueing theory) and the variance σ^2 of the service-time distribution are known, the *M/G/*1 queueing formulations provide closed-form solutions to all parameters of interest. The *M/G/*1 queueing formulation used for this problem would consist of the following standard queuing theory equations: the probability of being in state zero (e.g., no aircraft in the system) P_0 is

$$P_0 = 1-\rho, \quad \rho<1 \tag{29.3}$$

where the *traffic intensity* (also known as the *utilization factor*) ρ is defined as

$$\rho = \frac{\lambda}{s \cdot \mu} \tag{29.4}$$

and s is the number of servers (e.g., set to one in the example here). Queue lengths and waiting times are given by

$$L = L_q + \rho \tag{29.5}$$

$$L_q = \frac{\lambda^2 \cdot \sigma^2 + \rho^2}{2 \cdot (1-\rho)} \tag{29.6}$$

$$W = W_q + \frac{1}{\mu} \tag{29.7}$$

and

$$W_q = \frac{L_q}{\lambda} \tag{29.8}$$

If it is of interest to consider differing arrival rates, the effective arrival rate λ_e is then calculated using

$$\lambda_e = \sum_{x=0}^{\infty} \lambda_x \cdot P_x \tag{29.9}$$

Queueing theory provides a rigorous and accepted probabilistic methodology to measure the performance of waiting line systems and is often applied to manufacturing systems. Along with the sample queuing model given here, other queueing model formulations may be more applicable, as may be the case with deterministic models, other types of stochastic models, analysis of experimental data, or use of simulation software.

29.5 Benchmark Data for Measuring Unbalance

While designed for quantifying the level of balance, the A Priori data set of Chap. 10 is equally effective at providing a quantitative evaluation of the level of unbalance (along with other objectives); however, if the primary consideration is not the measure of balance, the results may not provide an adequate measure. In this case, the A Priori data set can be further modified to gather additional information, while still providing an optimal level of balance and optimal or near-optimal (but always known) values for the other metrics. For example, if the primary consideration was the evaluation of a heuristic's ability to remove hazardous parts as early (or as late) as possible, the parts designated as hazardous would consist of the entire latter half of each group of parts having the same A Priori data set part removal time. Alternatively, if the primary objective was demand, each group of parts having the same A Priori data set part removal time would contain parts having increasingly larger demand values. Finally, if using direction as the primary criteria, parts could be initialized with alternating directions. With these designs, perfect balance could still be achieved while attaining optimal placement for any of the other criteria, but it should be noted that only one of these extensions should be selected at any time to simplify the calculation of optimal values for the remaining two criteria.

References

Adenso-Díaz, B., Garcia-Carbajal, S., and Gupta, S. M. (2008). "A Path-Relinking Approach for a Bi-criteria Disassembly Sequencing Problem." *Computers & Operations Research*, 35, 3989–3997.

Agrawal, S. and Tiwari, M. K. (2006). "A Collaborative Ant Colony Algorithm to Stochastic Mixed-Model U-shaped Disassembly Line Balancing and Sequencing Problem." *International Journal of Production Research*, 46, 1405–1429.

Aho, A. V., Hopcroft, J. E., and Ullman, J. D. (1974). *The Design and Analysis of Computer Programs*, Addison-Wesley, Reading, Mass.

Altekín, F. T. (2005). *Profit Oriented Disassembly Line Balancing*, Ph.D. Dissertation, Department of Industrial Engineering, Middle East Technical University, Ankara, Turkey.

Altekín, F. T., Kandiller, L., and Ozdemirel, N. E. (2008). "Profit-Oriented Disassembly-Line Balancing." *International Journal of Production Research*, 46, 2675–2693.

Andonov, R., Poirriez, V., and Rajopadhye, S. (2000). "Unbounded Knapsack Problem: Dynamic Programming Revisited." *European Journal of Operational Research*, 123(2), 394–407.

Andrés, C., Lozano, S., and Adenso-Díaz, B. (2007). "Disassembly Sequence Planning in a Disassembly Cell Context." *Robotics and Computer-Integrated Manufacturing*, 23(6), 690–695.

Arola, D. F., Allen, L. E., Biddle, M. B., and Fisher, M. M. (1999). "Plastics Recovery from Electrical and Electronic Durable Goods: An Applied Technology and Economic Case Study." *Proceedings of the 1999 Society of Plastics Engineers Annual Recycling Conference*, Dearborn, Mich., 241–250.

Bader, C. D. (2004). "Where Will All That C&D Debris Go?" *MSW Management*, 14(5).

Barba-Gutíerrez, Y., Adenso-Díaz, B., and Gupta, S. M. (2008). "Lot Sizing in Reverse MRP for Scheduling Disassembly." *International Journal of Production Economics*, 111, 741–751.

Bautista, J. and Pereira, J. (2002). "Ant Algorithms for Assembly Line Balancing." *ANTS 2002, LNCS 2463*, M. Dorigo, et al., eds., Springer-Verlag, Berlin, Germany, 65–75.

Bierwirth, C., Mattfeld, D. C., and Kopfer, H. (1996). "On Permutation Representations for Scheduling Problems." *Parallel Problem Solving from Nature–PPSN IV, Lecture Notes in Computer Science*, H. M. Voigt, W. Ebeling, I. Rechenberg, and H.-P. Schwefel, eds., Springer-Verlag, Berlin, Germany, 310–318.

Boatman, J. K. (2004). "Build Me an Airplane: Cirrus Constant Changing." *AOPA Pilot*, 47(10), 90–97.

Bourjault, A. (1984). *Contribution à une Approache Méthodologique de l'assemblage Automatisé: Elaboration Automatique des Séquences Opératoires* (Contribution to a Systematic Approach of Automatic Assembly: Automatic Generation of Task Sequences). Ph.D. Dissertation, Université de Franche-Comté, Besançon, France. (In French.)

Bourjault, A. (1987). "Methodology of Assembly Operation: A New Approach." *Abstracts of the 2nd International Conference on Robotics and Factories of the Future*, San Diego, Calif., 34–45.

Boysen, N., Fliedner, M., and Scholl, A. (2008) "Assembly Line Balancing: Which Model to Use When?" *International Journal of Production Economics*, 111(2), 509–528.

Brander, P. and Forsberg, R. (2005). "Cyclic Lot Scheduling with Sequence-Dependent Set-Ups: A Heuristic for Disassembly Processes." *International Journal of Production Research*, 43, 295–310.

Brennan, L., Gupta, S. M., and Taleb, K. N. (1994). "Operations Planning Issues in an Assembly/Disassembly Environment." *International Journal of Operations and Production Management*, 14(9), 57–67.

Brown, A. S. (2009). "The Many Shades of Green." *Mechanical Engineering*, 131(1), 22–29.

Bukchin, J. and Tzur, M. (2000). "Design of a Flexible Assembly Line to Minimize Equipment Cost." *IIE Transactions*, 32, 585–598.

Charnes, A. and Cooper, W. W. (1961). *Management Models and Industrial Applications of Linear Programming*, John Wiley and Sons, New York, N.Y.

Charnes, A., Cooper, W. W., and Ferguson, R. O. (1955). "Optimal Estimation of Executive Compensation by Linear Programming." *Management Science*, 1(2), 138–151.

Chung, C. and Peng, Q. (2005). "An Integrated Approach to Selective-Disassembly Sequence Planning." *Robotics and Computer-Integrated Manufacturing*, 475–485.

Chung, C. and Peng, Q. (2006). "Evolutionary Sequence Planning for Selective Disassembly in De-manufacturing." *International Journal of Computer Integrated Manufacturing*, 19, 278–286.

Cook, S. A. (1971). "The Complexity of Theorem-Proving Procedures." *Proceedings 3rd Annual Association for Computing Machinery Symposium on Theory of Computing*, New York, N.Y., 151–158.

Cormen, T., Leiserson, C., Rivest, R., and Stein, C. (2001). *Introduction to Algorithms*, The MIT Press, Cambridge, Mass.

Dantzig, G. B. (1955). "Linear Programming under Uncertainty." *Management Science*, 1, 197–206.

Dantzig, G. B. (1957). "Discrete-Variable Extremum Problems." *Operations Research*, 5, 266–277.

Das, S. K. and Naik, S. (2002). "Process Planning for Product Disassembly." *International Journal of Production Research*, 40(6), 1335–1355.

Defense Acquisition University. (2009a). "Fundamentals of Systems Acquisition Management." *ACQ101 Notes*, Section 311, September, DAU Press, Fort Belvoir, Va.

Defense Acquisition University. (2009b). "Modeling and Simulation for Test and Evaluation." *CLE023 Notes*, Section 888, September, DAU Press, Fort Belvoir, Va.

Defense Acquisition University. (2010). "Production, Quality and Manufacturing Fundamentals." *PQM101 Notes*, Section 304, March, DAU Press, Fort Belvoir, Va.

Ding, L. -P., Feng, Y. -X., Tan, J. -R., and Gao, Y. -C. (2010). "A New Multi-objective Ant Colony Algorithm for Solving the Disassembly Line Balancing Problem." *International Journal of Advanced Manufacturing Technology*, 48(5-8), 761–771.

Dong, J. and Arndt, G. (2003). "A Review of Current Research on Disassembly Sequence Generation and Computer Aided Design for Disassembly." *Proceedings of the Institution of Mechanical Engineers, Part B: Journal of Engineering Manufacture*, 217, 299–312.

Dong, T., Zhang, L., Tong, R., and Dong, J. (2006). "A Hierarchical Approach to Disassembly Sequence Planning for Mechanical Product." *The International Journal of Advanced Manufacturing Technology*, 30, 507–520.

Dorigo, M., Di Caro, G., and Gambardella, L. M. (1999). "Ant Algorithms for Discrete Optimization." *Artificial Life*, 5(3), 137–172.

Dorigo, M., Maniezzo, V., and Colorni, A. (1996). "The Ant System: Optimization by a Colony of Cooperating Agents." *IEEE Transactions on Systems, Man, and Cybernetics–Part B*, 26(1), 1–13.

Duta, L., Filip, F. G., and Caciula, I. (2008a). "Real Time Balancing of Complex Disassembly Lines." *Proceedings of the 17th IFAC World Congress on Automatic Control*, Seoul, Korea, CD-ROM.

Duta, L., Filip, F. G., and Henrioud, J. M. (2002). "Automated Disassembly: Main Stage in Manufactured Products Recycling." *Proceedings of the 4th International Workshop on Computer Science and Information Technologies*, Paper No. 133, Patras, Greece, CD-ROM.

Duta, L., Filip, F. G., and Henrioud, J. M. (2005). "Applying Equal Piles Approach to Disassembly Line Balancing Problem." *Proceedings of the 16th IFAC World Congress*, Prague, Czech Republic, CD-ROM.

Duta, L., Filip, F. G., and Popescu, C. (2008b). "Evolutionary Programming in Disassembly Decision Making." *International Journal of Computers, Communications and Control*, 3, 282–286.

Elsayed, E. A. and Boucher, T. O. (1994). *Analysis and Control of Production Systems*, Prentice Hall, Upper Saddle River, N.J.

Erdos, G., Kis, T., and Xirouchakis, P. (2001). "Modelling and Evaluating Product End-of-Life Options." *International Journal of Production Research*, 39, 1203–1220.

Erel, E. and Gokcen, H. (1999). "Shortest-Route Formulation of Mixed-Model Assembly Line Balancing Problem," *European Journal of Operational Research*, 116, 194–204.

Franke, C., Basdere, B., Ciupek, M., and Seliger, S. (2006). "Remanufacturing of Mobile Phones–Capacity, Program and Facility Adaptation Planning." *Omega*, 34(6), 562–570.

Fleischmann, M., Bloemhof-Ruwaard, J. M., Dekker, R., van der Laan, E., van Nunen, J., and Van Wassenhove, L. N. (1997). "Quantitative Models for Reverse Logistics: A Review." *European Journal of Operational Research*, 103, 1–17.

Finch, B. J. (2008). *Operations Now*, 3rd ed., McGraw-Hill/Irwin, New York, N.Y.

Gao, M., Zhou, M. C., and Tang, Y. (2004). "Intelligent Decision Making in Disassembly Process Based on Fuzzy Reasoning Petri Nets." *IEEE Transactions on Systems, Man, and Cybernetics—Part B*, 34, 2029–2034.

Garey, M. and Johnson, D. (1979). *Computers and Intractability: A Guide to the Theory of NP Completeness*, W. H. Freeman and Company, San Francisco, Calif.

Garfinkel, R. S. and Nemhauser, G. L. (1972). *Integer Programming*, John Wiley & Sons, New York, N.Y.

Geoffrion, A. M. (1974). "Lagrangian Relaxation and Its Uses in Integer Programming." *Mathematical Programming Study*, 2, 82–114.

Giudice, F. and Fargione, G. (2007). "Disassembly Planning of Mechanical Systems for Service and Recovery: A Genetic Algorithms Based Approach." *Journal of Intelligent Manufacturing*, 18, 313–329.

Glover, F. (1989). "Tabu Search, Part I." *ORSA Journal of Computing*, 1(3), 190–206.

Glover, F. (1990). "Tabu Search, Part II." *ORSA Journal of Computing*, 2(1), 4–32.

Goksoy, E. (2010). *Disassembly Line Balancing Problem with Fixed Number of Workstations and Finite Supply*. Master's Thesis, Department of Industrial Engineering, Middle East Technical University, Ankara, Turkey.

Goldberg, D. E. (1989). *Genetic Algorithms in Search, Optimization and Machine Learning*, Kluwer Academic Publishers, Boston, Mass.

Gonzalez, B. and Adenso-Díaz, B. (2006). "A Scatter Search Approach to the Optimum Disassembly Sequence Problem." *Computers & Operations Research*, 33, 1776–1793.

Graedel, T. E. and Allenby, B. R. (1995). *Industrial Ecology*, Prentice Hall, Englewood Cliffs, N.J.

Grochowski, D. E. and Tang, Y. (2009). "A Machine Learning Approach for Optimal Disassembly Planning." *International Journal of Computer Integrated Manufacturing*, 22, 374–383.

Gualtieri, D. M. (2005). "Get the Lead Out." *Phi Kappa Phi Forum*, 85(2), 6–7.

Güngör, A. and Gupta, S. M. (1997). "An Evaluation Methodology for Disassembly Processes." *Computers and Industrial Engineering*, 33(1), 329–332.

Güngör, A. and Gupta, S. M. (1998). "Disassembly Sequence Planning for Products with Defective Parts in Product Recovery." *Computers and Industrial Engineering*, 35(1–2), 161–164.

Güngör, A. and Gupta, S. M. (1999a). "A Systematic Solution Approach to the Disassembly Line Balancing Problem." *Proceedings of the 25th International Conference on Computers and Industrial Engineering*, New Orleans, La., 70–73.

Güngör, A. and Gupta, S. M. (1999b). "Disassembly Line Balancing." *Proceedings of the 1999 Annual Meeting of the Northeast Decision Sciences Institute*, Newport, R.I., 193–195.

Güngör, A. and Gupta, S. M. (1999c). "Issues in Environmentally Conscious Manufacturing and Product Recovery: A Survey." *Computers and Industrial Engineering*, 36(4), 811–853.

Güngör, A. and Gupta, S. M. (2001a). "A Solution Approach to the Disassembly Line Problem in the Presence of Task Failures." *International Journal of Production Research*, 39(7), 1427–1467.

Güngör, A. and Gupta, S. M. (2001b). "Disassembly Sequence Plan Generation Using a Branch-and-Bound Algorithm." *International Journal of Production Research*, 39(3), 481–509.

Güngör, A. and Gupta, S. M. (2002). "Disassembly Line in Product Recovery." *International Journal of Production Research*, 40(11), 2569–2589.

Gupta, S. M. and Al-Turki, Y. A. Y. (1997). "An Algorithm to Dynamically Adjust the Number of Kanbans in a Stochastic Processing Times and Variable Demand Environment." *Production Planning and Control*, 8(2), 133–141.

Gupta, S. M. and Al-Turki, Y. A. Y. (1998a). "Adapting Just-in-Time Manufacturing Systems to Preventive Maintenance Interruptions." *Production Planning and Control*, 9(4), 349–359.

Gupta, S. M. and Al-Turki, Y. A. Y. (1998b). "The Effect of Sudden Material Handling System Breakdown on the Performance of a JIT System." *International Journal of Production Research*, 36(7), 1935–1960.

Gupta, S. M. and Güngör, A. (2001). "Product Recovery Using a Disassembly Line: Challenges and Solution." *Proceedings of the 2001 IEEE International Symposium on Electronics and the Environment*, Denver, Colo., 36–40.

Gupta, S. M. and McGovern, S. M. (2004). "Multi-objective Optimization in Disassembly Sequencing Problems." *Proceedings of the 2nd World Conference on Production & Operations Management and the 15th Annual Production & Operations Management Conference*, Cancun, Mexico, CD-ROM.

Gupta, S. M. and Taleb, K. (1994). "Scheduling Disassembly." *International Journal of Production Research*, 32(8), 1857–1866.

Gupta, S. M., Al-Turki, Y. A. Y., and Perry, R. F. (1999). "Flexible Kanban System." *International Journal of Operations and Production Management*, 19(10), 1065–1093.

Gupta, S. M., Erbis, E., and McGovern, S. M. (2004). "Disassembly Sequencing Problem: A Case Study of a Cell Phone." *Proceedings of the 2004 SPIE International Conference on Environmentally Conscious Manufacturing IV*, Philadelphia, Pa., 43–52.

Gupta, S. M., Imtanavanich, P., and Nakashima, K. (2009). "Using Neural Networks to Solve a Disassembly-to-Order Problem." *International Journal of Biomedical Soft Computing and Human Sciences (Special Issue on Total Operations Management)*, 15(1), 67–71.

Gutjahr, A. L. and Nemhauser, G. L. (1964). "An Algorithm for the Line Balancing Problem." *Management Science*, 11(2), 308–315.

Hackman, S. T., Magazine, M. J., and Wee, T. S. (1989). "Fast, Effective Algorithms for Simple Assembly Line Balancing Problems." *Operations Research*, 37(6), 916–924.

Held, M. and Karp, R. M. (1971). "The Traveling Salesman Problem and Minimum Spanning Trees: Part II." *Mathematical Programming*, 6, 62–88.

Heyman, D. P. (1977). "Optimal Disposal Policies for a Single-Item Inventory System with Returns." *Naval Research Logistics*, 24, 38–405.

Hindo, B. (2006). "Everything Old Is New Again." *Business Week*, (4002), 64–70.

Helgeson, W. B. and Birnie, D. P. (1961). "Assembly Line Balancing Using the Ranked Positional Weight Technique." *Journal of Industrial Engineering*, 12, 394–398.
Hillier, F. S. and Boling, R. W. (1966). "The Effect of Some Design Factors on the Efficiency of Production Lines with Variable Operation Times." *Journal of Industrial Engineering*, 17, 65–658.
Hillier, F. S. and Boling, R. W. (1979). "On the Optimal Allocation of Work in Symmetrically Unbalanced Production Line Systems with Variable Operation Times." *Management Science*, 25(8), 721–728.
Hillier, F. S. and Lieberman, G. J. (2005). *Introduction to Operations Research*, 8th ed., McGraw-Hill, New York, N.Y.
Holland, J. H. (1975). *Adaptation in Natural and Artificial Systems*, University of Michigan Press, Ann Arbor, Mich.
Hong, D. S. and Cho, H. S. (1997). "Generation of Robotic Assembly Sequences with Consideration of Line Balancing Using Simulated Annealing." *Robotica*, 15, 663–673.
Hopper, E. and Turton, B. C. H. (2000). "An Empirical Investigation of Meta-heuristic and Heuristic Algorithms for a 2D Packing Problem." *European Journal of Operational Research*, 128(1), 34–57.
Hopgood, A. A. (1993). *Knowledge-Based Systems for Engineers and Scientists*, CRC Press, Boca Raton, Fla.
Hu, T. C. (1969). *Integer Programming and Network Flows*, Addison-Wesley, Reading, Mass.
Hu, T. C. and Shing, M. T. (2002). *Combinatorial Algorithms*, Dover Publications, Mineola, N.Y.
Hui, W., Dong, X., and Guanghong, D. (2008). "A Genetic Algorithm for Product Disassembly Sequence Planning." *Neurocomputing*, 71, 2720–2726.
Ilgin, M. A. and Gupta, S. M. (2010). "Environmentally Conscious Manufacturing and Product Recovery (ECMPRO): A Review of the State of the Art." *Journal of Environmental Management*, 91(3), 563–591.
Ilgin, M. A. and Gupta, S. M. (in press a). "Comparison of Economic Benefits of Sensor Embedded Products and Conventional Products in a Multi-Product Disassembly Line." *Computers and Industrial Engineering*.
Ilgin, M. A. and Gupta, S. M. (in press b). "Evaluating the Impact of Sensor Embedded Products on the Performance of an Air Conditioner Disassembly Line." *International Journal of Advanced Manufacturing Technology*.
Ilgin, M. A. and Gupta, S. M. (in press c). "Performance Improvement Potential of Sensor Embedded Products in Environmental Supply Chains." *Resources, Conservation and Recycling*.
Ilgin, M. A. and Gupta, S. M. (in press d). "Recovery of Sensor Embedded Washing Machines Using a Multi-kanban Controlled Disassembly Line." *Robotics and Computer Integrated Manufacturing*.
Imtanavanich, P. and Gupta, S. M. (2006). "Calculating Disassembly Yields in a Multi-criteria Decision-Making Environment for a Disassembly to Order System." *Applications of Management Science, Vol. 12*, K. D. Lawrence, G. R. Reeves, and R. Klimberg, eds., Elsevier Science, North-Holland, Amsterdam, 109–125.
Inderfurth, K. and Langella, I. M. (2006). "Heuristics for Solving Disassemble-to-Order Problems with Stochastic Yields." *OR Spectrum*, 28(1), 73–99.
Inderfurth, K. and Langella, I. M. (2008). "Planning Disassembly for Remanufacture-to-Order Systems." *Environment Conscious Manufacturing*, S. M. Gupta and A. J. D. Lambert, eds., CRC Press, Boca Raton, Fla., 387–412.
Inderfurth, K. and van der Laan, E. (2001). "Leadtime Effects and Policy Improvement for Stochastic Inventory Control with Remanufacturing." *International Journal of Production Economics*, 71, 381–390.
Iori, M. (2003). *Metaheuristic Algorithms for Combinatorial Optimization Problems*, Ph.D. Dissertation, Department of Electronics, Computer Science, and Systems, University of Bologna, Bologna, Italy.
Isaacs, J. A. and Gupta, S. M. (1997). "A Decision Tool to Assess the Impact of Automobile Design on Disposal Strategies." *Journal of Industrial Ecology*, 1(4), 19–33.

Ishii, K., Eubanks, C. F., and Marco, P. D. (1994). "Design for Product Retirement and Material Life-Cycle." *Materials & Design*, 15(4), 225–233.

Johnson, M. R. and Wang, M. H. (1995). "Planning Product Disassembly for Material Recovery Opportunities." *International Journal of Production Research*, 33(11), 3119–3142.

Kaebernick, H., O'Shea, B., and Grewal, S. S. (2000). "A Method for Sequencing the Disassembly of Products." *CIRP Annals–Manufacturing Technology*, 49, 13–16.

Kang, J. G., Lee, D. H., and Xirouchakis, P. (2003). "Disassembly Sequencing with Imprecise Data: A Case Study." *International Journal of Industrial Engineering*, 10, 407–412.

Kara, S., Pornprasitpol, P., and Kaebernick, H. (2006). "Selective Disassembly Sequencing: A Methodology for the Disassembly of End-of-Life Products." *CIRP Annals–Manufacturing Technology*, 55(1), 37–40.

Karp, R. M. (1972). "Reducibility among Combinatorial Problems." *Complexity of Computer Computations*, R. E. Miller and J. W. Thatcher, eds., Plenum Press, New York, N.Y., 85–103.

Kazmierczak, K., Mathiassen, S. E., Forsman, M., and Winkel, J. (2005). "An Integrated Analysis of Ergonomics and Time Consumption in Swedish 'Craft-Type' Car Disassembly." *Applied Ergonomics*, 36(3), 263–273.

Kazmierczak, K., Neumann, W. P., and Winkel, J. (2007). "A Case Study of Serial-Flow Car Disassembly: Ergonomics, Productivity and Potential System Performance." *Human Factors and Ergonomics in Manufacturing*, 17(4), 331–351.

Kekre, S., Rao, U. S., Swaminathan, J. M., and Zhang, J. (2003). "Reconfiguring a Remanufacturing Line at Visteon, Mexico." *Interfaces*, 33(6), 30–43.

Kim, H. J., Ciupek, M., Buchholz, A. and Seliger, G. (2006a). "Adaptive Disassembly Sequence Control by Using Product and System Information." *Robotics and Computer-Integrated Manufacturing*, 22(3), 267–278.

Kim, H. J., Lee, D. H., and Xirouchakis, P. (2006b). "A Lagrangean Heuristic Algorithm for Disassembly Scheduling with Capacity Constraints." *Journal of the Operational Research Society*, 57, 1231–1240.

Kim, H. J., Lee, D. H., and Xirouchakis, P. (2006c). "Two-Phase Heuristic for Disassembly Scheduling with Multiple Product Types and Parts Commonality." *International Journal of Production Research*, 44, 195–212.

Kim, H. J., Lee, D. H., Xirouchakis, P., and Zust, R. (2003). "Disassembly Scheduling with Multiple Product Types." *CIRP Annals–Manufacturing Technology*, 52, 403–406.

Kim, J. G., Jeon, H. B., Kim, H. J., Lee, D. H., and Xirouchakis, P. (2006d). "Disassembly Scheduling with Capacity Constraints: Minimizing the Number of Products Disassembled." *Proceedings of the Institution of Mechanical Engineers—Part B—Engineering Manufacture*, 220, 1473–1481.

Kizilkaya, E. A. and Gupta, S. M. (2004). "Modeling Operational Behavior of a Disassembly Line." *Proceedings of the SPIE International Conference on Environmentally Conscious Manufacturing IV*, Philadelphia, Pa., 79–93.

Kizilkaya, E. A. and Gupta, S. M. (2005a). "Impact of Different Disassembly Line Balancing Algorithms on the Performance of Dynamic Kanban System for Disassembly Line." *Proceedings of the SPIE International Conference on Environmentally Conscious Manufacturing V*, Boston, Mass., 103–110.

Kizilkaya, E. A. and Gupta, S. M. (2005b). "Sudden Material Handling System Breakdown in a Disassembly Line." *Proceedings of the 2005 Northeast Decision Sciences Institute Conference*, Philadelphia, Pa., CD-ROM.

Koc, A., Sabuncuoglu, I., and Erel, E. (2009). "Two Exact Formulations for Disassembly Line Balancing Problems with Task Precedence Diagram Construction Using an AND/OR Graph." *IIE Transactions*, 41, 866–881.

Kongar, E. and Gupta, S. M. (2002a). "A Genetic Algorithm for Disassembly Process Planning." *Proceedings of the 2002 SPIE International Conference on Environmentally Conscious Manufacturing II*, Newton, Mass., 4569, 54–62.

Kongar, E. and Gupta, S. M. (2002b). "A Multi-criteria Decision Making Approach for Disassembly-to-Order Systems." *Journal of Electronics Manufacturing*, 11(2), 171–183.

Kongar, E. and Gupta, S. M. (2006a). "Disassembly to Order System under Uncertainty." *Omega*, 34, 550–561.
Kongar, E. and Gupta, S. M. (2006b). "Disassembly Sequencing Using Genetic Algorithm." *The International Journal of Advanced Manufacturing Technology*, 30, 497–506.
Kongar, E. and Gupta, S. M. (2009a). "A Multiple Objective Tabu Search Approach for End-of-Life Product Disassembly." *International Journal of Advanced Operations Management*, 1(2–3), 177–202.
Kongar, E. and Gupta, S. M. (2009b). "Solving the Disassembly-to-Order Problem Using Linear Physical Programming." *International Journal of Mathematics in Operational Research*, 1(4), 504–531.
Kongar, E., Gupta, S. M., and McGovern, S. M. (2003). "Use of Data Envelopment Analysis for Product Recovery." *Proceedings of the 2003 SPIE International Conference on Environmentally Conscious Manufacturing III*, Providence, R.I., 219–231.
Kotera, Y. and Sato, S. (1997). "An Integrated Recycling Process for Electric Home Appliances." *Mitsubishi Electric ADVANCE*, 80, 23–26.
Koza, J. R. (1992). *Genetic Programming: On the Programming of Computers by the Means of Natural Selection*, The MIT Press, Cambridge, Mass.
Kumar, S., Kumar, R., Shankar, R., and Tiwari, M. K. (2003). "Expert Enhanced Coloured Stochastic Petri Net and Its Application in Assembly/Disassembly." *International Journal of Production Research*, 41, 2727–2762.
Kuo, T. C. (2000). "Disassembly Sequence and Cost Analysis for Electromechanical Products." *Robotics and Computer-Integrated Manufacturing*, 16, 43–54.
Lambert, A. J. D. (1997). "Optimal Disassembly of Complex Products." *International Journal of Production Research*, 35, 2509–2524.
Lambert, A. J. D. (1999). "Linear Programming in Disassembly/Clustering Sequence Generation." *Computers and Industrial Engineering*, 36(4), 723–738.
Lambert, A. J. D. (2002). "Determining Optimum Disassembly Sequences in Electronic Equipment." *Computers and Industrial Engineering*, 43(3), 553–575.
Lambert, A. J. D. (2003). "Disassembly Sequencing: A Survey." *International Journal of Production Research*, 41(16), 3721–3759.
Lambert, A. J. D. (2006). "Exact Methods in Optimum Disassembly Sequence Search for Problems Subject to Sequence Dependent Costs." *Omega*, 34, 538–549.
Lambert, A. J. D. (2007). "Optimizing Disassembly Processes Subjected to Sequence-Dependent Cost." *Computers and Operations Research*, 34, 536–551.
Lambert, A. J. D. and Gupta, S. M. (2002). "Demand-Driven Disassembly Optimization for Electronic Products." *Journal of Electronics Manufacturing*, 11(2), 121–135.
Lambert, A. J. D. and Gupta, S. M. (2005). *Disassembly Modeling for Assembly, Maintenance, Reuse, and Recycling*, CRC Press, Boca Raton, Fla.
Lambert, A. J. D. and Gupta, S. M. (2008). "Methods for Optimum and Near Optimum Disassembly Sequencing." *International Journal of Production Research*, 46, 2845–2865.
Langella, I. M. (2007). "Heuristics for Demand-Driven Disassembly Planning." *Computers and Operations Research*, 34(2), 552–577.
Lapierre, S. D. and Ruiz, A. B. (2004). "Balancing Assembly Lines: An Industrial Case Study." *Journal of the Operational Research Society*, 55(6), 589–597.
Lapierre, S. D., Ruiz, A., and Soriano, P. (2006). "Balancing Assembly Lines with Tabu Search." *European Journal of Operational Research*, 168(3), 826–837.
Lawler, E. L., Lenstra, J. K., Rinnooy Kan, A. H. G., and Shmoys, D. B. (1985). *The Traveling Salesman Problem: A Guided Tour of Combinatorial Optimization*, John Wiley & Sons, New York, N.Y.
Lee, D.-H., Kang, J.-G., and Xirouchakis, P. (2001). "Disassembly Planning and Scheduling: Review and Further Research." *Journal of Engineering Manufacture*, 215(B5), 695–709.
Lee, D. H., Kim, H. J., Choi, G., and Xirouchakis, P. (2004). "Disassembly Scheduling: Integer Programming Models." *Proceedings of the Institution of Mechanical Engineers–Part B—Engineering Manufacture*, 218, 1357–1372.

Lee, D. H. and Xirouchakis, P. (2004). "A Two-Stage Heuristic for Disassembly Scheduling with Assembly Product Structure." *Journal of the Operational Research Society*, 55, 287–297.

Lee, D. H., Xirouchakis, P., and Zust, R. (2002). "Disassembly Scheduling with Capacity Constraints." *CIRP Annals–Manufacturing Technology*, 51, 387–390.

Li, J. R., Khoo, L. P., and Tor, S. B. (2005). "An Object-Oriented Intelligent Disassembly Sequence Planner for Maintenance." *Computers in Industry*, 56, 699–718.

Li, J. R., Khoo, L. P., and Tor, S. B. (2006). "Generation of Possible Multiple Components Disassembly Sequence for Maintenance Using a Disassembly Constraint Graph." *International Journal of Production Economics*, 102, 51–65.

Li, J. R., Wang, Q. H., Huang, P., and Shen, H. Z. (2010) "A Novel Connector-Knowledge Based Approach for Disassembly Precedence Constraint Generation." *International Journal of Advanced Manufacturing Technology*, 49(1–4), 293–304.

Mascle, C. and Balasoiu, B. A. (2003). "Algorithmic Selection of a Disassembly Sequence of a Component by a Wave Propagation Method." *Robotics and Computer Integrated Manufacturing*, 19, 439–448.

McGovern, S. M. and Gupta, S. M. (2003a). "2-Opt Heuristic for the Disassembly Line Balancing Problem." *Proceedings of the SPIE International Conference on Environmentally Conscious Manufacturing III*, Providence, R.I., 71–84.

McGovern, S. M. and Gupta, S. M. (2003b). "Greedy Algorithm for Disassembly Line Scheduling." *Proceedings of the 2003 IEEE International Conference on Systems, Man, and Cybernetics*, Washington, D.C., 1737–1744.

McGovern, S. M. and Gupta, S. M. (2004a). "Combinatorial Optimization Methods for Disassembly Line Balancing." *Proceedings of the 2004 SPIE International Conference on Environmentally Conscious Manufacturing IV*, Philadelphia, Pa., 53–66.

McGovern, S. M. and Gupta, S. M. (2004b). "Demanufacturing Strategy Based Upon Metaheuristics." *Proceedings of the 2004 Industrial Engineering Research Conference*, Houston, Tex., CD-ROM.

McGovern, S. M. and Gupta, S. M. (2004c). "Metaheuristic Technique for the Disassembly Line Balancing Problem." *Proceedings of the 2004 Northeast Decision Sciences Institute Conference*, Atlantic City, N.J., 223–225.

McGovern, S. M. and Gupta, S. M. (2004d). "Multi-criteria Ant System and Genetic Algorithm for End-of-Life Decision Making." *Proceedings of the 35th Annual Meeting of the Decision Sciences Institute*, Boston, Mass., 6371–6376.

McGovern, S. M. and Gupta, S. M. (2005a). "Local Search Heuristics and Greedy Algorithm for Balancing the Disassembly Line." *The International Journal of Operations and Quantitative Management*, 11(2), 91–114.

McGovern, S. M. and Gupta, S. M. (2005b). "Stochastic and Deterministic Combinatorial Optimization Solutions to an Electronic Product Disassembly Flow Shop." *Proceedings of the Northeast Decision Sciences Institute–34th Annual Meeting*, Philadelphia, Pa., CD-ROM.

McGovern, S. M. and Gupta, S. M. (2005c). "Uninformed and Probabilistic Distributed Agent Combinatorial Searches for the Unary NP-Complete Disassembly Line Balancing Problem." *Proceedings of the 2005 SPIE International Conference on Environmentally Conscious Manufacturing V*, Boston, Mass., 5997-12, 81–92.

McGovern, S. M. and Gupta, S. M. (2006a). "Ant Colony Optimization for Disassembly Sequencing with Multiple Objectives." *The International Journal of Advanced Manufacturing Technology*, 30(5–6), 481–496.

McGovern, S. M. and Gupta, S. M. (2006b). "Computational Complexity of a Reverse Manufacturing Line." *Proceedings of the 2006 SPIE International Conference on Environmentally Conscious Manufacturing VI*, Boston, Mass., 1–12.

McGovern, S. M. and Gupta, S. M. (2006c). "Deterministic Hybrid and Stochastic Combinatorial Optimization Treatments of an Electronic Product Disassembly Line." *Applications of Management Science, Vol. 12*, K. D. Lawrence, G. R. Reeves, and R. Klimberg, eds., Elsevier Science, North-Holland, Amsterdam, 175–197.

McGovern, S. M. and Gupta, S. M. (2006d). "Performance Metrics for End-of-Life Product Processing." *Proceedings of the 17th Annual Production & Operations Management Conference*, Boston, Mass., CD-ROM.

McGovern, S. M. and Gupta, S. M. (2007a). "A Balancing Method and Genetic Algorithm for Disassembly Line Balancing." *European Journal of Operational Research*, 179(3), 692–708.
McGovern, S. M. and Gupta, S. M. (2007b). "Benchmark Data Set for Evaluation of Line Balancing Algorithms." *Proceedings of the IFAC Workshop on Intelligent Assembly and Disassembly*, Alicante, Spain, 54–59.
McGovern, S. M. and Gupta, S. M. (2007c). "Combinatorial Optimization Analysis of the Unary NP-Complete Disassembly Line Balancing Problem." *International Journal of Production Research*, 45(18–19), 4485–4511.
McGovern, S. M. and Gupta, S. M. (2008a). "Deterministic Search Algorithm for Sequencing and Scheduling Problems." *Meta-heuristics for Scheduling in Industrial and Manufacturing Applications, Studies in Computational Intelligence Vol. 128*, F. Xhafa and A. Abraham, eds., Springer-Verlag London Ltd., London, UK, 105–124.
McGovern, S. M. and Gupta, S. M. (2008b). "Disassembly Line Balancing." *Environment Conscious Manufacturing*, S. M. Gupta and A. J. D. Lambert, eds., CRC Press, Boca Raton, Fla., 235–310.
McGovern, S. M. and Gupta, S. M. (2008c). "Lexicographic Goal Programming and Assessment Tools for a Combinatorial Production Problem." *Multi-objective Optimization in Computational Intelligence: Theory and Practice*, L. T. Bui and S. Alam, eds., Idea Group Incorporated, Hershey, Pa, 151–188.
McGovern, S. M. and Gupta, S. M. (in press). "Metrics and Experimental Data for Assessing Unbalanced Disassembly Lines." *International Journal of Manufacturing Technology and Management*.
McGovern, S. M., Gupta, S. M., and Kamarthi, S. V. (2003). "Solving Disassembly Sequence Planning Problems Using Combinatorial Optimization." *Proceedings of the 2003 Northeast Decision Sciences Institute Conference*, Providence, R.I., 178–180.
McGovern, S. M., Gupta, S. M., and Nakashima, K. (2004). "Multi-criteria Optimization for Non-linear End of Lifecycle Models." *Proceedings of the Sixth Conference on EcoBalance*, Tsukuba, Japan, 201–204.
McMullen, P. R. and Frazier, G. V. (1998). "Using Simulated Annealing to Solve a Multiobjective Assembly Line Balancing Problem with Parallel Workstations." *International Journal of Production Research*, 36(10), 2717–2741.
McMullen, P. R. and Tarasewich, P. (2003). "Using Ant Techniques to Solve the Assembly Line Balancing Problem." *IIE Transactions*, 35, 605–617.
Meacham, A., Uzsoy, R., and Venkatadri, U. (1999). "Optimal Disassembly Configurations for Single and Multiple Products." *Journal of Manufacturing Systems*, 18, 311–322.
Merkle, D. and Middendorf, M. (2000). "An Ant Algorithm with a New Pheromone Evaluation Rule for Total Tardiness Problems." *Proceedings of Real-World Applications of Evolutionary Computing, EvoWorkshops 2000: EvoSTIM*, Edinburgh, Scotland, 287–296.
Metz, R. (2005). "Out with the Old Phone, in with the Cash." *The New York Times*, July 7, C10.
Miller, I. and Freund, J. E. (1985). *Probability and Statistics for Engineers*, Prentice Hall, Englewood Cliffs, N.J.
Moore, K. E. and Gupta, S. M. (1996). "Petri Net Models of Flexible and Automated Manufacturing Systems: A Survey." *International Journal of Production Research*, 34(11), 3001–3035.
Moore, K. E., Güngör, A., and Gupta, S. M. (1998). "A Petri Net Approach to Disassembly Process Planning." *Computers and Industrial Engineering*, 35, 165–168.
Moore, K. E., Güngör, A., and Gupta, S. M. (2001). "Petri Net Approach to Disassembly Process Planning for Products with Complex AND/OR Precedence Relationships." *European Journal of Operational Research*, 135(2), 428–449.
Nahmias, S. (2009). *Production and Operations Analysis*, 6th ed., McGraw-Hill/Irwin, New York, N.Y.
Navin-Chandra, D. (1994). "The Recovery Problem in Product Design." *Journal of Engineering Design*, 5(1), 65–86.

Opalić, M., Kljajin, M., and Vučković, K. (2010). "Disassembly Layout in WEEE Recycling Process." *Strojarstvo: Journal for Theory and Application in Mechanical Engineering*, 52(1), 51–58.

Opalić, M., Vučković, K., and Panić, N. (2004). "Consumer Electronics Disassembly Line Layout." *Polimeri*, 25(1–2), 20–22.

O'Shea, B., Kaebernick, H., Grewal, S. S., Perlewitz, H., Müller, K., and Seliger, G. (1999). "Method for Automatic Tool Selection for Disassembly Planning." *Assembly Automation*, 19(1), 47–54.

Osman, I. H. (2004). "Metaheuristics: Models, Design and Analysis." *Proceedings of the Fifth Asia Pacific Industrial Engineering and Management Systems Conference*, Gold Coast, Australia, 1.2.1–1.2.16.

Osman, I. H. and Laporte, G. (1996). "Metaheuristics: A Bibliography." *Annals of Operations Research*, 63, 513–623.

Pan, L. and Zeid, I. (2001). "A Knowledge Base for Indexing and Retrieving Disassembly Plans." *Journal of Intelligent Manufacturing*, 12, 77–94.

Pearce, J. A. (2009). "The Profit-Making Allure of Product Reconstruction." *MIT Sloan Management Review*, 50(3), 59–65.

Papadimitriou, C. H. and Steiglitz, K. (1998). *Combinatorial Optimization: Algorithms and Complexity*, Dover Publications, Mineola, N.Y.

Pinedo, M. (2002). *Scheduling Theory, Algorithms and Systems*, Prentice-Hall, Upper Saddle River, N.J.

Ponnambalam, S. G., Aravindan, P., and Naidu, G. M. (1999). "A Comparative Evaluation of Assembly Line Balancing Heuristics." *The International Journal of Advanced Manufacturing Technology*, 15, 577–586.

Prakash, A. and Tiwari, M. K. (2005). "Solving a Disassembly Line Balancing Problem with Task Failure Using a Psychoclonal Algorithm." *Proceedings of the ASME 2005 International Design Engineering Technical Conferences & Computers and Information in Engineering Conference,* Long Beach, Calif., CD-ROM.

Rabin, M. O. (1976). "Probabilistic Algorithms." *Algorithms and Complexity: New Directions and Recent Results*, J. F. Traub, ed., Academic Press, New York, N.Y., 21–39.

Rai, R., Rai, V., Tiwari, M. K., and Allada, V. (2002). "Disassembly Sequence Generation: A Petri Net Based Heuristic Approach." *International Journal of Production Research*, 40, 3183–3198.

Ranky, P. G., Subramanyam, M., Caudill, R. J., Limaye, K., and Alli, N. (2003). "Dynamic Scheduling and Line Balancing Methods, and Software Tools for Lean and Reconfigurable Disassembly Cells and Lines." *Proceedings of the IEEE International Symposium on Electronics and the Environment*, 234–239.

Reingold, E. M., Nievergeld, J., and Deo, N. (1977). *Combinatorial Algorithms: Theory and Practice*, Prentice-Hall, Englewood Cliffs, N.J.

Reveliotis, S. A. (2007). "Uncertainty Management in Optimal Disassembly Planning Through Learning-Based Strategies." *IIE Transactions*, 39, 645–658.

Rios, P. J. and Stuart, J. A. (2004). "Scheduling Selective Disassembly for Plastics Recovery in an Electronics Recycling Center." *IEEE Transactions on Electronics Packaging Manufacturing*, 27, 187–197.

Ripley, T. (2008). "Recycling Airliners." *Aerospace Testing International*, March, 6.

Rosen, K. H. (1999). *Discrete Mathematics and Its Applications*, McGraw-Hill, Boston, Mass.

Sarin, S. C., Sherali, H. D., and Bhootra, A. (2006). "A Precedence-Constrained Asymmetric Traveling Salesman Model for Disassembly Optimization." *IIE Transactions*, 38, 223–237.

Scholl, A. (1995). *Balancing and Sequencing of Assembly Lines*, Physica-Verlag, Heidelberg, Germany.

Scholz-Reiter, B., Scharke, H., and Hucht, A. (1999). "Flexible Robot-Based Disassembly Cell for Obsolete TV-Sets and Monitors." *Robotics and Computer-Integrated Manufacturing*, 15(3), 247–255.

Schultmann, F. and Rentz, O. (2001). "Environment-Oriented Project Scheduling for the Dismantling of Buildings." *OR Spektrum*, 23, 51–78.

Sen, S. and Higle, J. L. (1999). "An Introductory Tutorial on Stochastic Linear Programming Models." *Interfaces*, 29, 33–61.

Seo, K. K., Park, J. H., and Jang, D. S. (2001). "Optimal Disassembly Sequence Using Genetic Algorithms Considering Economic and Environmental Aspects." *The International Journal of Advanced Manufacturing Technology*, 18, 371–380.

Shimizu, Y., Tsuji, K., and Nomura, M. (2007). "Optimal Disassembly Sequence Generation Using a Genetic Programming." *International Journal of Production Research*, 45, 4537–4554.

Silverman, F. and Carter, J. (2003). "Choosing a Station Loading Rule in Assembly Line Design." *Journal of the Academy of Business and Economics*, March, 131–138.

Singh, A. K., Tiwari, M. K., and Mukhopadhyay, S. K. (2003). "Modelling and Planning of the Disassembly Processes Using an Enhanced Expert Petri Net." *International Journal of Production Research*, 41, 3761–3792.

Sodhi, M. S. and Reimer, B. (2001). "Models for Recycling Electronics End-of-Life Products." *OR Spektrum*, 23, 97–115.

Srinivasan, H., Figueroa, R., and Gadh, R. (1999). "Selective Disassembly for Virtual Prototyping as Applied to De-manufacturing." *Robotics and Computer-Integrated Manufacturing*, 15(3) 231–245.

Stape, A. L. (2004). "In Computers, It's out with the Old–But Where Do They Go?" *The Providence Journal*, January 6, G1–G3.

Stecke, K. E. and Solberg, J. J. (1985). "The Optimality of Unbalancing Both Workloads and Machine Group Sizes in Closed Queueing Networks of Multiserver Queues." *Operations Research*, 33(4), 882–910.

Steuer, R. E. (1989). *Multiple Criteria Optimization: Theory, Computation, and Application*, Krieger, Melbourne, Fla.

Stuart, J. A. and Christina, V. (2003). "New Metrics and Scheduling Rules for Disassembly and Bulk Recycling." *IEEE Transactions on Electronics Packaging Manufacturing*, 26, 133–140.

Subramani, A. K. and Dewhurst, P. (1991). "Automatic Generation of Product Disassembly Sequence." *Annals of the CIRP*, 40(1), 115–118.

Suresh, G., Vinod, V. V., and Sahu, S. (1996). "A Genetic Algorithm for Assembly Line Balancing." *Production Planning and Control*, 7(1), 38–46.

Takahashi, K. and Nakamura, N. (1999). "Reacting JIT Ordering Systems to the Unstable Changes in Demand." *International Journal of Production Research*, 37(10), 2293–2313.

Takahashi, K. and Nakamura, N. (2000). "Reactive Logistics in a JIT Environment." *Production Planning & Control*, 11(1), 20–31.

Takahashi, K. and Nakamura, N. (2002). "Decentralized Reactive Kanban System." *European Journal of Operational Research*, 139(2), 262–276.

Takahashi, K., Morikawa, K., and Nakamura, N. (2004). "Reactive JIT Ordering System for Changes in the Mean and Variance of Demand." *International Journal of Production Economics*, 92(2), 181–196.

Taleb, K. N. and Gupta, S. M. (1997). "Disassembly of Multiple Product Structures." *Computers & Industrial Engineering*, 32, 949–961.

Taleb, K. N., Gupta, S. M., and Brennan, L. (1997). "Disassembly of Complex Products with Parts and Materials Commonality." *Production Planning and Control*, 8(3), 255–269.

Tang, Y. and Zhou, M. -C. (2006). "A Systematic Approach to Design and Operation of Disassembly Lines." *IEEE Transactions on Automation Science and Engineering*, 3, 324–329.

Tang, Y., Zhou, M. -C., and Caudill, R. (2001a). "A Systematic Approach to Disassembly Line Design." *Proceedings of the 2001 IEEE International Symposium on Electronics and the Environment*, Denver, Colo., 173–178.

Tang, Y., Zhou, M. -C., and Caudill, R. (2001b). "An Integrated Approach to Disassembly Planning and Demanufacturing Operation." *IEEE Transactions on Robotics and Automation*, 17(6), 773–784.

Tang, Y., Zhou, M. -C., and Gao, M. (2006). "Fuzzy-Petri-Net Based Disassembly Planning Considering Human Factors." *IEEE Transactions on Systems, Man and Cybernetics*, 36, 718–726.

Tapping, D. (2003). *The Lean Pocket Guide*, MCS Media, Chelsea, Mich.

Tardif, V. and Maaseidvaag, L. (2001) "An Adaptive Approach to Controlling Kanban Systems." *European Journal of Operational Research*, 132(2), 411–424.

Thilakawardana, D., Driscoll, J., and Deacon, G. (2003a). "A Forward-Loading Heuristic Procedure for Single Model Assembly Line Balancing." *Proceedings of the 17th International Conference on Production Research*, Blacksburg, Va., CD-ROM.

Thilakawardana, D., Driscoll, J., and Deacon, G. (2003b). "Assembly Line Work Assignment Using a Front Loading Genetic Algorithm." *Proceedings of the 17th International Conference on Production Research*, Blacksburg, Va., CD-ROM.

Tiacci, L., Saetta, S., and Martini, A. (2003). "A Methodology to Reduce Data Collection in Lean Simulation Modeling for the Assembly Line Balancing Problem." *Proceedings of Summer Computer Simulation Conference 2003*, Montreal, Canada, 841–846.

Tiwari, M. K., Sinha, N., Kumar, S., Rai, R., and Mukhopadhyay, S. K. (2001). "A Petri Net Based Approach to Determine the Disassembly Strategy of a Product." *International Journal of Production Research*, 40(5), 1113–1129.

Toffel, M. W. (2002). "End-of-Life Product Recovery: Drivers, Prior Research, and Future Directions." *Proceedings of the INSEAD Conference on European Electronics Take-Back Legislation: Impacts on Business Strategy and Global Trade*, Fontainebleau, France, CD-ROM.

Torres, F., Gil, P., Puente, S. T., Pomares, J., and Aracil, R. (2004). "Automatic PC Disassembly for Component Recovery." *International Journal of Advanced Manufacturing Technology*, 23(1–2), 39–46.

Torres, F., Puente, S. T., and Aracil, R. (2003). "Disassembly Planning Based on Precedence Relations among Assemblies." *The International Journal of Advanced Manufacturing Technology*, 21, 317–327.

Tovey, C. A. (2002). "Tutorial on Computational Complexity." *Interfaces*, 32(3), 30–61.

Tripathi, M., Agrawal, S., Pandey, M. K., Shankar, R., and Tiwari, M. K. (2009). "Real World Disassembly Modeling and Sequencing Problem: Optimization by Algorithm of Self-Guided Ants (ASGA)." *Robotics and Computer-Integrated Manufacturing*, 25, 483–496.

Udomsawat, G. and Gupta, S. M. (2005a). "Multi-kanban in Disassembly Line with Component Discriminating Demand." *Proceedings of the 2005 Northeast Decision Sciences Institute Conference*, Philadelphia, Pa., CD-ROM.

Udomsawat, G. and Gupta, S. M. (2005b). "Multi-kanban Mechanism for Appliance Disassembly." *Proceedings of the SPIE International Conference on Environmentally Conscious Manufacturing V*, Boston, Mass., 30–41.

Udomsawat, G. and Gupta, S. M. (2005c). "The Effect of Sudden Server Breakdown on the Performance of a Disassembly Line." *Proceedings of the SPIE International Conference on Environmentally Conscious Manufacturing V*, Boston, Mass., 93–102.

Udomsawat, G. and Gupta, S. M. (2006). "Controlling Disassembly Line with Multi-kanban System." *Proceedings of the SPIE International Conference on Environmentally Conscious Manufacturing VI*, Boston, Mass., 42–53.

Udomsawat, G. and Gupta, S. M. (2008). "Multikanban System for Disassembly Line." *Environment Conscious Manufacturing*, S. M. Gupta and A. J. D. Lambert, eds., CRC Press, Boca Raton, Fla., 311–330.

Udomsawat, G., Gupta, S. M. and Al-Turki, Y.A.Y. (2003a). "Multi-kanban Model for Disassembly Line with Demand Fluctuation." *Proceedings of the SPIE International Conference on Environmentally Conscious Manufacturing III*, Providence, R.I., 85–93.

Udomsawat, G., Gupta, S. M., and Kamarthi, S. V. (2003b). "A Multi-kanban Model for Disassembly." *Proceedings of the 2003 Northeast Decision Sciences Institute Conference*, Providence, R.I., 169–171.

Uhlmann, E., Elbing, F., and Dittberner, J. (2004) "Innovative Manufacturing Technologies for the Disassembly of Consumer Goods." *Proceedings of the Institution of Mechanical Engineers, Part B: Journal of Engineering Manufacture*, 218(9), 1039–1046.

Veerakamolmal, P. and Gupta, S. M. (1998). "Optimal Analysis of Lot-Size Balancing for Multiproducts Selective Disassembly." *International Journal of Flexible Automation and Integrated Manufacturing*, 6(3&4), 245–269.

Veerakamolmal, P. and Gupta, S. M. (1999). "Analysis of Design Efficiency for the Disassembly of Modular Electronic Products." *Journal of Electronics Manufacturing*, 9(1), 79–95.

Veerakamolmal, P. and Gupta, S. M. (2002). "A Case-Based Reasoning Approach for Automating Disassembly Process Planning." *Journal of Intelligent Manufacturing*, 13, 47–60.

Vujosevic, R., Raskar, T., Yetukuri, N. V., Jothishankar, M. C., and Juang, S. H. (1995). "Simulation, Animation, and Analysis of Design Assembly for Maintainability Analysis." *International Journal of Production Research*, 33(11), 2999–3022.

Walsh, B. (2009). "E-Waste Not. How–and Why–We Should Make Sure Our Old Cell Phones, TVs and PCs Get Dismantled Properly." *Time*, January 19, 49–50.

Wang, J. F., Lui, J. H., Li, S. Q., and Zhong, Y. F. (2003). "Intelligent Selective Disassembly Using the Ant Colony Algorithm." *Artificial Intelligence for Engineering Design, Analysis and Manufacturing*, 17, 325–333.

Willems, B., Dewulf, W., and Dulflou, J. (2004). "End-of-Life Strategy Selection: A Linear Programming Approach to Manage Innovations in Product Design." *Proceedings of the 11th International CIRP Life Cycle Engineering Seminar on Product Life Cycle–Quality Management*, Belgrade, Serbia, 35–43.

Yano, C. A. and Lee, H. L. (1995). "Lot Sizing with Random Yields: A Review." *Operations Research*, 43(2), 311–334.

Zeid, I., Gupta, S. M., and Bardasz, T. (1997). "A Case-Based Reasoning Approach to Planning for Disassembly." *Journal of Intelligent Manufacturing*, 8(2), 97–106.

Zha, X. F. and Lim, S. Y. E. (2000). "Assembly/Disassembly Task Planning and Simulation Using Expert Petri Nets." *International Journal of Production Research*, 38, 3639–3676.

Zussman, E. and Zhou, M. (1999). "A Methodology for Modeling and Adaptive Planning of Disassembly Processes." *IEEE Transactions on Robotics and Automation*, 15, 190–194.

Zussman, E. and Zhou, M. C. (2000). "Design and Implementation of an Adaptive Process Planner for Disassembly Processes." *IEEE Transactions on Robotics and Automation*, 16, 171–179.

Appendix: Acronyms

This section provides a list of the acronyms developed and referred to throughout this text. The full terminology for each acronym is provided.

2-opt	2-optimal
3-opt	3-optimal
5M	machinery, method, material, man, and measurement
ACO	ant colony optimization
AEHC	adjacent element hill climbing
CAD	computer-aided design
CONWIP	constant work in progress
CPU	central processing unit
CR	critical ratio
DLBP	DISASSEMBLY LINE BALANCING PROBLEM
DTO	disassembly to order
ECMPRO	environmentally conscious manufacturing and product recovery
EDD	earliest due date
EOQ	economic order quantity (also, the economic lot size model)
EPA	Environmental Protection Agency
EU	European Union
FCFS	first come first served
FIFO	first in first out
FFD	first fit decreasing
GA	genetic algorithm
GHz	gigahertz
GP	goal programming

H-K	hunter-killer
JIT	just-in-time
k-opt	*k*-optimal
LP	linear programming
MCDM	multicriteria decision making
MRP	material requirements planning
NP	nondeterministic polynomial
P	polynomial
PC	personal computer
PPX	precedence preservative crossover
prec	precedence constraints
prmu	permutation
RFP	request for proposal
SALB-I	simple assembly-line-balancing problem type I
SALB-II	simple assembly-line-balancing problem type II
SAT	SATISFIABILITY
SMED	single minute exchange of dies
SPT	shortest processing time
TARMAC AEROSAVE	Tarbes Advanced Recycling & Maintenance Aircraft Company
TSP	TRAVELING SALESPERSON PROBLEM
U.S.	United States
V&V	verification and validation
WEEE	Proposal for a Directive on Waste Electrical and Electronic Equipment
WIP	work in progress

Author Index

A

Abraham, A., 357
Adenso-Díaz, B., 314, 315, 349, 351
Agrawal, S., 278, 349, 360
Aho, A. V., 85, 349
Alam, S., 357
Allada, V., 358
Allen, L. E., 349
Allenby, B. R., 289, 351
Alli, N., 358
Altekín, F. T., 59, 68, 276, 338, 341, 349
Al-Turki, Y. A. Y., 327, 328, 352, 360
Andonov, R., 321, 349
Andrés, C., 315, 349
Aracil, R., 360
Aravindan, P., 358
Arndt, G., 314, 350
Arola, D. F., 294, 341, 349

B

Bader, C. D., 48, 349
Balasoiu, B. A., 315, 356
Barba-Gutíerrez, Y., 316, 349
Bardasz, T., 361
Basdere, B., 351
Bautista, J., 51, 349
Bhootra, A., 358
Biddle, M. B., 349
Bierwirth, C., 166, 349
Birnie, D. P., 28, 49, 276, 353
Bloemhof-Ruwaard, J. M., 351
Boatman, J. K., 133, 349
Boling, R. W., 344, 353
Boucher, T. O., 49, 70, 97, 103, 114, 276, 277, 278, 279, 351
Bourjault, A., 37, 62, 63, 64, 65, 66, 67, 68, 349, 350
Boysen, N., 52, 350
Brander, P., 316, 350
Brennan, L., 5, 48, 350, 359
Brown, A. S., 33, 350
Buchholz, A., 354
Bui, L. T., 357
Bukchin, J., 52, 350

C

Caciula, I., 351
Carter, J., 50, 359
Caudill, R., 358, 359
Charnes, A., 153, 350
Cho, H. S., 50, 286, 353
Choi, G., 355
Christina, V., 316, 359
Chung, C., 314, 315, 350
Ciupek, M., 351, 354
Colorni, A., 350
Cook, S. A., 82, 84, 350
Cooper, W. W., 153, 350
Cormen, T., 7, 59, 123, 350

D

Dantzig, G. B., 86, 321, 350
Das, S. K., 54, 350
Deacon, G., 360
Defense Acquisition University, 19, 35, 156, 295, 350
Dekker, R., 351
Deo, N., 358
Dewhurst, P., 290, 359
Dewulf, W., 361
Di Caro, G., 350
Ding, L. -P., 198, 350
Dittberner, J., 360
Dong, J., 314, 350
Dong, T., 75, 350
Dong, X., 353
Dorigo, M., 60, 121, 122, 123, 181, 182, 184, 186, 187, 198, 200, 349, 350
Driscoll, J., 360
Dulflou, J., 361
Duta, L., 57, 59, 276, 314, 351

E

Elbing, F., 360
Elsayed, E. A., 49, 70, 97, 103, 114, 276, 277, 278, 279, 351
Erbis, E., 352
Erdos, G., 315, 351
Erel, E., 50, 351, 354
Eubanks, C. F., 354

F

Fargione, G., 314, 351
Feng, Y. -X., 350
Ferguson, R. O., 350
Figueroa, R., 359
Filip, F. G., 351
Finch, B. J., 23, 24, 25, 38, 300, 301, 303, 351
Fisher, M. M., 349
Fleischmann, M., 324, 351
Fliedner, M., 350
Forsberg, R., 316, 350
Forsman, M., 354
Franke, C., 59, 351
Frazier, G. V., 51, 357
Freund, J. E., 59, 143, 357

G

Gadh, R., 359
Gambardella, L. M., 350
Gao, M., 54, 351, 359
Gao, Y. -C., 50
Garcia-Carbajal, S., 349
Garey, M., 7, 8, 59, 60, 77, 78, 79, 80, 81, 84, 87, 88, 90, 111, 113, 114, 123, 351
Garfinkel, R. S., 88, 351
Geoffrion, A. M., 88, 351
Gil, P., 360
Giudice, F., 314, 351
Glover, F., 200, 351
Gokcen, H., 50, 351
Goksoy, E., 341, 351
Goldberg, D. E., 60, 351
Gonzalez, B., 314, 351
Graedel, T. E., 289, 351
Grewal, S. S., 354, 358
Grochowski, D. E., 54, 351
Gualtieri, D. M., 47, 134, 136, 137, 351
Guanghong, D., 353
Güngör, A., 6, 37, 38, 48, 55, 56, 59, 96, 134, 275, 276, 314, 315, 351, 352, 357
Gupta, S. M., 6, 8, 37, 38, 48, 51, 53, 54, 55, 56, 57, 58, 59, 61, 62, 63, 65, 67, 68, 70, 75, 96, 97, 98, 103, 105, 106, 107, 111, 113, 114, 125, 134, 136, 137, 140, 148, 159, 166, 170, 171, 182, 187, 189, 199, 214, 215, 228, 243, 244, 263, 275, 276, 286, 290, 293, 294, 314, 315, 316, 318, 319, 327, 328, 334, 344, 349, 350, 351, 352, 353, 354, 355, 356, 357, 359, 360, 361
Gutjahr, A. L., 50, 55, 352

H

Hackman, S. T., 50, 352
Held, M., 88, 352
Helgeson, W. B., 28, 276, 353
Henrioud, J. M., 351
Heyman, D. P., 324, 352
Higle, J. L., 321, 358
Hillier, F. S., 31, 318, 344, 346, 347, 353
Hindo, B., 34, 352
Holland, J. H., 60, 120, 353
Hong, D. S., 50, 286, 353
Hopcroft, J. E., 349
Hopgood, A. A., 124, 215, 353
Hopper, E., 60, 353
Hu, T. C., 87, 88, 244, 353
Huang, P., 356
Hucht, A., 358
Hui, W., 314, 353

I

Ilgin, M. A., 48, 328, 353
Imtanavanich, P., 319, 352, 353
Inderfurth, K., 319, 321, 323, 353
Iori, M., 60, 353
Isaacs, J. A., 290, 353
Ishii, K., 289, 354

J

Jang, D. S., 359
Jeon, H. B., 354
Johnson, D., 7, 8, 59, 60, 77, 78, 79, 80, 81, 84, 87, 88, 90, 111, 113, 114, 123, 217, 232, 351
Johnson, M. R., 290, 354
Jothishankar, M. C., 361
Juang, S. H., 361

K

Kaebernick, H., 74, 354, 358
Kamarthi, S. V., 357, 360
Kandiller, L., 349
Kang, J. G., 315, 354, 355
Kara, S., 286, 354
Karp, R. M., 77, 78, 82, 83, 84, 88, 111, 352, 354
Kazmierczak, K., 286, 354
Kekre, S., 59, 354
Khoo, L. P., 356
Kim, H. J., 303, 315, 316, 354, 355
Kim, J. G., 316, 354
Kis, T., 351
Kizilkaya, E. A., 327, 354
Klimberg, R., 353, 356
Kljajin, M., 358
Koc, A., 75, 354
Kongar, E., 54, 57, 59, 136, 314, 319, 354, 355

Kopfer, H., 349
Kotera, Y., 49, 355
Koza, J. R., 60, 169, 355
Kumar, R., 355
Kumar, S., 54, 355, 360
Kuo, T. C., 286, 314, 355

L

Lambert, A. J. D., 8, 37, 52, 56, 57, 58, 59, 61, 62, 63, 65, 67, 68, 70, 74, 286, 294, 314, 315, 318, 319, 353, 355, 357, 360
Langella, I. M., 319, 321, 323, 353, 355
Lapierre, S. D., 50, 51, 286, 355
Laporte, G., 60, 358
Lawler, E. L., 123, 128, 217, 228, 232, 308, 355
Lawrence, K. D., 353, 356
Lee, D. H., 52, 316, 354, 355, 356
Lee, H. L., 321, 361
Leiserson, C., 350
Lenstra, J. K., 355
Li, J. R., 68, 74, 109, 286, 314, 356
Li, S. Q., 361
Lieberman, G. J., 31, 318, 346, 347, 353
Lim, S. Y. E., 54, 361
Limaye, K., 358
Lozano, S., 349
Lui, J. H., 361

M

Maaseidvaag, L., 327, 360
Magazine, M. J., 352
Maniezzo, V., 350
Marco, P. D., 354
Martini, A., 360
Mascle, C., 315, 356
Mathiassen, S. E., 354
Mattfeld, D. C., 349
McGovern, S., 37, 59, 70, 97, 98, 103, 105, 106, 107, 111, 113, 114, 121, 125, 134, 136, 140, 148, 159, 166, 170, 182, 187, 199, 214, 215, 228, 243, 244, 263, 344, 352, 355, 356, 357
McMullen, P. R., 51, 357
Meacham, A., 316, 357
Merkle, D., 229, 357
Metz, R., 138, 357
Middendorf, M., 229, 357
Miller, I., 59, 143, 357
Miller, R. E., 354
Moore, K. E., 33, 51, 54, 75, 314, 357
Morikawa, K., 359
Mukhopadhyay, S. K., 359, 360
Müller, K., 358

N

Nahmias, S., 24, 28, 303, 307, 313, 318, 325, 357
Naidu, G. M., 358
Naik, S., 54, 350
Nakamura, N., 327, 359
Nakashima, K., 352, 357
Navin-Chandra, D., 290, 357
Nemhauser, G. L., 50, 55, 88, 351, 352
Neumann, W. P., 354
Nievergeld, J., 358
Nomura, M., 359

O

O'Shea, B., 57, 75, 286, 354, 358
Opalić, M., 303, 358
Osman, I. H., 60, 358
Ozdemirel, N. E., 349

P

Pan, L., 33, 315, 358
Pandey, M. K., 360
Panic, N., 358
Papadimitriou, C. H., 60, 77, 79, 89, 113, 119, 123, 149, 217, 232, 358
Park, J. H., 359
Pearce, J. A., 35, 337, 338, 358
Peng, Q., 314, 315, 350
Pereira, J., 51, 349
Perlewitz, H., 358
Perry, R. F., 352
Pinedo, M., 31, 50, 97, 307, 358
Poirriez, V., 349
Pomares, J., 360
Ponnambalam, S. G., 50, 358
Popescu, C., 351
Pornprasitpol, P., 354
Prakash, A., 59, 358
Puente, S. T., 360

R

Rabin, M. O., 90, 358
Rai, R., 54, 358, 360
Rai, V., 358
Rajopadhye, S., 349
Ranky, P. G., 303, 358
Rao, U. S., 354
Raskar, T., 361
Reeves, G. R., 353, 356
Reimer, B., 48, 359
Reingold, E. M., 86, 358
Rentz, O., 48, 358
Reveliotis, S. A., 314, 358
Rinnooy Kan, A. H. G., 355
Rios, P. J., 316, 358
Ripley, T., 35, 358
Rivest, R., 350

Rosen, K. H., 8, 59, 80, 150, 164, 176, 196, 201, 358
Ruiz, A., 51, 286, 355

S

Sabuncuoglu, I., 354
Saetta, S., 360
Sahu, S., 359
Sarin, S. C., 315, 358
Sato, S., 49, 355
Scharke, H., 358
Scholl, A., 50, 350, 358
Scholz-Reiter, B., 286, 303, 358
Schultmann, F., 48, 358
Seliger, G., 354, 358
Seliger, S., 351
Sen, S., 321, 358
Seo, K. K., 314, 359
Shankar, R., 355, 360
Shen, H. Z., 356
Sherali, H. D., 358
Shimizu, Y., 314, 359
Shing, M. T., 87, 244, 353
Shmoys, D. B., 355
Silverman, F., 50, 359
Singh, A. K., 54, 359
Sinha, N., 360
Sodhi, M. S., 48, 359
Solberg, J. J., 344, 359
Soriano, P., 355
Srinivasan, H., 315, 359
Stape, A. L., 136, 359
Stecke, K. E., 344, 359
Steiglitz, K., 60, 77, 79, 89, 113, 119, 149, 217, 358
Stein, C., 350
Steuer, R. E., 153, 154, 359
Stuart, J. A., 316, 358, 359
Subramani, A. K., 290, 359
Subramanyam, M., 358
Suresh, G., 50, 359
Swaminathan, J. M., 354

T

Takahashi, K., 327, 328, 359
Taleb, K., 53, 315, 316, 350, 352, 359
Tan, J. -R., 350
Tang, Y., 53, 54, 276, 351, 359
Tapping, D., 24, 27, 359
Tarasewich, P., 51, 357
Tardif, V., 327, 360
Thatcher, J. W., 354
Thilakawardana, D., 50, 360
Tiacci, L., 51, 278, 279, 360
Tiwari, M. K., 54, 59, 278, 349, 355, 358, 359, 360
Toffel, M. W., 48, 360
Tong, R., 350
Tor, S. B., 356
Torres, F., 57, 290, 315, 360
Tovey, C. A., 59, 77, 81, 87, 88, 89, 90, 360
Traub, J. F., 358
Tripathi, M., 14, 360
Tsuji, K., 359
Turton, B. C. H., 60, 353
Tzur, M., 52, 350

U

Udomsawat, G., 328, 334, 360
Uhlmann, E., 286, 360
Ullman, J. D., 349
Uzsoy, R., 357

V

van der Laan, E., 323, 351, 353
van Nunen, J., 351
Van Wassenhove, L. N., 351
Veerakamolmal, P., 58, 293, 294, 315, 361
Venkatadri, U., 57
Vinod, V. V., 359
Vučković, K., 358
Vujosevic, R., 290, 361

W

Walsh, B., 33, 361
Wang, J. F., 286, 361
Wang, M. H., 290, 354
Wang, Q. H., 356
Wee, T. S., 352
Willems, B., 341, 361
Winkel, J., 354

X

Xhafa, F., 357
Xirouchakis, P., 316, 351, 354, 355, 356

Y

Yano, C. A., 321, 361
Yetukuri, N. V., 361

Z

Zeid, I., 33, 41, 315, 358, 361
Zha, X. F., 54, 361
Zhang, J., 354
Zhang, L., 350
Zhong, Y. F., 361
Zhou, M. C., 54, 276, 351, 359, 361
Zussman, E., 54, 361
Zust, R., 354, 356

Subject Index

2

2-opt, 124, 227

5

5M, 20

A

ACO, 6, 122, 181
adjacent element hill climbing. *See* AEHC
AEHC, 124, 213, 291
aerospace industry, 35, 133, 284, 346
Aker's graphical procedure, 308
algorithm, 27, 79, 118, 129, 149, 160, 167, 170, 184, 201, 216, 230, 245, 276, 285, 287, 303, 310, 312, 318
ant colony optimization. *See* ACO
approximation, 89
artificial part. *See* artificial task
artificial task, 68, 76
artificial vertex. *See* artificial task
assembly line, 22, 37, 49
assembly tree, 64
asymptotic notation, 79, 150, 161, 166, 182, 187, 203, 217, 231, 271
automobile industry, 48, 133, 212, 286, 290, 325
average tardiness, 310

B

balance, 97, 103, 124, 154, 184, 187, 203, 213, 264, 293, 304, 308, 343, 348
batches, 326
before life, 3
benchmark data sets, 140
best-fit, 87
BIN PACKING, 7, 87, 88, 112, 114, 121, 123, 124, 186, 198, 227, 244, 287
binary NP-complete, 86
birth-and-death process, 346
blind search, 119
breakeven analysis, 301
breath-first search, 118
British Museum search, 118
brute-force search, 118
bubble sort, 201
buffer, 330
buffer stock, 326
buffers, 31, 309, 344

C

calling population, 344
candidate solution, 119
canister, 326
carrying cost. *See* holding cost
cellular layout, 23, 303
Chicago slaughterhouses. *See* meatpacking industry
chromosome, 122, 165
closed half-space, 153
cloud computing. *See* distributed computing
cluster computing. *See* distributed computing
cluster graph, 57
clustering, 56
coefficient of determination, 155
coherent subassemblies, 57
combinatorial optimization, 5, 79, 118, 119, 149, 263, 271, 279
combinatoric, 118
completion time. *See* flow time
complexity theory, 81
component disassembly optimization, 58
computational complexity, 78
computer, 38, 80, 91, 121, 134, 136, 163, 291, 308
connection diagram, 63
connection state diagram, 66
co-NP-complete, 89
constrained connection diagram, 67
contacting matrix, 109
containers. *See* canister

continuous review, 318
convex, 153
convex combination, 153
CONWIP, 326
CR, 308, 312
critical ratio. *See* CR
crossover, 122, 166, 167, 169, 264
cycle, 183, 186, 187, 264
cycle time, 105, 133, 293, 304

D

data, 91, 127, 128, 132, 133, 140, 149, 151, 155, 168, 244, 252, 263, 264, 278, 279, 286, 308, 348
decision, 81, 90, 301, 321, 343
decision problem, 82, 87
decision tree, 301
decision version, 6, 117
decision-tree analysis, 301
defective part, 39
delivery lead/lag time, 318
delta skip, 130, 252, 264
demand, 4, 43, 106, 284, 287, 293, 304, 317, 319, 326, 329
depth-first search, 118, 160
descriptive solution, 158, 344
design, 284, 287, 289, 297
design for disassembly index, 58
destructive disassembly, 95
detachable subassemblies, 58
deterministic search, 118
digraph. *See* directed graph
directed graph, 61, 70
direction, 4, 107, 286, 293, 304
disassembly, 3, 33
disassembly bill of materials, 54
disassembly constraint graph, 68
disassembly graph, 56, 319
DISASSEMBLY LINE BALANCING PROBLEM. *See* DLBP
disassembly planning, 52
disassembly precedence graph, 66
disassembly precedence matrix, 55, 109
disassembly process plan, 54
disassembly scheduling, 52
disassembly sequence plan, 54
disassembly sequencing, 52
disassembly tree, 64
disassembly-to-order. *See* DTO
discrete optimization. *See* combinatorial optimization
distributed autocatalytic process, 122
distributed computing, 126, 180, 261, 280
distributed intelligent agent, 118
DLBP, 4, 96, 98, 99, 102, 111, 117, 133, 151, 264, 275
dominance, 153
DTO, 13, 318
due dates, 309
dynamic stochastic sequencing and scheduling, 313

E

earliest due date. *See* EDD
economic order quantity model. *See* EOQ
EDD, 308, 310, 311
efficacy, 98
efficacy index, 154, 271, 290
electronics industry, 47, 136, 137, 138, 286, 289, 293, 303, 329, 341
end-of-life, 4
EOQ, 284, 287, 317, 318
evaporation, 123, 182, 187, 264
exhaustive search, 4, 6, 118, 119, 121, 159, 246
exponential, 89
exponential time, 78, 88
exponential-time algorithm, 80
external subassembly, 331

F

facilities, 27, 284, 287, 295
facilities engineering, 295
facility layout, 23, 300, 302
facility location, 285, 300
factorial complexity, 78, 164
FCFS, 308, 310
feasible, 5, 102, 118, 153
FFD, 87, 121, 123, 199, 203, 308
FIFO, 97, 345
first come first served. *See* FCFS
first in first out. *See* FIFO
first-fit, 87
first-fit-decreasing. *See* FFD
fitness, 122, 166, 167
flow shop, 22, 31, 78, 97, 113, 309, 327, 344
flow time, 308, 309

G

GA, 6, 120, 122, 165, 211
gene, 122, 165
generation, 165, 168, 264
genetic algorithm. *See* GA
geometric constraints, 64, 286
goal programming, 153, 319
gradient, 124, 127, 153
greedy. *See* greedy algorithm
greedy algorithm, 7, 119, 123, 184, 189, 198, 199, 228, 291
grid computing. *See* distributed computing

H

Hamiltonian, 79, 84, 90, 112, 228
hazardous, 4, 105, 200, 293, 304, 309

heuristic, 89, 90, 96, 118, 119, 151, 213, 227, 243, 263, 276, 287, 308, 341
hill climbing, 7, 119, 124
H-K, 125, 243
holding area, 326
holding cost, 317
hot start, 120, 166, 261, 280
hybrid, 118, 120, 129, 213, 227, 261, 291
hyperplane, 153

I

idle time, 97, 102
immediate update first fit, 49
Industrial Revolution, 23, 31
informed search, 119
instance, 79, 82, 97, 118, 133, 149, 161, 170, 187, 203, 217, 232, 244, 252, 264, 279, 308
internal subassembly, 331
intractable, 80, 82
inventory theory, 286, 287, 315, 317, 325

J

JIT, 14, 157, 284, 286, 325
job shop, 31, 278, 309
Johnson's algorithm, 308
just-in-time. *See* JIT

K

kanban, 285, 325, 326
kanban, part, 331
kanban, subassembly, 331
Kendall notation, 346
KNAPSACK, 80, 86, 88, 112, 285, 287, 321

L

lateness, 308, 310
Lawler's algorithm, 308
lean, 24, 27, 325
lexicographic, 129, 153, 202
line balancing, 97
line design, 27, 284, 285, 287, 295, 298, 302, 304
line efficiency, 103
line layout. *See* line design
local search, 89, 119

M

makespan, 97, 308, 309, 310
Markov chain, 346
material requirements planning. *See* MRP
maximum lateness, 308
MCDM, 98, 102, 113, 151, 153, 154, 279, 280, 343
mean flow time, 308, 310
meatpacking industry, 23, 31, 37
merge sort, 211
metaheuristic, 118, 120, 165, 181, 263, 276, 308
mixed-model line, 50, 213, 303, 328
model, 96, 98, 102, 108, 156, 284, 303, 318, 319, 321, 338, 344, 345, 348
modeling, 156, 284
Moore's algorithm, 308, 310
MRP, 53, 58, 284, 287, 315, 318, 327
multicriteria. *See* MCDM
multifactor rating, 301
multikanban system, 328
multiple optimum extreme points, 143, 152, 163, 174, 206, 246, 273
MULTIPROCESSOR SCHEDULING, 86, 112, 113
mutation, 122, 167, 170, 264

N

naïve search, 119
nearest neighbor, 123
neighborhood, 89, 214
next-fit, 244, 247, 308
no-free-lunch theorem, 277
nondestructive disassembly, 95
nondeterministic, 78
nondeterministic algorithm, 83
nondeterministic computer, 82
nondeterministic polynomial, 78
normalizing , 154
NP, 78, 82, 83, 84, 86, 87, 113
NP-complete, 81, 82, 86, 87, 88, 96, 111, 140, 285, 286, 303, 307, 308, 320, 341
NP-complete in the strong sense. *See* unary NP-complete
NP-completeness proof, 84
NP-hard, 81, 86, 87, 114, 320, 341
n-tuple, 97, 124, 127, 150, 182, 186, 346
number of tardy jobs, 309
number of workstations, 104

O

offspring , 165, 166
one-tape deterministic Turing machine, 78, 80
operator balance chart, 27
optimization problem, 87
optimization version, 81, 114
ordering cycle, 318
overflow item, 330

P

P, 78, 81
paced line, 16, 22, 38, 95, 98, 344

parallel computing. *See* distributed computing
parent, 122
Pareto optimum, 154
PARTITION, 80, 84, 86, 87, 88, 111, 112, 114
periodic review, 318
permutation, 5, 97, 118, 129, 131, 160, 224
permutation schedule, 308
Petri net, 51, 75, 276
pheromone, 123, 182, 187, 264
planning horizon, 318
polyhedral, 153
polynomial, 78
polynomial complexity, 196
polynomial time, 79, 85, 86
polynomial time reduction, 82
polynomial transformation, 84
polynomial-time algorithm, 80
polytope, 153
pool, 122, 165
population, 165, 167, 168, 264
post. *See* canister
PPX, 166
precedence, 96, 167, 184, 243, 309
precedence constraints, 5, 64, 112
precedence preservative crossover. *See* PPX
predictive model, 301
prescriptive solution, 158, 344
probabilistic, 30, 89, 198, 277, 278, 344
probability, 183
probability theory, 143, 145, 155
problem, 79, 118
process-oriented layout, 23
producibility, 20
product reconstruction, 337
product recovery, 3, 48, 95
production centers, 326
production cost. *See* purchase price
production lead time, 326
product-oriented layout, 23
pseudocode. *See* algorithm
pseudopolynomial time algorithm, 86, 114, 321
PSPACE-complete, 90
pull, 326, 330
purchase price, 317
push, 328, 334

Q

quasi-component, 63
queue, 345
queueing system, 345
queueing theory, 15, 278, 284, 286, 287, 344
quick sort, 211

R

ranked positional weight, 28
reasonable encoding, 81
reasonable machine, 80, 164
recourse models, 321
recycling, 3, 95, 337
reduced tree, 65
refurbishing, 337
regression, 151, 155, 271
regular item, 330
remanufacturing, 3, 95, 284, 337
request for proposal. *See* RFP
revenue, 15, 285, 286, 287, 301, 337
RFP, 297
rule. *See* algorithm
runtime, 164, 176, 194, 210, 224, 239, 259, 269, 272

S

scheduling, 96, 113, 284, 285, 307, 315
search algorithm, 118
search space, 118
sequence, 97, 186, 304
sequencing, 96, 284, 285, 307, 313, 338
service discipline, 345
service facility, 345
service layout, 24, 303
set of natural numbers, 105, 130
set of positive integers, 127
setup costs, 318
shortage costs, 317
shortest processing time. *See* SPT
simulation, 156, 285, 286, 287, 328, 334, 348
simulator, 156
single minute exchange of dies. *See* SMED
skip size, 127, 129, 130, 245, 253, 264
slaughterhouses. *See* meatpacking industry
SMED, 325, 327
smoothness index, 103
software, 91, 134, 151, 155, 252, 280, 302, 303, 341
space complexity, 150
SPT, 308, 310, 311
SS policy model, 318
state, 345
static stochastic sequencing and scheduling, 313
station time, 98, 102
steady-state operation, 345
stochastic, 90, 165, 181, 264, 277, 278, 284, 287, 321, 348
stock replenishment, 318
stopping criteria, 119, 129
subassembly state diagram, 66

T

takt, 27
tardiness, 308, 309, 310
technical constraints, 68
time complexity, 80, 91, 151, 176, 196, 210, 224, 240, 244, 247, 261, 271, 285
time complexity function, 80
topological constraints, 63, 286
trail, 183, 186
transformed AND/OR graph, 75
transient period, 345
TRAVELING SALESPERSON PROBLEM. *See* TSP
Turing network model, 319
TSP, 79, 81, 83, 119, 121, 123, 140, 184, 198, 227, 285, 315
Turing machine, 78, 80

U

unary NP-complete, 86, 113
unbalanced lines, 285, 286, 287, 343
undirected graph, 61
uninformed search, 119
unpaced line, 38, 278, 303, 344
useful period, 3

V

V&V, 157
validation, 157
verification, 157
virtual component. *See* virtual part
virtual part, 8, 63, 98

W

weak search, 119
weighting, 28, 49, 55, 79, 98, 127, 151, 184, 198, 279, 301, 313, 321
workstation, 78, 95, 99, 104, 214, 293, 304, 305, 329, 344

Y

yes-no version, 81

ACCESS Engineering
NEW BOOKS
Search AccessEngineering
Search
Search All Content
Search Dictionary

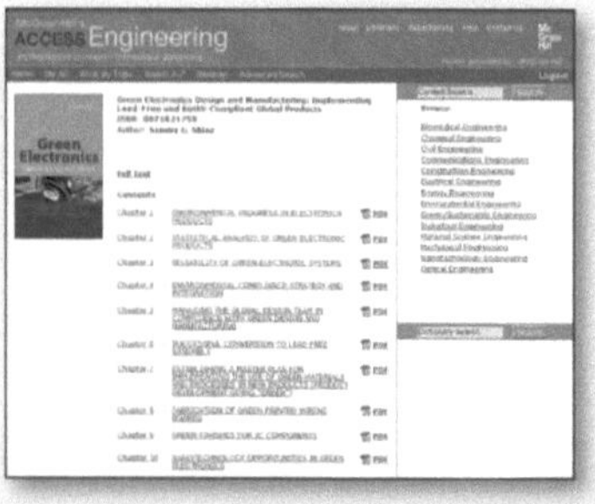
ACCESS Engineering
Green Electronics

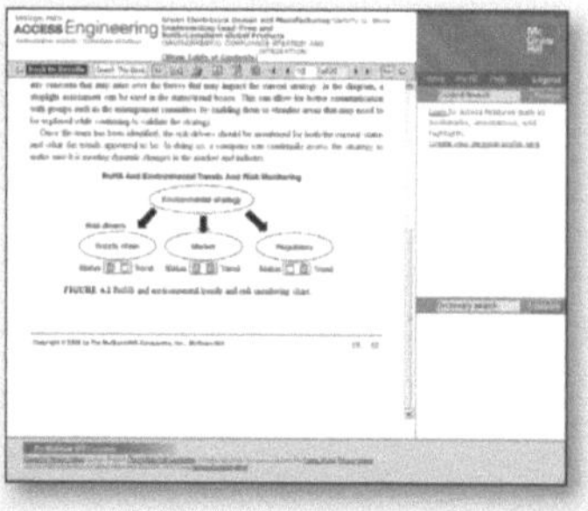
ACCESS Engineering

Printed by Libri Plureos GmbH in Hamburg,
Germany

9 780071 622875